WITHDRAWN

Biomechanics of Training and Testing

Jean-Benoit Morin · Pierre Samozino
Editors

Biomechanics of Training and Testing

Innovative Concepts and Simple Field Methods

Editors
Jean-Benoit Morin
Faculty of Sport Sciences
University of Nice
Nice
France

Pierre Samozino
Laboratoire Inter-universitaire de Biologie
 de la Motricité
Université de Savoie Mont Blanc
Chambéry
France

ISBN 978-3-319-05632-6   ISBN 978-3-319-05633-3   (eBook)
https://doi.org/10.1007/978-3-319-05633-3

Library of Congress Control Number: 2017958616

© Springer International Publishing AG 2018
This work is subject to copyright. All rights are reserved by the Publisher, whether the whole or part of the material is concerned, specifically the rights of translation, reprinting, reuse of illustrations, recitation, broadcasting, reproduction on microfilms or in any other physical way, and transmission or information storage and retrieval, electronic adaptation, computer software, or by similar or dissimilar methodology now known or hereafter developed.
The use of general descriptive names, registered names, trademarks, service marks, etc. in this publication does not imply, even in the absence of a specific statement, that such names are exempt from the relevant protective laws and regulations and therefore free for general use.
The publisher, the authors and the editors are safe to assume that the advice and information in this book are believed to be true and accurate at the date of publication. Neither the publisher nor the authors or the editors give a warranty, express or implied, with respect to the material contained herein or for any errors or omissions that may have been made. The publisher remains neutral with regard to jurisdictional claims in published maps and institutional affiliations.

Printed on acid-free paper

This Springer imprint is published by Springer Nature
The registered company is Springer International Publishing AG
The registered company address is: Gewerbestrasse 11, 6330 Cham, Switzerland

Acknowledgements

The Editors are very grateful to the authors who participated to the writing of this book, for their time and dedication to this project. We also thank Prof. Alain Belli for teaching us how to generate "simple methods" for locomotion and sports biomechanical analysis, during our time as Masters and Ph.D. students under his supervision.

Several colleagues significantly contributed to this book and the underlying experimental work. In particular Mr. Régis Bonnefoy, who designed, engineered and fixed many devices and sensors used during our experimental studies, and our colleague and friend Pedro Jimenez-Reyes who significantly contributed to several studies presented in this book. We also thank Caroline Giroux, Cameron Josse, Carlos Balsalobre, Johan Cassirame, Peter Weyand and Ken Clark for the specific support they provided for typical practical application examples or figure designs.

None of the research presented in this book could have been produced without the collaboration and trust of all the students, athletes and their coaches/manager who volunteered to participate to our tests, sometimes with maximal, repeated efforts, but always with a high motivation, compliance and understanding. Over the years, the application of the methods presented in this book and their improvement have been systematically made possible through their participation, constructive feedback and questioning.

Finally, we would like to thank our families for their constant support, patience and love.

Contents

1 **Introduction** 1
  Jean-Benoit Morin and Pierre Samozino

Part I Cycling

2 **Maximal Force-Velocity and Power-Velocity Characteristics
  in Cycling: Assessment and Relevance** 7
  Sylvain Dorel

3 **Mechanical Effectiveness and Coordination: New Insights
  into Sprint Cycling Performance** 33
  Sylvain Dorel

Part II Ballistic Movements of Upper and Lower Limbs

4 **A Simple Method for Measuring Lower Limb Force, Velocity
  and Power Capabilities During Jumping** 65
  Pierre Samozino

5 **Optimal Force-Velocity Profile in Ballistic Push-off:
  Measurement and Relationship with Performance** 97
  Pierre Samozino

6 **A Simple Method for Measuring Lower Limb Stiffness
  in Hopping** 121
  Teddy Caderby and Georges Dalleau

7 **A Simple Method for Measuring Force, Velocity, Power
  and Force-Velocity Profile of Upper Limbs** 139
  Abderrahmane Rahmani, Baptiste Morel and Pierre Samozino

Part III Running

8 A Simple Method for Measuring Lower Limb Stiffness During Running 165
Jean-Benoit Morin

9 A Simple Method for Determining Foot Strike Pattern During Running 195
Marlene Giandolini

10 The Measurement of Sprint Mechanics Using Instrumented Treadmills 211
Jean-Benoit Morin, Scott R. Brown and Matthew R. Cross

11 A Simple Method for Measuring Force, Velocity and Power Capabilities and Mechanical Effectiveness During Sprint Running 237
Pierre Samozino

12 The Energy Cost of Sprint Running and the Energy Balance of Current World Records from 100 to 5000 m 269
Pietro E. di Prampero and Cristian Osgnach

13 Metabolic Power and Oxygen Consumption in Soccer: Facts and Theories 299
Cristian Osgnach and Pietro E. di Prampero

Chapter 1
Introduction

Jean-Benoit Morin and Pierre Samozino

Everything should be made as simple as possible, but not simpler.

Albert Einstein (1879–1955)

Although it is a "young" scientific discipline, locomotion and sport biomechanics has taken an important place in the daily routine of many practitioners of sports training, medicine and rehabilitation. It allows both a better understanding of human locomotion and performance and a better design of sports training and injury prevention programs. In these processes, the testing of athletes is crucial, and the quality and quantity of variables analysed will directly influence the effectiveness of coaches, physiotherapists and other practitioners' interventions.

This book presents a state of the art of innovative methods, and for most of them, gives direct and practical insights into how practitioners may benefit from using them in their everyday practice. It also details how to interpret the data measured, and the underlying neuromuscular and biomechanical factors related to sport performances.

Written and edited by the same researchers who proposed and validated these methods and concepts, the aim of this book is to both present innovative methods and concepts for an effective and accurate training and testing process (most of them being based on very simple technology and data processing methods), and discuss the associated underlying knowledge. Before presenting in details the theoretical basis and practical applications of these methods and concepts in the

---

J.-B. Morin (✉)
Laboratory of Human Motricity, Education Sport and Health,
Université Côte D'Azur, 261 Route de Grenoble, 06205 Nice, France
e-mail: jean-benoit.morin@unice.fr

P. Samozino (✉)
Laboratoire Inter-universitaire de Biologie de la Motricité, Université de Savoie Mont Blanc,
Campus Scientifique, 73000 Le Bourget du Lac, Chambéry, France
e-mail: pierre.samozino@univ-smb.fr

following chapters, this introduction section will focus on the specificities of the overall approach the authors of this book used as sport scientists to bring some new insights in human performances.

1.1 Optimizing Sport Performance Is like Cooking

A good dish is the result of the optimal combination of different ingredients. A head cooking chief chooses the best ingredients and mixes them in the optimal way. Similar processes happen in sports. Performance is a complex integration of different qualities, abilities and skills. The head cooking chief is the coach, or the strength and conditioning coach if we focus on physical qualities. He has the genius of training to mix at best the different ingredients required to reach high levels of performance. To improve athlete's performance, one needs to know the different ingredients well, and how they can interact, to achieve the best mix possible. Both of the latter can come from empirical experience, but also, from evidence and data brought by sport sciences. Sport scientists do not aim to propose take-away recipes to sport practitioners, but only to bring some insights about ingredients, effects of their combinations and how to accurately taste/test them. This book presents some of these "ingredients" related to running, jumping, throwing and cycling performance, notably innovative methods and concepts to test and quantify some of these ingredients for each athlete, most of these methods being easily usable out of laboratories.

In the same way that only one ingredient cannot be not responsible of the success of a delicious dish, the performance in sport does not depend on only one or two factors. However, to better understand the effect of one specific physical, technical, psychological or tactical quality on the final performance, the sport scientist is forced to isolate each of them and to study their effect on only one part of the performance. This does not mean that he neglects the other factors also contributing to performance, but increasing the knowledge about a specific factor necessary goes by playing with this factor and considering the others as stable (*ceteris paribus*). For instance, explosive movements and sprint accelerations are key factors in soccer. While a good sprinter would not be necessarily a good soccer player since many other specific qualities and skills are needed, a soccer player who jumps higher, accelerates more and runs faster than his opponent, all other qualities being equal, will take a certain advantage in the game. So, increasing the understanding and evaluation of individual capabilities determining explosive performances is of great interest, yet not sufficient, to optimize soccer performance. In a delicious dish, each ingredient is indispensable, even not enough to explain the final flavor. This book will present theoretical and practical insights about biomechanical factors determining the ability to run or pedal faster, to jump higher or to throw further, which can be interesting to improve performance in some sports, keeping in mind that they are only some ingredients of the success. Sport practitioners should be aware about these factors and how to evaluate and train them, but they have to integrate and associate them at best with the other ingredients involved in the targeted performance. Coaches are and remain the head cooking chiefs and sport sciences an indispensable tool.

1.2 See the Big Picture First

You can't see the forest for the tree

When aiming at understanding and contributing to improve sport performance, a scientific approach going from macroscopic to microscopic levels is of great interest. It consists in starting the analyses from the performance itself, its different integrative biomechanical factors (when focusing on physical qualities) to then study the biological or neurophysiological underlying mechanisms. This allows a clear understanding, in a logical order, of the relationships between performance, the mechanical requirements of the underlying tasks and the associated mechanical outputs, the various athlete's individual intrinsic qualities involved and in fine, the corresponding biological features. In the field of applied sport sciences, this implies to use a back-and-forth approach between fields of practice and laboratory. Most of time, the initial basic questions come from sport practitioners on the field regarding what they need to better know to improve performance. Some of these interrogations require laboratory approaches using standardized experimental protocols, biomechanical or physiological models, mathematics, physics, and statistics. This inevitably puts some distance between research and the actual performance on the field, but this makes possible to find some answers which have then direct practical applications for performance optimization and training. This book will thus present some theoretical approaches, mainly based on biomechanical models, which bring some new insights contributing, at least in part, to answer practical issues for sport practitioners. These theoretical answers are associated to validation by comparison to experimental data and practical applications supporting their relevance and interest in training and testing.

1.3 Simple Models, Simple Methods

The simpler the model, the clearer it is which of its characteristics are essential to the observed effect

Alexander (2003)

The originality of the biomechanical models presented in this book and associated to the above-mentioned macroscopic approaches, is to correctly explain human performance from the fewest variables possible. This philosophy of such models is well illustrated by the words of Robert McNeill Alexander in an interview presented in the "Questions and Answers" section of Current Biology journal in

2006 (Alexander 2006): "*Use simple mathematical models for clarifying arguments and generating hypotheses. Don't try to make your model as complex as the animal it represents: you will never succeed, and the effort may be counterproductive because it is often not apparent which features of a complex model are responsible for the effects it shows. On the other hand, if a model is simple enough, you can tell what caused the effect. I have found optimization models particularly useful — models that seek the best possible structure or behaviour. For example, if a model tells me that a particular pattern of behaviour is the best possible in given circumstances, and if real animals do something quite different, that suggests that I may have failed to understand the issues at stake*". Such biomechanical models do not aim at representing all the biological structures forming human body, but rather at characterizing in the simplest way possible[1] the actual mechanical behavior of the different part of the body acting in sport performance. These models allow sport scientists and practitioners to better identify and understand the different integrative biomechanical factors affecting human performances (dynamics approach). When using these models in the opposite way, i.e. the model's input is the performance and the outputs are the underlying mechanical properties (inverse dynamics approach), they can be used to develop simple methods to evaluate some individual mechanical properties without any specific dynamometers or other laboratory devices. This book will present both ways to use such biomechanical models. First, it will present some models that led to some original concepts to better understand the biomechanical factors affecting sport performances. Second, this book will present simple methods to assess, easily and out of labs, mechanical properties of an athlete's neuromuscular system or biomechanical features of human locomotion. Theoretical background, validation against "gold standard" methods and practical applications in training will be detailed in the following chapters.

References

Alexander RM (2003) Modelling approaches in biomechanics. Philos Trans R Soc Lond B Biol Sci 358(1437):1429–1435

Alexander RM (2006) R. McNeill Alexander. Curr Biol 16(14):R519–R520

---

[1]The simplest as possible in respect of the initial question at which the model is used to answer.

Part I
Cycling

Chapter 2
Maximal Force-Velocity and Power-Velocity Characteristics in Cycling: Assessment and Relevance

Sylvain Dorel

Abstract Cycling is a "common" task considered relatively intuitive and hence easy to perform by everybody. Although pedaling represents a typical multi-joint movement characterized by several degrees of freedom, in contrast with running it can be regarded as less complex since the fixed trajectory of the pedals and the mechanical coupling between both legs constrain lower-extremity movements to a higher degree. In this chapter we will define a macroscopic model to measure maximal cycling power output and typical measurements involved in stationary ergometer conditions and also on the field in Sect. 2.5. We will give some reference data and discuss how best to interpret the force, power and velocity indexes obtained from this testing procedure. Force-velocity and power-velocity relationships allow reliable assessment of maximal power capabilities and its two force and velocity components in cycling. Provided that the proper methodology and advice presented in this chapter are used, the different indexes extracted from these relationships give very useful information on (i) evaluating lower-limb muscle function, (ii) monitoring specific and general strength training effects and (iii) interpreting and characterizing muscle involvement in the efforts performed on the field within the training or competition context. Use of a stationary ergometer is now straightforward and particularly suited for each individuals and hence remains the reference method to use, yet portable power meters are now accessible and will become more and more suitable.

S. Dorel (✉)
Faculty of Sport Sciences, Laboratory "Movement, Interactions, Performance", University of Nantes, 25 bis, Boulevard Guy Mollet, EA 4334, BP 72206, Nantes, 44322 cedex 3, France
e-mail: Sylvain.Dorel@univ-nantes.fr

2.1 Introduction

Cycling is a "common" task considered relatively intuitive and hence easy to perform by everybody. After the French engineers Michaux and Lallement added the pedals, the pedaling machine (i.e., the first "bicycle") appeared at the beginning of the 20th century with approximately the same general design as today's bicycle. Currently, it is more and more widely used in daily life (e.g., to go to work) as well as for recreational, fitness, sport or rehabilitation activity. From a biomechanical point of view, cycling consists in propelling (or "driving") a system (cyclist and its bicycle or an inertia flywheel) using alternative actions of both legs on a crank system in rotation. Although pedaling represents a typical multi-joint movement characterized by several degrees of freedom, in contrast with running it can be regarded as less complex since the fixed trajectory of the pedals and the mechanical coupling between both legs constrain lower- extremity movements to a greater extent.

Historically, the maximal power produced during sprint cycling has been considered an indirect measurement of "maximal anaerobic power"; that is, ability related to the rate of energy turnover from anaerobic metabolism (specially related to the phosphagen pathway). The last three decades have shown it is now well established that this ability is more representative of the global muscle function of the lower limbs and mechanical muscle properties. Hence explosive cycling movement is greatly determined by the maximal force and power capabilities of the muscles involved which directly depend on the well-known force-velocity relationship (Hill 1938). As a consequence, at the macroscopic level, it is well established that the total maximal power produced during cycling is similarly very well characterized by both force-velocity and a power-velocity relationships. Although these relationships reflect a range of other neuro-mechanical properties (see Chap. 3), they allow the determination of useful indexes of global power, force and velocity abilities that provide interesting simple tools to evaluate athletes and at the same time monitor specific and general strength training effects.

In this chapter we will define a macroscopic model to measure maximal cycling power output and typical measurements found under stationary ergometer conditions and also on the field in Sect. 2.5. We will provide some reference data and discuss the best way to interpret the force, power and velocity indexes obtained from this testing procedure. Additionally, we make a special attempt to provide the main practical tips to ensure validity and precision in the determinations of the different indexes in order to encourage a good interpretation of the results in the context of training and sport performance.

2.2 Measurement of Mechanical Output (Force, Velocity and Power) During Sprint Pedaling

The system involved in riding an ergometer is generally represented by a crankset and the associated wheel in rotation on which external forces are applied. Then the "external power" produced by the subject can be considered part of the power output used to overcome external resistances (i.e., different frictions and the weight or inertia of the system) or linked to the changes in total mechanical energy of this system (i.e., kinetic energy in this case). This external power (Ps) is therefore directly related to the torque generated at the level of the crank axis (T_{Feff}, which only depends on the perpendicular component of the total force F_{eff} produced on the pedals) and the crank angular velocity (ω, in rad·s^{-1}, Eq. 2.1). In practice, Ps can also be calculated considering the effective force (F_{eff}, in N, by dividing torque by the crank length in m) and velocity (V, in m·s^{-1}) at the pedal level (Eq. 2.2).

$$Ps = T_{Feff} \cdot \omega \qquad (2.1)$$

$$Ps = F_{eff} \cdot V \qquad (2.2)$$

Many methods have been developed in the last 30 years to calculate the two components of Ps (Driss and Vandewalle 2013) and two main modalities were used: isokinetic (i.e., control of pedaling rate and direct measurement of force with sensors in the crank or pedal; Mc Cartney et al. 1983b; Sargeant et al. 1981) and the more common isoinertial condition (i.e., control of the external resistance and inertia parameters of the flywheel and measurement of its velocity and acceleration; Dorel et al. 2005; Hautier et al. 1996; Martin et al. 1997; Vandewalle et al. 1987). All these methodologies use the same general equations of movement for measuring the torque produced by the subject (T_{Feff}, Eq. 2.3) and then the power (Ps, Eq. 2.4):

$$T_{Feff} = I \cdot \alpha + T_{Fric} \qquad (2.3)$$

$$Ps = I \cdot \alpha \cdot \omega + T_{Fric} \cdot \omega \qquad (2.4)$$

where T_{Fric} represents the sum of the resistive torques applied on the system in rotation (friction or magnetic resistance in most cases); I is the moment of inertia of this system (in kg.m^2); and α is its angular acceleration (in rad·s^{-2}). Historically this equation has been used on friction-loaded cycle ergometers (e.g., Monark) and then considered the flywheel as the system in rotation. Instead of the torque and angular velocity at the crank, the values are in this case expressed referring to the flywheel axis. Ultimately, in studies using the Monark ergometer (Hautier et al. 1996; Morin and Belli 2004; Vandewalle et al. 1987), force and velocity are expressed in components corresponding to the forces applied at the periphery of the flywheel (in N or kg equivalent) and the velocity of the point of application of these forces (v, in m·s^{-1}). So Eq. 2.4 becomes:

$$Ps = I/R^2 \cdot a \cdot v + F_{Fric} \cdot v \qquad (2.5)$$

where R represents the radius of the flywheel (in m), a the tangential acceleration (in m·s^{-2}) and F_{fric} the friction force applied by the belt (generally measured by a strain gauge) or the magnetic resistance (in N). Note that this last processing methodology is reliable and convenient depending on the equipment resources. However, taking into account the variability of the ergometer's characteristics (radius of the flywheel, gear ratio...), it is not really useful to compare values of force between the studies and the ergometers (e.g., the maximal force index, see below). On the same issue, it is important to keep in mind that a slight but non-negligible difference is observed between the power measured at the flywheel level and that measured directly by strain gauges at the crank or at the shoe-pedal interface, depending on the losses induced by frictions in the chain, sprockets and the different rotation axis (Driss and Vandewalle 2013).

During a sprint exercise, it is possible to monitor the force and then power output produced provided that (i) the friction force applied and the velocity and acceleration of the flywheel (or the crank) are precisely measured and that (ii) the moment of inertia of this flywheel is known. The time course of force, velocity and power during a single sprint of few seconds can be tracked, depending on the sampling rate. Finally, the mean value on each pedal stroke or each complete cycle allows to plot the crank velocity-time and power-time relationships (Fig. 2.1a–b), and also the evolution of force and power in function of the velocity (generally expressed as pedaling rate, Fig. 2.1c). These typical relationships are detailed in the next part.

This methodology was largely applied in the last twenty years and remains very useful when using friction-inertial-loaded (Arsac et al. 1996; Dorel et al. 2003) or only inertial-loaded (Martin et al. 1997) cycle ergometers. However other methods were historically proposed and can still be used (see the two detailed reviews for more information, (Cross et al. 2017; Driss and Vandewalle 2013)). Indeed, the first approach to assess the force-velocity in cycling was based on simplification by considering only the second part of the Eq. 2.5 (i.e. the power required to overcome the braking force) omitting the power required to accelerate the flywheel inertia (Vandewalle et al. 1987). The power estimated by this methodology is valid only when the acceleration is equal to zero, which is solely the case when the peak velocity is reached during sprint performed in isoinertial condition. Then a single couple of force-velocity (i.e. braking force-peak velocity) can be assessed per sprint. It is necessary to repeat a series of six to ten sprints against progressively increasing braking forces, to record different peak velocities and finally plot a force-velocity relationship as well (Fig. 2.1c). Theoretically this approach remains reliable to assess the maximal force- and power-velocity relationships and you can see on Fig. 2.1c the good agreement between the F-V relationship resulted from "peak velocity" method and the relationship resulted from the "reference" method (i.e. Eq. 2.5). The advantage was that the material is very simple: the braking forces are predetermined by known weights and only the peak flywheel velocity should be measured. However, the repetition of the 8–10 sprints can induce a significant

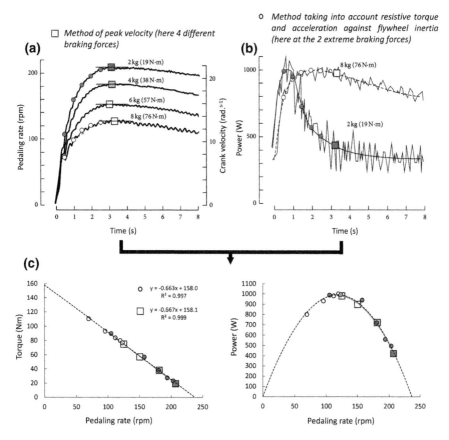

Fig. 2.1 a Typical example of the time-course of the pedaling rate (or crank velocity) during a force-velocity test on a Monark cycle ergometer: 4 short sprints against 4 braking forces (in kg at the circumference of the flywheel and equivalent resistive torque in Nm at the crank axis). Squares correspond to the peak values and circles to the mean values on each complete cycle before. **b** Time-power curves for the two extreme loads with the raw and filtered data (grey lines) and the same representation of the mean values on the five first complete crank cycles and at the peak velocities. **c** Mean value of torque, pedaling rate and power on these complete cycles (first 3 s of each sprint) can be used to plot torque- and power-velocity relationships. Note the good match in this example between the shape of these relationships for the two methods: one that considers only the peak velocity values of the 4 distinct sprints of panel a (squares: power corresponds to the product of peak velocity and braking force) and the other that considers all the cycles of the 2 sprints of panel b (circle: power measurement also takes into account the force to accelerate the flywheel inertia during the first phase of the sprint). Adapted from Driss and Vandewalle (2013)

fatigue throughout the test session and the protocol assumes an absence of fatigue at the time of the peak velocity occurrence which can be questionable in some cases (see the next parts). Note that although there are still few people which propose to use this methodology (i.e. only the power against the braking force) to estimate power also during the acceleration phase, this approach should be logically ruled out due to its invalidity (Morin and Belli 2004).

Finally, with the development of a new generation of commercialized electronic cycle ergometers (e.g., SRM^{TM}, Lode Excalibur Sport^{TM}), especially from the last fifteen years, they are increasingly used. With these devices, the estimation of the resistive torque (braking or magnetic) is no longer required because of the direct measurement by strain gauges of the force at the crank or the pedal level (Buttelli et al. 1996; Capmal and Vandewalle 1997; Dorel et al. 2010). Moreover, the setting of these ergometers has been particularly improved to be better adapted to the size and muscle capacities of all athletes (Dorel et al. 2012; Gardner et al. 2007). For this reason, the next parts of the chapter presenting in detail the methodology and interpretation of force- and power-velocity capabilities will often use data recorded with this type of ergometry.

As for all biomechanical models aiming at simplifying the complex multi-joint movement to facilitate its interpretation, this macroscopic model of external power output measurement in cycling implies several assumptions:

- From a mechanical standpoint cycling remains a double task: (i) moving the leg segments in such a way that the foot moves on a circular trajectory; (ii) producing torque at the crank levels. The work produced by the muscles is then transformed into mechanical work at the crank level but also used to move the leg segments (Driss and Vandewalle 2013; Kautz and Hull 1993). Therefore, it is important to keep in mind that pedal force measured at the shoe-pedal interface can be decomposed into a muscular component due directly to the intersegmental net joint torques and a non-muscular component due to gravitational and inertial effects of the segments.
- This cyclic movement of the lower extremity requires a specific coordination of several lower-limb muscles with a lot of co-activation between synergists as well as antagonist muscles (see next chapter, Dorel et al. 2012; Hug and Dorel 2009). In this line, despite a maximal involvement, the maximal activation level is not verified for all the muscles and the cyclic characteristics of the task induce that mechanical output is also governed by excitation-relaxation kinetics (Neptune and Kautz 2001).
- This movement is classically considered by the scientific and coaching communities as a useful evaluation of the human dynamic muscle function of the lower limb extensor muscles in a "concentric" mode. However, some muscles could theoretically act in different modalities of contraction (particularly the bi-articular muscles), and also operate in different range of their force-length relationship.

Note that all of these points will be further examined and discussed in the next chapter (see Chap. 3 for more information).

2.3 Maximal Force- and Power-Velocity Relationships in Cycling

2.3.1 Testing and Processing

Sprint exercise on cycle ergometer is widely used to evaluate force and power characteristics of the lower limbs. These muscle capacities can be accurately measured on a cycle ergometer, using the well-known "force-velocity" test that consists of performing 3 brief all-out exercises of 5-s duration against difference resistances. The corresponding resistive torques generally applied are 0, 0.5–0.8, and 1–1.8 $Nm.kg^{-1}$ body mass depending on the body mass and level of expertise. Actually, these values of resistance are indicative and may be adapted in line with the general principle of allowing subjects to attain a large range of pedaling rates over the cumulative three bouts. Classically the value of force, velocity and power are averaged on a period corresponding to a full crank cycle (or a half cycle, i.e. one downstroke) and then 4 to 10–12 couples of force and velocity values are obtained from each sprint, allowing modeling the relationships with a large number of points (Fig. 2.2).

Whatever the approach to measure the mechanical data, the velocity is preferentially expressed in pedaling rate at the crank level (in rpm). Then force-velocity relationship obtained during pedaling is classically presented as a maximal torque (or effective force)-pedaling rate relationship which is very well fitted by a linear regression model (Dorel et al. 2003, 2005; Driss and Vandewalle 2013; Driss et al. 2002; Hautier et al. 1996; Hintzy et al. 1999; Vandewalle et al. 1987). The linear relationship obtained (Eq. 2.6) enables assessment of V_{max} and F_{max}, which have the dimensions of maximal pedaling rate at the zero force axis and the effective force corresponding to a zero pedaling rate, respectively (Fig. 2.2; Eq. 2.7).

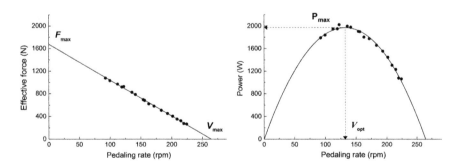

Fig. 2.2 Typical force-velocity (left) and power-velocity (right) relationships of an elite track cyclist, fitted using linear and quadratic regressions, respectively and considering mean crank cycle values obtained from three sprints of 4-s duration. Determination of maximal power (P_{max}), optimal pedaling rate (V_{opt}), maximal pedaling rate (V_{max}), and maximal effective force (F_{max})

$$F = \alpha V + F_{max} \tag{2.6}$$

$$V_{max} = -F_{max} \cdot \alpha^{-1} \tag{2.7}$$

Maximal power generation is described by a polynomial (2nd order,) power-velocity relationship (Eq. 2.8) with a maximum value (P_{max}) reached at an optimal cycling rate (V_{opt}) and hence at an optimal force (F_{opt}).

$$P = aV^2 + bV + c \tag{2.8}$$

P_{max} can be determined directly basing on the F-V relationship with the following equations:

$$P_{max} = V_{opt} \cdot F_{opt}, \quad P_{max} = 0.5 V_{max} \cdot 0.5 F_{max}, \quad P_{max} = 0.25 \cdot V_{max} \cdot F_{max} \tag{2.9a–c}$$

where V_{opt} and F_{opt} are expressed in official unit (rad·s^{-1} and Nm at the crank axis or m·s^{-1} and N at the pedal, respectively).

V_{opt} and P_{max} can be determined based on Eq. 2.8 (P-V relationship) with the following equation:

$$V_{opt} = \frac{-b}{2a} \tag{2.10}$$

$$P_{max} = a \cdot \left(\frac{-b}{2a}\right)^2 + b \cdot \left(\frac{-b}{2a}\right) + c \tag{2.11}$$

Practically, the relation between the power and velocity is fundamental because it means that (i) an athlete can reach the actual maximal value of power (P_{max}) only at an optimal trade-off between force and velocity (Eq. 2.9a–c), and (ii) the maximal power-generating capacity dramatically decreases when velocity significantly moves away below or above this V_{opt} value (Fig. 2.2). Interestingly, the more the target power is below P_{max} value (i.e. submaximal values under the curve), the more the possibilities to produce this power on a large range of different pedaling rates (and hence force) exist. On the training and testing viewpoint, the power-velocity relationship implies that P_{max} can be reached around to the small range of pedaling rates corresponded to almost V_{opt} value. It means that the higher V_{opt} the more the athletes should pedal at a high cadence to be able to reach their maximal power (Fig. 2.3). Then even if two athletes with different V_{opt} have a large difference in their absolute P_{max} (e.g, 12 rpm and ~200 W in the example of Fig. 2.3), this discrepancy in power production could be very lower or even disappear on the field if the effort is performed at specific pedaling rate advantageous for the athlete with the higher V_{opt} (e.g., at almost 180 rpm on Fig. 2.3).

However, it is important to keep in mind that due the mechanical fundamental laws, even if the mechanical constraints (e.g., effects of gravity, rolling resistance,

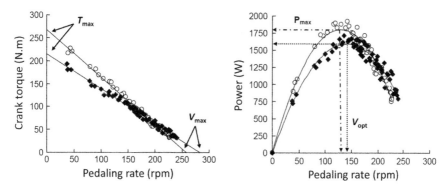

Fig. 2.3 Determination of the torque- and power-velocity relationships of two typical elite sprinters in cycling with large differences in torque and velocity capabilities: one (white circle) with very high value of maximal torque (T_{max}: ~270 Nm) giving him a higher value of maximal power (P_{max}: 1800 W), the other (black diamond) with extremely high value of maximal pedaling rate (V_{max}: 285 rpm) giving him a high P_{max} also (despite lower: 1650 W). Observe the greater discrepancy between power capabilities of both athletes at the pedaling lower than V_{opt} (60–70 to 130 rpm) and the lower discrepancy at higher pedaling rates and even an absence of difference at pedaling rates around 180–200 rpm

air friction, moment of inertia of the system) can directly influence the speed of the system on the field (i.e. the cyclist + bicycle), they do not directly act on the capacity to reach maximal power output per se if the pedaling rate can be adjusted. Indeed, contrary to running for example, it is easier to change the ratio between speed and pedaling rate by using a gear ratio system on the field. Consequently, the athlete by changing the gear ratio can adjust his pedaling rate to be close to V_{opt} conditions and then to be able to reach his maximal power output in very different mechanical situations (speed, external resistances…etc…).

2.3.2 Meaning of the Indexes Extracted from the Relationships

The force- and power-velocity relationships allow the determination of several useful parameters (i.e. indexes) for normative evaluation as well as training monitoring. One often tries to link these indexes with performance factors in different explosives disciplines requiring power of the lower limbs (Dorel et al. 2005; Morin et al. 2002) and to describe the alteration of these indexes with training or with age to better understand the loss of power capacity and functional performance in elderly population (Bonnefoy et al. 1998; Martin et al. 2000). Then, it is important to state on the interpretation of the different parameters and how each can account to the lower limb muscle function.

Optimal and maximal pedaling rate. Due to the linear model, both indexes correspond to the same global velocity capability (i.e. $V_{opt} = 0.5\ V_{max}$). The pedaling rate being directly linked to the pedal speed it influences indirectly the angular velocity of the main joints (especially the hip and the knee) and hence is considered to influence also the muscle shortening velocity. Therefore, V_{opt} is often considered as the condition where the majority of the muscles involved are shortening near their optimal velocity (Sargeant 1994; Zatsiorsky 2008). In this line it is interesting to note that V_{opt} values reported in subjects specialists in "explosive" performances (and hence presumably characterized by a higher percentage of fast twitch muscle fibers) are classically higher than those reported in endurance athletes (Buttelli et al. 1996; Davies et al. 1984; Dorel et al. 2003, 2005; Gardner et al. 2007; Hintzy et al. 1999; Sargeant et al. 1981; Vandewalle et al. 1987). The optimal pedaling rate range between almost 90–100 rpm for extremely low values up to 140–145 rpm for extremely high values. Interestingly, this link between the proportion of fast twitch fibers and V_{opt} was experimentally demonstrated in two subjects (McCartney et al. 1983a) and ten healthy specifically trained subjects (Hautier et al. 1996) confirming that this high proportion of type II fibers, at least on the knee extensor muscle, may be one factor associated with a high pedaling rate for maximal power. As depicted on Fig. 2.4, the fact that V_{opt} is gradually impaired with advancing aging (Bonnefoy et al. 1998; Martin et al. 2000) can be confidently related to the well-known deterioration of the maximal unloaded shortening velocity of muscle fibers with aging (Power et al. 2016) and further suggest that fiber type distribution influences, at least partly, the V_{opt} ability.

Nevertheless due to the characteristics of the movement, other factors clearly influence the maximal pedaling rate. Among them, we can point specifically the muscle coordination and the activation dynamics (see next chapter for more details). As cycle frequency increases, the capacity of nervous system to activate and deactivate the muscles increase in importance, and then have a significant

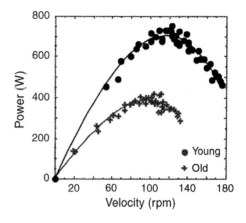

Fig. 2.4 Typical power-velocity of an young and an old individual illustrating the great difference in maximal power due to both a decrease in maximal force and velocity capabilities (note the decrease in optimal velocity, Bonnefoy et al. (1998))

influence on force production additionally to the intrinsic force- and power-velocity characteristics of the muscles (Bobbert et al. 2015; Caiozzo and Baldwin 1997; Neptune and Kautz 2001; van Soest and Casius 2000). Moreover, the coordination between the muscles seems to have a significant impact on the capacity to continue to produce a high force at pedaling rate higher than V_{opt} (Samozino et al. 2007) and is even more critical at extreme pedaling rate (Dorel et al. 2014).

Optimal and maximal effective force (or crank torque). Due to the linear model, both indexes corresponds the same global qualities of force (i.e. F_{opt} = 0.5 F_{max}). F_{max} is the theoretical maximal "isometric" force produced on the pedals by both legs. This data is not frequently reported in the literature compared to P_{max} and often expressed in different units (braking force at the flywheel, crank torque or effective force applied on the pedals) making difficult to compare the values. By referring to data directly observed or basing on the P_{max} and V_{opt} values we can state that F_{max} values expressed in torque range between almost 115–120 Nm for extremely low values (unpublished data: recreational or endurance road cyclists or triathletes) up to 300–320 Nm for extremely higher values (unpublished data: world-class BMX and track sprint male cyclists). This index is clearly considered as reflecting the maximal force ability of the main lower limb extensor muscles and then as a good indicator of maximal strength. In this line, significant relationships have been demonstrated between F_{max} and different indexes of specific maximal isometric or isokinetic peak torque (i.e. between 0 and $240° \ s^{-1}$) of the knee extensors on single-joint ergometer when data were expressed in absolute units or normalized to the quadriceps mass (Driss et al. 2002). Moreover, F_{opt} and F_{max} were significantly related to thigh muscle area or volume determined from tomodensitometry (Linossier et al. 1996; McCartney et al. 1983a). Finally, a similar link was also reported with the lean leg volume (estimated by anthropometry using Jones and Pearson's technique Jones and Pearson 1969) in a relative homogeneous population of elite track sprint cyclists (r = 0.77, p < 0.01, Dorel et al. 2005).

Maximal power. Theoretically maximal power depends on both force and velocity capabilities. P_{max} values measured in very different populations range between 500 600 W/10 W·kg^{-1} and 2100 W/22 W·kg^{-1} (Arsac et al. 1996; Dorel et al. 2005; Driss and Vandewalle 2013; Gardner et al. 2007; Hintzy et al. 1999; Martin et al. 1997; Vandewalle et al. 1987) up to almost extremely high values of 2400–2500 W and 25–26 W·kg^{-1} (unpublished data on world-class BMX and track sprint male cyclists). These extreme values are logically recorded on subjects exhibiting low (almost 100 rpm) and high V_{opt} (almost 135–140 rpm) respectively (in relation with the proportion of fast twitch fibers). In this line it has been reported significant relationship between P_{max} and V_{opt} in quite heterogeneous population (Arsac et al. 1996; Hintzy et al. 1999). However, the $\sim$40% difference in "velocity" capability is dramatically lower compared to the $\sim$300% difference in P_{max}. It should therefore keep in mind that despite importance of velocity, force capability is definitively the main key factor that explains such great differences in P_{max} between subjects. In this line, no relationship was reported between P_{max} and V_{opt} in homogeneous elite track cycling sprinters while a strong relationship (r = 0.92, p < 0.001) was observed with F_{max} (Dorel et al. 2005). In the same way, a lot of

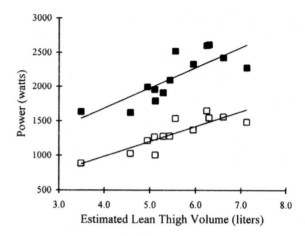

Fig. 2.5 The relationship between maximal power measured during a force-velocity test and the estimated lean thigh volume (n = 13): peak power obtained during the cycle in black, mean power produced over the complete cycle in white, Martin et al. (1997)

studies highlighted direct correlation (Fig. 2.5) between P_{max} and indices of muscle mass or lean leg of thigh volume (Linossier et al. 1996; Martin et al. 1997; Pearson et al. 2006) and interestingly some of them reported the same statistical significance for the relationship between strength indexes of knee extensors and P_{max} than that obtained with F_{max} ability (Driss et al. 2002).

What about the improvement possibilities by training? Based on the aforementioned literature and a longitudinal follow-up of elite track cyclists in the last 15 years with the French Federation of Cycling, it is reasonable to think that strength capacities are the best candidate for improvement power at both short and long terms. It is in agreement with the impressive increase of muscle mass and the maximal force measured in strength and conditioning movements (e.g., squat exercise) observed on the athletes throughout their career. Moreover, it is also corroborated by the concomitant alteration classically observed in maximal force and power in cycling and the muscle mass and force indexes of the athletes in periods of detraining or reprise of training. In the same time, the alteration of V_{opt} or V_{max} are note really noticeable (between these periods or throughout the career). The question is: does it means that velocity capabilities are not important and cannot be improved? Three arguments suggest that the question is not so obvious and the answer is likely no. First, the extremely highly powerful athletes are also those that exhibit the highest V_{opt} values (e.g. between 135 up to 145 rpm for the best 4–5 elite sprint track and BMX cyclists performing all at the highest international level; unpublished personal data). Basically, compared to an athlete A with P_{max} = 2000 W and V_{opt} = 130 rpm an athlete B with the same maximal force but with a V_{opt} = 140 rpm directly benefits from almost additional 150 W at P_{max}. Second, even rarely, new young adult athletes can sometimes show non negligible increase of V_{opt} ($\sim$10 rpm) in early years of training; perhaps linked to an improvement of muscle coordination. Thirdly, although the gain of V_{opt} are limited (certainly mainly due to the influence of heredity on the muscle typology), we know that possible change in a range of 5–8 rpm can appear over the time (as a result of a velocity-specific training block or related to a period of very high level of expertise during the career). For all these reasons, it appears that velocity capabilities are non-negligible and that testing V_{max} remains interesting and should still be considered for talent identification and development (Tofari et al. 2016).

2.4 Methodological Consideration and Practical Advices

2.4.1 Period of Averaging to Draw F-V or P-V Relationships and Duration of the Sprint

Mean cycle versus peak instantaneous values. As previously described, the phase/period for averaging values on cycle ergometer classically corresponds to a full crank cycle because matching with the period of the cyclic movement (i.e. during which each muscles are activated on one phase and deactivate on another). As movement is done by both legs in antiphase it is also often proposed to average the data only on one half cycle which corresponds to only one pedal stroke (i.e. the downstroke of one leg + upstroke of the contralateral leg) while a full cycle corresponds to two pedal strokes (i.e. downstroke and upstroke of each legs in antiphase). Note that if the different measurement sensors and the acquisition system allow to record quasi instantaneous values it is then possible to describe the torque profile inside each pedal cycle (see the next chapter). As a consequence, in rare cases (specially in former studies) the peak "instantaneous" values of force and power reach in each pedaling cycle are reported (Beelen and Sargeant 1991; McCartney et al. 1983a; Sargeant et al. 1981), these values being clearly not comparable with the classical mean values on cycle (i.e. almost 50% higher).

Sprint duration and influence of the occurrence of early fatigue. Practically, the choice of the period of time (or number of cycles) to be included for each bout is an important question to draw fatigue-free force- and power-velocity relationships. During a maximal sprint performed at a constant pedaling rate (to avoid the effect of the pedaling rate confounding factor, Gardner et al. 2009; Tomas et al. 2010) the power output can be maintained during a very short period of almost 4–5 s depending on velocity (higher pedaling rate inducing a higher decrease) and certainly the athletes' individual profile (endurance vs. sprint athletes) (Beelen and Sargeant 1991; Dorel et al. 2003; McCartney et al. 1983a). It is therefore important to remove frome the analysis all the values for which fatigue potentially already occurs at the end of the sprint: after 3–4 s for low pedaling rates (50–120 rpm) and maximally after 3 s for high pedaling rates (120–250 rpm). On the other hand, it is possible to include more data (4–5 or 5–6 s) keeping in mind that the P_{max} and associated V_{max} resulted from the relationships would then represent a slight different capability (e.g., 5–8%) also already accounting a fatigue resistance ability (Fig. 2.6).

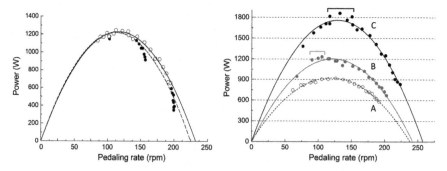

Fig. 2.6 Left: Effects of taking into account the power values produced during the first 3 s (white) versus the first 5 s (white + black) of the three sprints performed during F-V testing session. Despite an absence of a great influence on the maximal power estimation, black points illustrate an effect of early fatigue for this individual leading to a significant underestimation of the maximal velocity and the power at the high pedaling rates (and likely an overestimation of the maximal force, not illustrated here). **Right**: typical example of P-V relationships for 3 subjects illustrating the importance of visual inspection of data in addition to the coefficient of determination of the model (R^2). Maximal power (P_{max}), optimal velocity (V_{opt}) extracted from the model, and mean of the three higher power values (P_{Peak}, bracket) and the associated velocity (V_{Ppeak}) for each subject are as follows: (A) P_{max} = 912 W, V_{opt} = 119 rpm; R^2 = 0.977; P_{peak} = 900 W at V_{Ppeak} = 121 rpm; (B) P_{max} = 1182 W, V_{opt} = 120 rpm; R^2 = 0.968; P_{peak} = 1195 W at V_{Ppeak} = 105 rpm; (C) P_{max} = 1760 W, V_{opt} = 129 rpm; R^2 = 0.982; P_{peak} = 1828 W at V_{Ppeak} = 133 rpm. See details of interpretation in the text

2.4.2 Quality of the F-V and P-V Models: "Calculated" Versus "True" Data

As a whole, the reliability of the force-velocity test is well established (Jaafar et al. 2015) and we can be even more confident for data obtained on athletes. Nevertheless, to go a step further in the use of the indexes, some advices can be serve. The power of the linear and quadratic models to estimate useful values of P_{max} and V_{opt} and others indexes logically depends on the number of points, the capacity to obtain points on a large range of pedaling rates below and above V_{opt} and the coefficient of determination (R^2). One can consider that the latter should be at least equal to 0.80–0.85 but it might be better to obtain a coefficient higher than 0.9–0.95 to enable valid assessments. Figure 2.6 illustrates three typical examples of P-V relationships for which the model fits the data very well (i.e. extremely high R^2 values: from 0.968 to 0.982) and the P_{max} values nicely represent the differences in the power capacity of the three subjects. Beyond that, the visual inspection of data in respect to the fitted curve remains important to avoid some over-interpretation. The gold standard should correspond to the typical example A for which both P_{max} and V_{opt} extracted from the model exactly correspond to the real values obtained. For the example B, while P_{max} remains very reliable (only 1.1% of difference between P_{max} and the mean of the three higher power values: P_{Peak}), it appears that V_{opt} determined by the model (120 rpm) is partly overestimated in the sense that P_{peak} is almost reached at a 12.5% lower pedaling rate (around 105 rpm). That means that the athlete B is actually not prevented from producing his maximal power for pedaling rates around 100 rpm as it is predicted by the quadratic model. For the athlete C, P_{max} assessed by the model is almost 50 W lower than P_{peak} which corresponds to a non-negligible difference of almost 3%.

Consequently, in the context of the longitudinal follow-up of training it can be advised to primary consider P_{max}, F_{max} and V_{opt} to better quantify gains in force and velocity capacities in response to the training period. In the same time, in the cases where modeling has the aforementioned limitations it is advisable to refer to real values for some practical application: (i) for choosing the suitable gear ratio to adapt pedaling rate with the goal to maximize the power during a sprint on the field (for example for the cyclist B) or (ii) for comparing and better interpreting the peak power obtained using powermeter on the field in comparison with the maximal power obtained during the stationary cycle ergometer testing procedure (for example for the cyclist C). Finally, note that P-V model often used a more constrained equation by setting the y-intercept as 0 (i.e. constant $c = 0$, in all the figures presented in this chapter). This process can slightly influence the capacity to fit the true data but is interesting to extrapolate more realistic power at extreme cadences (low or high). So, that is important to control for inter and intra-individual comparisons.

2.4.3 Main Factors to Control that may influence Maximal Power Output

Different parameters can influence the force and/or the velocity produced on the pedals and the question is to know whether it might alter the maximal power capability. Among them, the seat height and the crank length, the nature of the shoe-pedal interface and the use of a standing versus seating position are the main recognized factors but they do not act on the same manner and with the same extent on the power output. Rather than presenting a detail analysis of these factors, the purpose here is just to insist on their main effect, in order to (i) select them appropriately and (ii) control these parameters in both the context of longitudinal follow-up of training and evaluation/detection of athletes.

Ergometer setup: seat height and crank length. Seat configuration (seat tube angle and height) can theoretically influence the pedal power by altering the lower limbs kinematics (specifically the range of motion of each joints) and ultimately the force produced by each muscle groups. However, by simulation (Rankin and Neptune 2010) it was demonstrated that the influence of seat tube angle is actually limited (i.e. almost 1% for a wide range of 65°–110°) when considering only the classical cycling configuration (i.e. not the recumbent cycling). For these authors, the seat height has a greater impact and they proposed an optimal value corresponded to 102% of greater trochanter height (and an average knee flexion angle of 101.7° with min = 59.6° at TDC and max = 153.6° at TDC). Actually and despite the more and more popularity of the bike-fitting consulting activity, the setting of these parameters is not obvious. The problem is that too many parameters interact to be able to conclude about an "optimization": the anthropometric (segment lengths), the comfort (depending on the duration of the effort, the joint flexibility, etc.), the alteration of upper body posture and hence the aerodynamic resistances, the degree of freedom allowed by the natural displacement of the hip on the saddle. Moreover, it should be examined in concert with the choice of the crank length since they can together alter the kinematics and then the range for force generation by each joint based on the force-length relationship. Overall, different trials (e.g. force-velocity test) should be done to experimentally confirm that a position is better for each individual.

By itself the crank length does not drastically influence the maximal power (Martin and Spirduso 2001) because even if it changes the link between the force and the velocity, P_{max} is not deteriorated in a large range of crank lengths (i.e., 145–195 mm). Although a standard crank length (170–175 mm) can therefore be confidently used in adults without concern of decreasing maximal power, an optimal crank length of 20% of leg length was still suggested by these authors. It is interesting for specific population such as very short athletes or the children and adolescents (Martin et al. 2002). One practical consequence however is that the optimal pedaling rate increases with the decrease of crank length. When different bicycles equipped with different crank sizes are used by the athletes on the field, it is therefore interesting to take into account this influence to adapt pedaling rate and gear ratio. Basing on the previous studies and the testing database of elite sprint cyclists (personal unpublished data), we can confidently state that a change in 10 mm of the crank length corresponds to almost an alteration of 4–5 rpm in V_{opt}.

Pedal and toe-clips. It was early demonstrated that during the upstroke phase of the pedal, athletes can actively pull on the pedal to partially transmit additional force to the crank (Beelen et al. 1994; Dorel et al. 2010; Martin and Brown 2009). As a consequence, presence of toe-clips with straps or clipless pedals necessary induce significant change in the maximal values obtained during a torque-velocity test (Capmal and Vandewalle 1997). Hence an almost 10–20% difference has been reported in P_{max} or F_{max} in the last study depending on the shoe-pedal interface, and we can reasonably think that it also significantly influences the value of V_{max} (or V_{opt}). As we will discuss in the next chapter, the relative contribution of the muscles involved in the flexion phase should not be neglected especially during the sprint compared to the submaximal exercise (Dorel et al. 2010; Driss and Vandewalle 2013; Elmer et al. 2011). If the use of flat pedals would theoretically be interesting to better isolate on the work of the lower limb extensors only, it is not advised for two reasons: it does not avoid a partial non-controlled contribution of the contralateral leg during the flexion phase (at least the partial effect of its weight) and it remains very challenging to follow the pedal trajectory in this phase, especially at very high pedaling rates.

Effect of body position: seating/standing. Standing position when sprinting allows to produce a higher maximal power output compared to seating (Driss and Vandewalle 2013; Hug et al. 2011; Reiser et al. 2002). This is mainly explained by the effects of a higher participation of the body weight over the pedals, an additional recruiting of the upper limb and trunk musculature (Turpin et al. 2017), along with some positive adjustment of muscle coordination such like a higher or longer recruitment of some muscles throughout the pedaling cycle (hamstring during the extension-flexion transition, quadriceps and gluteus maximus during the second part of the pushing phase; (Hug et al. 2011) and personal unpublished data). In this context, further studies are still needed to better elucidate all the biomechanical and neuromuscular factors of influence. If some studies reported a gain of about 8–15% (Driss and Vandewalle 2013; Reiser et al. 2002; Vandewalle et al. 1988) it is not easy to state on a 'typical' mean value regarding the amount of the benefit. Indeed, the benefit can be very different between the subjects and sometimes very poor (e.g., 50 W = +2.2%, on an elite male sprint cyclist with 2250 W of P_{max} in seating position) or very important (e.g. 160 W = 16.7% on an elite female sprint cyclist with 960 W of P_{max} in seating position; unpublished personal data). It is often accompanied by difference in the optimal velocity which are often slightly lower (0–10 rpm) in standing position; however, it is again difficult to draw up a general rule because few subjects can also exhibit a slight higher V_{opt} in standing position (e.g., elite BMX or sprint track riders). Maybe specific skills and segment length can also partially explain this interindividual variability in the seating/standing difference observed regarding the maximal power and velocity abilities. Both values can have the same practical interest if the athlete rides in both positions on the field, but it is then more appropriate (and very recommended) to test cyclist in the same position in laboratory and field conditions to gain a better overview of his capabilities.

2.5 Field Measurement in Ecological Condition

As detailed above, the force-velocity test on cycle ergometer is specific of the cycling performance and then is a largely accepted method to describe and predict the mechanical behavior of the cyclist in sprint condition on the field. However, in addition to this classical reference method, the use of additional "field" data bring real benefits for both coaches and sports scientists:

- firstly, little changes in body configuration can occur in field condition (especially lateral oscillation of the cyclist and his bicycle commonly observed when sprinting) and hence may slightly influence the power produced (Faria et al. 2005);
- consistent with that, this would be particularly right at very high level of force during a sprint starting performance and then the classical stationary force-velocity test does not perfectly evaluate the real-life practice specifically

for conditions at extremely high force and low cadence (i.e., lower than 70 rpm);
- additionally, power measurement allow today to better characterize the effort on the field, and to better identify and control the muscular quality worked during training sessions (e.g., force or power predominance);
- finally, beyond all of that, the external resistance and mechanical constraints applied on the cyclist on the field are strongly different compared to the stationary force-velocity test and could be very diversified. Although that should not impair the maximal power capability of the subject (except if optimal velocity condition is not verified), it largely influences the relative contribution of each resistance/constraint on the power demand and hence factors of performance such like the maximal speed or the acceleration of the athlete.

2.5.1 Mathematical Model of Sprint Cycling

Since the first work of di Prampero et al. (1979), different mathematical model were developed to estimate the power produced on the field by the cyclist. Historically, the challenge was really to determine the power demand basing on the different applied external resistances and the variation in the mechanical energy in order to infer a reliable estimation of the muscular power produced by the athlete. Whatever the model proposed (Martin et al. 2006, 1998; Olds 2001; Olds et al. 1993) that corresponds to the power required to overcome air resistance and rolling resistance and the power required to change the kinetic energy and/or the potential energy of the system (Eq. 2.12):

$$P = C_d A \cdot \frac{1}{2} \cdot \rho \cdot V^3 + \mu . F_n . V + \frac{\Delta E_p}{\Delta t} + \frac{\Delta E_k}{\Delta t} \qquad (2.12)$$

with $C_d A$ represents a coefficient including both the effective frontal area and the drag coefficient of the system (cyclist + bicycle); ρ is the air density, μ is a global coefficient of friction, F_n is the normal force on the surface due to the weight of the system, V, the velocity of the system cyclist + bicycle (in absence of wind) and ΔE_p and ΔE_k the variation on the period of interest (Δt) of potential and kinetic energy, respectively.

When looking in detail this equation it appears not simple to use this model to evaluate maximal power of the cyclist in routine on the field. Indeed, measuring (or reliably estimating) the different coefficients requires complex, expensive and less accessible methodologies especially for the aerodynamic coefficient (e.g., wind tunnel), which often represents the most important factor due to the importance of the air resistance with the increased speed. For that, it is possible to use field-derived values for modeling $C_d A$ and μ coefficients (thanks to testing sessions by means of a portable powermeter system, see the next part) and to apply the model to accurately estimate power during a maximal sprint cycling (Martin et al. 2006). Beyond this apparent difficulty, this modeling approach has significant practical applications for training and testing:

- for better characterizing the relative contribution of each external constraints in the total power demand during a specific exercise (e.g., the power demand of increasing the kinematic energy and concomitant power supply of decreasing potential energy from the top of the turn during the acceleration phase of a 200 m flying start performance on the track);
- for assessing aerodynamics issues and simulating how subtle changes in all the constant included in the model (e.g., riding position, mass, tire pressure, configuration of the track or the road, environmental condition, etc.) and power of the subject itself will influence cycling speed and in fine the sprint performance;
- for avoiding misinterpretation and shortcut about the link between very high speed and the amount of power output. For illustrating with a numerical example, the model can estimate that an athlete of 80 kg sprinting at a mean speed of 64.5 km.h^{-1} on a 5-s period (on the flat and in absence of wind and drafting) produce an additional power of almost 200 W when the exact initial and final instantaneous speeds on this period are 64 km.h^{-1} and 65 km.h^{-1}, respectively (i.e. very slight acceleration < 0.06.ms^{-2}) than then in inverse condition (i.e. 65 and 64 km h^{-1}, respectively: slight deceleration). This is not obvious to imagine that without modeling.

In line with the last point, the measurement interval on which all the kinematic data can be measured practically (Δt: period of averaging the velocity, and on which the variation of velocity and of the vertical position, h, are detected) directly affects the capacity of model to account for rapid changes in power production (which is crucial in sprint cycling). However the new technologies available to measure the displacement speed (e.g., commercial high speed video, see Chaps. 4 and 5, and also the portable systems, see the next part of this chapter) are promising opportunities to obtain a precise speed profile of the athlete during a competition or training/testing session. Finally, the condition on the field can be organized to simplify the model by limiting the number of variables that can contribute. For practical testing session, it is easy to perform a sprint in starting-block condition to have very low value of speed and on the flat to omit the first and the fourth components of Eq. 2.12, respectively (Bertucci et al. 2005). This approach is interesting to have a field estimation of maximal force but it is more critical for maximal power assessment if V_{opt} is not reached which would be probable if speed is limited to 20 km h^{-1}. The assessment of force- and power-velocity and their associated indexes on the field requires therefore additional equipment.

2.5.2 Direct Measurement with Portable System

As on cycle ergometer, new technologies available from the last decade allowed the emergence of several portable devices to reliably measure the power output on the field. One main issue of the existing commercial powermeter systems (SRM^{TM}, Powertap^{TM}, Stages^{TM}, Garmin vector^{TM}, Look keo power^{TM}...) for determining maximal power during explosive efforts remains their sensitivity and the sampling rate. Some of them now fulfil these essential conditions: they have been validated and store data at least at 1 Hz. That represents the minimum value to ensure a sufficient temporal resolution during explosive exercise to estimate a realistic maximal value of power and associated torque and pedaling rate. As a consequence, these tools begins to be used and raise interesting possibilities for training and testing in sport performance (Gardner et al. 2009, 2007; Martin et al. 2007, 2006; Menaspa et al. 2015a, b).

Valuable inputs regarding the sprint performance. As aforementioned, recording the power and pedaling rate with these portable devices allow a better understanding of the characteristics of the sprint effort done on the field whatever the discipline (track, road, BMX). During training sessions, it is a means to better describe the power-velocity demand of a specific training session and in turn for coaches a possibility to adjust the speed and gear ratio in order to preferentially target a muscle quality to work (e.g. force-power vs. power-velocity). It can provide a rapid post-exercise feedback on the involvement of the athletes. For long-lasting sprint exercise (i.e. 10–30 s duration) the analysis gives additional information on the capacity to resist to fatigue which is also a very valuable physiology-related quality. Indeed the dissociation of the effect of the pedaling rate confounding factor (on the time-course of power) represents a real benefit to isolate the influence of fatigue but also to characterize the interaction between these parameters (Gardner et al. 2009; Tomas et al. 2010).

A very important advantage of these devices is they can be used in a competition context. As for a specific training session the time course of the power and pedaling rate during a race can be plotted (Fig. 2.7). For the example of the 200 m flying start performance depicted in details on this figure, it permits to identify the different phases of the effort and to characterize each of them in term of force-velocity-power demand and fatigue. It could be a very useful input to also understand the influence of external variables such like the configuration of the track (height, width and friction feature) and in turn to adjust the gear ratio but also the individual pacing strategy to attempt to optimize the power production in one or all the phases and hence the speed performance. All of these applications require to express the data referring to the F-V and P-V relationships measured in the laboratory condition (i.e. reference method) but we can wonder whether using powermeter may also be an option to directly assess these relationships on the field.

Assessment of F-V and P-V relationships. Theoretically, nothing prevents to determine the global force- and power-velocity relationships with these tools. In this line, a very good agreement was logically reported between "laboratory" and

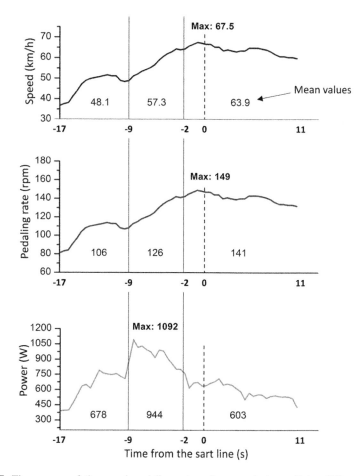

Fig. 2.7 Time-course of the speed, pedaling rate and power during a flying 200-m time trial measured during a world cup competition (female athlete, SRM^TM pro system at 2 Hz). The chronometric portion begins at the start line (time 0: dotted line) and the distance was performed in 11.3 s after this line. Three phases can be considered with the associated mean value obtained on each: an initial acceleration phase (on the left, from −17 to −9 s before the start line), a maximal acceleration phase (on the middle, from −9 to −2 s) and the final phase where the athlete attempts to maintain maximal effort for the remainder of the 200-m (on the right, from −2 to the end of the effort). Here for example, the second phase (7 s) during which the power has to be maximized was performed at a mean cadence (126 rpm) corresponded to 104% of V_{opt} and a mean power (944 W) corresponded to 84% of P_{max} (measured with the cycle ergometer reference method: Lode Excalibut Sport^TM, in standing position because this phase is performed in this configuration on the field). Note in this typical example the very high values of pedaling rate in the third chronometric phase (Max = 123% and mean = 117% of V_{opt} for this athlete) which could be detrimental for maintaining a high level of power during this phase. Max corresponds to the peak value measured on 1 s interval

"field" conditions regarding the force- and power-velocity relationships assessment (Gardner et al. 2007). Despite some limitations (lack of data around and above the optimal pedaling rate to draw the relationship) this study makes us confident on the opportunity to determine reliable F-V and P-V relationships and the associated force and velocity indexes. An interesting attempt was recently proposed in BMX (Debraux et al. 2013) again with some limitations in terms of range of velocities (i.e. lack of points below V_{opt} this time, due to only one sprint configuration was used to plot the curve). To this end, 3–5 sprints must be performed in different conditions (speed and gear ratio to alter the amount of external resistance) in order to combine a sufficient number of force-velocity data couples and on a large range of pedaling rates (from 70–80 rpm up to 180–190 rpm). The justification is to ensure that pedaling rates around V_{opt} are reached during the effort ant that both ascending and descending parts of the P-V relationship can be fitted with true data. Moreover, the same practical precautions than in laboratory should be taken and particularly the duration of the sprint to avoid potential effect of fatigue and then error in descending portion of the power-velocity relationship.

Overall, both field tools and cycle ergometers are generally limited to record data below almost 50 rpm, at least at this time. We will see in the next chapter that it is possible with more sophisticated pedal dynamometer to better investigate the specificity of the exercises performed at extremely high level of force (compared to the F_{max} index extrapolated from the F-V relationship). Indeed, these efforts are performed in a very specific standing position with lateral oscillations leading likely to adjustment in terms of pedaling technique and coordination.

2.6 Conclusion

Force- and power-velocity relationships allow the reliable assessment of maximal power capabilities and its two force and velocity components in cycling. Provided that proper methodology and advices presented in this chapter are used, the different indexes (F_{max}, V_{max}, P_{max}, V_{opt}) extracted from these relationships give very useful information (i) to evaluate the lower limb muscle function, (ii) to monitor the specific and general strength training effects and (iii) to interpret and characterize the muscle involvement in the efforts performed on the field in training or competition context. The method on stationary ergometer is now straightforward and particularly adapted to the setting of the individuals and hence remains the reference method to use, whereas portable power-meters are accessible and will become more and more suitable.

References

Arsac LM, Belli A, Lacour JR (1996) Muscle function during brief maximal exercise: accurate measurements on a friction-loaded cycle ergometer. Eur J Appl Physiol Occup Physiol 74(1–2):100–106

Beelen A, Sargeant AJ (1991) Effect of fatigue on maximal power output at different contraction velocities in humans. J Appl Physiol (1985) 71(6):2332–2337

Beelen A, Sargeant AJ, Wijkhuizen F (1994) Measurement of directional force and power during human submaximal and maximal isokinetic exercise. Eur J Appl Physiol Occup Physiol 68(2):177–181

Bertucci W, Taiar R, Grappe F (2005) Differences between sprint tests under laboratory and actual cycling conditions. J Sports Med Phys Fitness 45(3):277–283

Bobbert MF, Casius LJ, van Soest AJ (2015) On the relationship between pedal force and crank angular velocity in sprint cycling. Med Sci Sports Exerc. https://doi.org/10.1249/mss.0000000000000845

Bonnefoy M, Kostka T, Arsac LM, Berthouze SE, Lacour JR (1998) Peak anaerobic power in elderly men. Eur J Appl Physiol Occup Physiol 77(1–2):182–188

Buttelli O, Vandewalle H, Peres G (1996) The relationship between maximal power and maximal torque-velocity using an electronic ergometer. Eur J Appl Physiol Occup Physiol 73(5):479–483

Caiozzo VJ, Baldwin KM (1997) Determinants of work produced by skeletal muscle: potential limitations of activation and relaxation. Am J Physiol 273(3 Pt 1):C1049–C1056

Capmal S, Vandewalle H (1997) Torque-velocity relationship during cycle ergometer sprints with and without toe clips. Eur J Appl Physiol Occup Physiol 76(4):375–379

Cross MR, Brughelli M, Samozino P, Morin JB (2017) Methods of power-force-velocity profiling during sprint running: a narrative review. Sports Med 47(7):1255–1269. https://doi.org/10.1007/s40279-016-0653-3

Davies CT, Wemyss-Holden J, Young K (1984) Measurement of short term power output: comparison between cycling and jumping. Ergonomics 27(3):285–296. https://doi.org/10.1080/00140138408963490

Debraux P, Manolova AV, Soudain-Pineau M, Hourde C, Bertucci W (2013) Maximal torque and power pedaling rate relationships for high level BMX riders in field tests, vol 1 and 2

di Prampero PE, Cortili G, Mognoni P, Saibene F (1979) Equation of motion of a cyclist. J Appl Physiol Respir Environ Exerc Physiol 47(1):201–206

Dorel S, Bourdin M, Van Praagh E, Lacour JR, Hautier CA (2003) Influence of two pedalling rate conditions on mechanical output and physiological responses during all-out intermittent exercise. Eur J Appl Physiol 89(2):157–165. https://doi.org/10.1007/s00421-002-0764-4

Dorel S, Hautier CA, Rambaud O, Rouffet D, Van Praagh E, Lacour JR, Bourdin M (2005) Torque and power-velocity relationships in cycling: relevance to track sprint performance in world-class cyclists. Int J Sports Med 26(9):739–746. https://doi.org/10.1055/s-2004-830493

Dorel S, Couturier A, Lacour JR, Vandewalle H, Hautier C, Hug F (2010) Force-velocity relationship in cycling revisited: benefit of two-dimensional pedal forces analysis. Med Sci Sports Exerc 42(6):1174–1183. https://doi.org/10.1249/MSS.0b013e3181c91f35

Dorel S, Guilhem G, Couturier A, Hug F (2012) Adjustment of muscle coordination during an all-out sprint cycling task. Med Sci Sports Exerc 44(11).2154–2164. https://doi.org/10.1249/MSS.0b013e3182625423

Dorel S, Couturier A, Hug F (2014) Maintain a proper muscle coordination is critical to achieve maximal velocity during a pedaling cyclic task. In: XXth congress of international society of electromyography and kinesiology, Rome

Driss T, Vandewalle H (2013) The measurement of maximal (anaerobic) power output on a cycle ergometer: a critical review. Biomed Res Int 2013:589361. https://doi.org/10.1155/2013/589361

Driss T, Vandewalle H, Le Chevalier JM, Monod H (2002) Force-velocity relationship on a cycle ergometer and knee-extensor strength indices. Can J Appl Physiol 27(3):250–262

Elmer SJ, Barratt PR, Korff T, Martin JC (2011) Joint-specific power production during submaximal and maximal cycling. Med Sci Sports Exerc 43(10):1940–1947. https://doi.org/10.1249/MSS.0b013e31821b00c5

Faria EW, Parker DL, Faria IE (2005) The science of cycling: factors affecting performance—Part 2. Sports Med 35(4):313–337

Gardner AS, Martin JC, Martin DT, Barras M, Jenkins DG (2007) Maximal torque- and power-pedaling rate relationships for elite sprint cyclists in laboratory and field tests. Eur J Appl Physiol 101(3):287–292. https://doi.org/10.1007/s00421-007-0498-4

Gardner AS, Martin DT, Jenkins DG, Dyer I, Van Eiden J, Barras M, Martin JC (2009) Velocity-specific fatigue: quantifying fatigue during variable velocity cycling. Med Sci Sports Exerc 41(4):904–911. https://doi.org/10.1249/MSS.0b013e318190c2cc

Hautier CA, Linossier MT, Belli A, Lacour JR, Arsac LM (1996) Optimal velocity for maximal power production in non-isokinetic cycling is related to muscle fibre type composition. Eur J Appl Physiol Occup Physiol 74(1–2):114–118

Hill AV (1938) The heat of shortening and the dynamic constants of muscle. Proc Royal Soc London Ser B—Biol Sci 126(843):136–195. https://doi.org/10.1098/rspb.1938.0050

Hintzy F, Belli A, Grappe F, Rouillon JD (1999) Optimal pedalling velocity characteristics during maximal and submaximal cycling in humans. Eur J Appl Physiol Occup Physiol 79(5):426–432

Hug F, Dorel S (2009) Electromyographic analysis of pedaling: a review. J Electromyogr Kinesiol 19(2):182–198. https://doi.org/10.1016/j.jelekin.2007.10.010 (S1050-6411(07)00181-2 [pii])

Hug F, Turpin NA, Couturier A, Dorel S (2011) Consistency of muscle synergies during pedaling across different mechanical constraints. J Neurophysiol 106(1):91–103. https://doi.org/10.1152/jn.01096.2010 (jn.01096.2010 [pii])

Jaafar H, Attiogbe E, Rouis M, Vandewalle H, Driss T (2015) Reliability of force-velocity tests in cycling and cranking exercises in men and women. Biomed Res Int 2015:954780. https://doi.org/10.1155/2015/954780

Jones PR, Pearson J (1969) Anthropometric determination of leg fat and muscle plus bone volumes in young male and female adults. J Physiol 204(2):63P–66P

Kautz SA, Hull ML (1993) A theoretical basis for interpreting the force applied to the pedal in cycling. J Biomech 26(2):155–165

Linossier MT, Dormois D, Fouquet R, Geyssant A, Denis C (1996) Use of the force-velocity test to determine the optimal braking force for a sprint exercise on a friction-loaded cycle ergometer. Eur J Appl Physiol Occup Physiol 74(5):420–427

Martin JC, Brown NA (2009) Joint-specific power production and fatigue during maximal cycling. J Biomech 42(4):474–479. https://doi.org/10.1016/j.jbiomech.2008.11.015 (S0021-9290(08)00601-5 [pii])

Martin JC, Spirduso WW (2001) Determinants of maximal cycling power: crank length, pedaling rate and pedal speed. Eur J Appl Physiol 84(5):413–418

Martin JC, Wagner BM, Coyle EF (1997) Inertial-load method determines maximal cycling power in a single exercise bout. Med Sci Sports Exerc 29(11):1505–1512

Martin JC, Milliken DL, Cobb JE, McFadden KL, Coggan AR (1998) Validation of a mathematical model for road cycling power. J Appl Biomech 14(3):276–291. https://doi.org/10.1123/jab.14.3.276

Martin JC, Farrar RP, Wagner BM, Spirduso WW (2000) Maximal power across the lifespan. J Gerontol A Biol Sci Med Sci 55(6):M311–M316

Martin JC, Malina RM, Spirduso WW (2002) Effects of crank length on maximal cycling power and optimal pedaling rate of boys aged 8–11 years. Eur J Appl Physiol 86(3):215–217

Martin JC, Gardner AS, Barras M, Martin DT (2006) Modeling sprint cycling using field-derived parameters and forward integration. Med Sci Sports Exerc 38(3):592–597. https://doi.org/10.1249/01.mss.0000193560.34022.04

Martin JC, Davidson CJ, Pardyjak ER (2007) Understanding sprint-cycling performance: the integration of muscle power, resistance, and modeling. Int J Sports Physiol Perform 2(1):5–21

McCartney N, Heigenhauser GJ, Jones NL (1983a) Power output and fatigue of human muscle in maximal cycling exercise. J Appl Physiol Respir Environ Exerc Physiol 55(1 Pt 1):218–224

McCartney N, Heigenhauser GJ, Sargeant AJ, Jones NL (1983b) A constant-velocity cycle ergometer for the study of dynamic muscle function. J Appl Physiol Respir Environ Exerc Physiol 55(1 Pt 1):212–217

Menaspa P, Martin DT, Victor J, Abbiss CR (2015a) Maximal sprint power in road cyclists after variable and nonvariable high-intensity exercise. J Strength Cond Res 29(11):3156–3161. https://doi.org/10.1519/jsc.0000000000000972

Menaspa P, Quod M, Martin DT, Peiffer JJ, Abbiss CR (2015b) Physical demands of sprinting in professional road cycling. Int J Sports Med 36(13):1058–1062. https://doi.org/10.1055/s-0035-1554697

Morin JB, Belli A (2004) A simple method for measurement of maximal downstroke power on friction-loaded cycle ergometer. J Biomech 37(1):141–145

Morin JB, Hintzy F, Belli A, Grappe F (2002) Force-velocity relationships and sprint running performances in trained athletes. Sci Sports 17(2):78–85

Neptune RR, Kautz SA (2001) Muscle activation and deactivation dynamics: the governing properties in fast cyclical human movement performance? Exerc Sport Sci Rev 29(2):76–80

Olds T (2001) Modelling human locomotion: applications to cycling. Sports Med 31(7):497–509

Olds TS, Norton KI, Craig NP (1993) Mathematical model of cycling performance. J Appl Physiol (1985) 75(2):730–737

Pearson SJ, Cobbold M, Orrell RW, Harridge SD (2006) Power output and muscle myosin heavy chain composition in young and elderly men. Med Sci Sports Exerc 38(9):1601–1607. https://doi.org/10.1249/01.mss.0000227638.75983.9d

Power GA, Minozzo FC, Spendiff S, Filion ME, Konokhova Y, Purves-Smith MF, Pion C, Aubertin-Leheudre M, Morais JA, Herzog W, Hepple RT, Taivassalo T, Rassier DE (2016) Reduction in single muscle fiber rate of force development with aging is not attenuated in world class older masters athletes. Am J Physiol Cell Physiol 310(4):C318–C327. https://doi.org/10.1152/ajpcell.00289.2015

Rankin JW, Neptune RR (2010) The influence of seat configuration on maximal average crank power during pedaling: a simulation study. J Appl Biomech 26(4):493–500

Reiser RF 2nd, Maines JM, Eisenmann JC, Wilkinson JG (2002) Standing and seated Wingate protocols in human cycling. A comparison of standard parameters. Eur J Appl Physiol 88(1–2):152–157. https://doi.org/10.1007/s00421-002-0694-1

Samozino P, Horvais N, Hintzy F (2007) Why does power output decrease at high pedaling rates during sprint cycling? Med Sci Sports Exerc 39(4):680–687. https://doi.org/10.1249/MSS.0b013e3180315246

Sargeant AJ (1994) Human power output and muscle fatigue. Int J Sports Med 15(3):116–121. https://doi.org/10.1055/s-2007-1021031

Sargeant AJ, Hoinville E, Young A (1981) Maximum leg force and power output during short-term dynamic exercise. J Appl Physiol 51(5):1175–1182

Tofari PJ, Cormack SJ, Ebert TR, Gardner AS, Kemp JG (2016) Comparison of ergometer- and track-based testing in junior track-sprint cyclists. Implications for talent identification and development. J Sports Sci 1–7. https://doi.org/10.1080/02640414.2016.1243795

Tomas A, Ross EZ, Martin JC (2010) Fatigue during maximal sprint cycling: unique role of cumulative contraction cycles. Med Sci Sports Exerc 42(7):1364–1369. https://doi.org/10.1249/MSS.0b013e3181cae2ce

Turpin NA, Costes A, Moretto P, Watier B (2017) Upper limb and trunk muscle activity patterns during seated and standing cycling. J Sports Sci 35(6):557–564. https://doi.org/10.1080/02640414.2016.1179777

van Soest O, Casius LJ (2000) Which factors determine the optimal pedaling rate in sprint cycling? Med Sci Sports Exerc 32(11):1927–1934

Vandewalle H, Peres G, Heller J, Panel J, Monod H (1987) Force-velocity relationship and maximal power on a cycle ergometer. Correlation with the height of a vertical jump. Eur J Appl Physiol Occup Physiol 56(6):650–656

Vandewalle H, Pérès G, Heller J, Monod H (1988) Intérêts et limites des relations force-vitesse chez l'homme. Sci et Motricité 4:38–46

Zatsiorsky VM (2008) Biomechanics of strength and strength training. In: Strength and power in sport. Blackwell Science Ltd, pp 439–487. https://doi.org/10.1002/9780470757215.ch23

Chapter 3
Mechanical Effectiveness and Coordination: New Insights into Sprint Cycling Performance

Sylvain Dorel

Abstract The pedaling task remains a multijoint task with biomechanical constraints (e.g., circular trajectory of the pedal) requiring specific coordination of the lower-limb muscles. This chapter attempts to provide an overview of how aspects related to pedaling technique and muscle coordination partly account for maximal cycling power capability (i.e., sprint exercise). Our aim in this chapter is (i) to define the typical concepts involved, (ii) discuss the practical information provided (with some examples) and (iii) highlight the main messages for optimizing coordination and sprint performance. Provided that the dynamometer used is capable of dissociating the force produced by each leg with sufficient temporal resolution and measuring the orientation of force, it is possible to analyze the pedal force profile throughout the cycle and the capacity of the athlete to effectively orientate this force. Using inverse dynamics (including calculation of both muscular and non-muscular components) and measuring EMG activity our interpretation can clearly be taken to a new level by better characterizing the involvement of each muscle group. Despite their inherent limitations, these data represent useful information about the force-generating capacity of big muscle groups (extensors and flexors) as well as about pedaling technique, especially for exercise performed at low and intermediate pedaling rates around which maximal power output is produced. It is possible in practice to identify any potential weakness in the contribution of specific muscle groups (extensor and flexors of the hip, knee and ankle) and finally to characterize muscle coordination and its key role as a limiting factor of maximal power.

S. Dorel (✉)
Faculty of Sport Sciences Laboratory "Movement, Interactions, Performance" (EA 4334), University of Nantes, Nantes, France
e-mail: Sylvain.Dorel@univ-nantes.fr

3.1 Introduction

As mentioned in the previous chapter, pedaling could be considered intuitive and hence not really dependent on complex control by the central nervous system. On the other hand, this remains a multijoint task with biomechanical constraints (e.g., circular trajectory of the pedal) requiring specific coordination of the lower-limb muscles. The torque generated over a full crank revolution is therefore not homogeneous and analysis of the torque profile may indirectly give some important insights into the role of the main functional phases and associated muscle groups. Additionally, it is important to remember that the effective force (i.e., that acts perpendicular to the bicycle crank and thus produces a torque driving the pedal around in its circle) represents only one component of the total force produced at the shoe–pedal interface. The pedaling technique can then be characterized by the ability to effectively orientate the total force on the pedal. This technical aspect of pedaling is often evaluated using an index of mechanical effectiveness (IE), defined as the ratio of the effective force to the total force exerted at the shoe–pedal interface during a complete cycle (LaFortune and Cavanagh 1983; Sanderson 1991). It is often suggested here that substantial differences exist between subjects regarding their power generation techniques. Among coaches, specialists or scientists many still believe and propose that this "pedaling effectiveness" would be improved by using some innovative training "methods." Actually, there is no experimental evidence of a clear link between this parameter and endurance-cycling performance (i.e., level of expertise, training status, muscular efficiency or metabolic demand; Bini et al. 2013; Korff et al. 2007). Nevertheless, when power output increases up to a very high level, the potential relationship between mechanical effectiveness and performance becomes patently clear and some recent information is now available regarding the significance of this index during explosive cycling exercise.

Additionally, direct measurement of the electromyographic (EMG) activity of the main muscles (see, for a review, Hug and Dorel 2009) and estimation of the torque produced at the level of the three main joints (Hull and Jorge 1985) can provide some deeper understanding of the involvement and functional role of each muscle group. Until recently, the literature has been concerned with submaximal exercise, but in the last decade some significant findings have been reported regarding sprint cycling. As a whole they offer a better understanding of the force, velocity and power indexes derived from the force–velocity cycling test. This chapter attempts to provide an overview of how aspects related to pedaling technique and muscle coordination also partly account for maximal cycling power capability (i.e., sprint exercise). We will propose an approach which to our mind provides a deeper understanding of the neuromechanical construction of sprint cycling by combining three types of analysis: 2D pedal force, joint-specific power and EMG activity measurements. Our aim in this chapter is (i) to define the typical concepts involved, (ii) discuss the practical information provided (with some examples) and (iii) highlight the main messages for optimizing coordination and sprint performance.

3.2 Torque Profile and Concept of Mechanical Effectiveness

3.2.1 Production of Power over the Pedaling Cycle

Interpretation of the force–velocity test considering the different phases. Compared to submaximal exercise, the increase in mean total force produced during sprint cycling (i.e., over a complete crank cycle) leads to an increase in pedal force not only during the downstroke phase but also during the upstroke phase. Then, whereas a slight negative contribution is commonly observed during this second phase for submaximal exercise, the torque profile during sprint is characterized by a positive impulse (Beelen et al. 1994; Dorel et al. 2010). This means that the contribution of the flexor muscles involved in the pull-up action may largely increase under sprint conditions. As a consequence, although the force–velocity test presented in the previous chapter is classically used to assess the muscle function of the extensor muscles (and particularly the quadriceps), the actual participation of these muscle groups compared with the flexor muscles has to be fully evaluated.

To investigate the contribution of each functional cycle phase, we focused on the force produced in different parts of the pedaling cycle during a classical force–velocity test performed on a specific cycle ergometer (Lode Excalibur Sport^{TM}; Dorel et al. 2010). Mean values of force and power corresponding to four functional angular sectors of the pedaling cycle were calculated: sector 1, 330°–30° (top); sector 2, 30°–150° (downstroke); sector 3, 150°–210° (bottom); and sector 4, 210°–330° (upstroke), with 0° corresponding to the highest position of the pedal. From a functional standpoint, sectors 1 and 2 correspond to the sectors around the top dead center (TDC: flexion-to-extension transition) and the main propulsive downstroke phases, and sectors 3 and 4 correspond to the sectors around bottom dead center (BDC: extension-to-flexion transition) and the propulsive upstroke phases (Fig. 3.1). The results confirmed that force–velocity and power–velocity characteristics in cycling mainly reflect the function of muscles responsible for pushing on the pedal during the extension phase (sectors 1 + 2, hip and knee extensors). Moreover, they showed an important positive contribution from the upstroke phase especially at low and moderate pedaling rates (Fig. 3.1). Indeed, when pedaling at maximal power (P_{max}), almost 19% of the total power on the entire cycle is produced during the flexion phase of the lower limb (almost corresponding to sectors 3 + 4). Additionally, significant relationships also exist between the indexes of maximal force and power produced in the upstroke phase and those obtained on the complete revolution. Overall, these findings clearly indicate that data from all-out

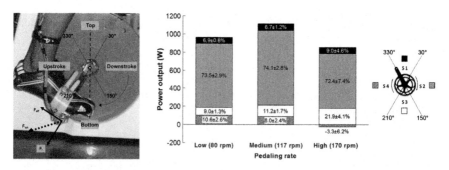

Fig. 3.1 *Left* Representation of the 2D forces applied on the pedal (sagittal plan), and the four angular sectors considered to decompose the pedaling cycle. Total force (F_{tot}) produced at the shoe–pedal interface ant its two components: the effective force (F_{eff}) acting perpendicularly to the crank and F_I, acting along the crank. *Right* Relative contribution to the total power output (in %) produced during each of the four functional sectors in three pedaling conditions of the force–velocity relationship (below, around and above the optimal pedaling rate: V_{opt}, respectively). Adapted from Dorel et al. (2010)

cycling exercise also reflect the muscular capability of flexor muscles involved during the pedal upstroke (i.e., knee and hip flexors).

Practical implications for interpreting and monitoring the force–velocity characteristics. In line with the previous paragraph, measurement of the contributions of both the flexion and extension phases appears really pertinent to be included in cycling force–velocity testing procedures. Moreover, pedaling is an alternative cyclic task characterized by an antiphase action of both lower limbs, meaning that when one is pushing on the pedal (i.e., downstroke), the other is recovering or pulling up on the second pedal (i.e., upstroke). Due to the mechanical coupling between both cranks at the bottom bracket, the net crank torque results from these simultaneous actions on the two pedals. Consequently, it should be kept in mind that it is necessary to measure the torque/force with dissociated sensors on each crank or pedal to be able to estimate the contribution of each lower limb and the possible imbalance between both (Fig. 3.2). The opportunity to measure the mechanical output for each lower limb separately allows not only (i) evaluation of the global contribution of each leg but also (ii) interpretation of potential asymmetry by isolating the part of the cycle where this difference occurs. In the typical example depicted in Fig. 3.2b, we can conclude that during this submaximal exercise the right leg produces more force than the left and that concerns both the downstroke and the upstroke phases.

From the practical point of view of training, it is interesting to describe this information during a classical force–velocity (F–V) test. Measuring the torque with sensors placed either in both pedals or both cranks the global F–V relationship can be expanded into two subrelationships which provide additional information on muscle capabilities. First, by simplifying the cycle to two main phases,

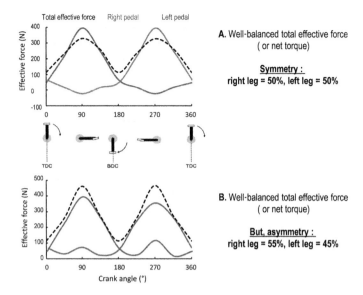

Fig. 3.2 Two theoretical examples of the effective force profiles produced on each pedal and the corresponding net (total) effective force (i.e., crank torque equivalent) during a submaximal cycling exercise. Note the similarity in the shape and peak values obtained during each stroke for total effective force profiles in both examples, which does not imply that a significant imbalance cannot appear between each lower limb in their relative contribution to the power in the two main phases of the cycle (downstroke and upstroke)

force–velocity relationships on the downstroke and upstroke can be determined to better identify the force and velocity capabilities of the extensor and the flexor muscles, respectively. Second, a global force–velocity relationship can be determined individually for each leg and ultimately expressed in its two components: the downstroke and upstroke phases. A typical report of F–V testing session on an elite sprint track cyclist describing these results is presented in Fig. 3.3. In addition to the global force–velocity indexes (Panel a), we can rapidly observe for this example: (i) a maximal force produced during the upstroke amounting to 21% of the total force (Panel b), (ii) global asymmetry with the left leg exhibiting higher force and velocity capacities resulting in higher power produced by this leg of almost 10% (Panel c) and finally (iii) the explanation that this asymmetry mainly results from the higher force, velocity and hence power produced by this leg in the upstroke (Panel d), meaning that it is primarily related to the properties of the flexor muscles such as the psoas and the hamstrings. As a whole, there is a greater likelihood that this approach would be very useful for monitoring performance in sprint cycling and orienting a strength training program.

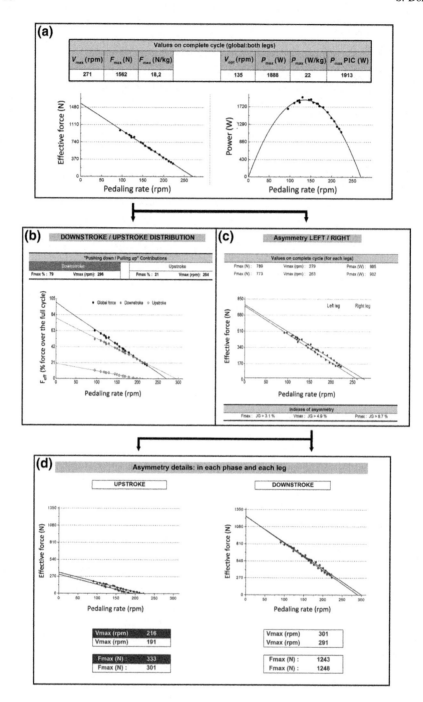

◄**Fig. 3.3** Typical detailed report of a force–velocity test performed on a cycle ergometer equipped with strain gauges in both cranks for an elite sprint cyclist. **a** Classical force–velocity and power–velocity relationships and the associated force, power and velocity indexes: V_{max}, maximal pedaling rate (rpm); V_{opt}, optimal pedaling rate (rpm); F_{max}, maximal force (N and N kg^{-1} of body mass); P_{max}, maximal power (W and W kg^{-1} of body mass); and P_{max} PIC, peak power corresponding to the mean of the three cycles with the highest values. **b** Relative contribution of the downstroke (0–180°) and the upstroke (180–360°) represented as two distinct global force–velocity relationships (ant their F_{max} and V_{max}) expressed in percentage of the value obtained on the full cycle. **c** force–velocity relationships and F_{max}, V_{max} and P_{max} measured on each leg and the consequence in terms of asymmetry. **d** Analysis of the contribution of each leg referring to the two functional phases (expressed as force–velocity relationships also) to better evaluate the potential imbalance between both legs in terms of pushing action in the downstroke and pulling action in the upstroke

3.2.2 Mechanical Effectiveness: The Orientation of Pedal Force

The concept of mechanical effectiveness in cycling. One characteristic of the biomechanics of force application during cycling is that the propulsive crank torque is proportional to the effective force, which represents only one component of the total force produced at the shoe–pedal interface. Throughout a pedaling cycle, not only the magnitude but also the orientation of the force vector on the pedal is altered. Thus, in order to account for this ability to effectively orient the force on the pedal (i.e., so that more of the resultant force participates in the propulsive action), the concept of mechanical effectiveness was introduced and an index of mechanical effectiveness (IE) proposed almost 30 years ago (Ericson and Nisell 1988; Sanderson 1991). As far as the sagittal plan is concerned, it is defined as the ratio (expressed from 0 to 1 or in %) between the magnitude of the effective force component (F_{eff}) produced perpendicularly to the crank and the magnitude of the resultant total force (F_{tot}): IE(%) = ($F_{eff} \cdot F_{tot}^{-1}$) × 100 (Fig. 3.4). Note that this index is sometimes called the index of mechanical efficiency, which is not really appropriate because of the risk of confusing muscular efficiency corresponding to the ratio between mechanical output and metabolic demand (energy consumption).

Note that different pedal dynamometers have been described in the literature as historically restricted to laboratory use (Davis and Hull 1981), but recently some specific prototypes (Drouet et al. 2008) or commercial devices (I-Crankset-2, Sensix™, Poitiers) allow both a research and training application on the field. Briefly, a series of strain gauges and optical encoders to record both the pedal angle to the crank and the crank to the vertical allow by means of trigonometry to calculate F_{tot} as well as its orientation and hence to determine F_{eff} and IE (Fig. 3.4). Hence, IE has been logically proposed as an indicator of cycling skill (see detailed review of Bini et al. 2013), which leads us to discuss its significance regarding force–velocity and power–velocity capabilities in the following sections.

Force–velocity relationship and mechanical effectiveness. Regarding explosive sprint cycling, information on the index of mechanical effectiveness has

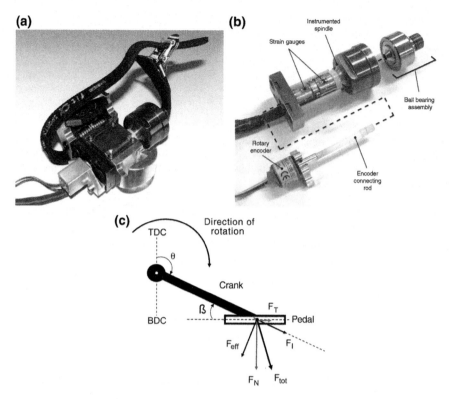

Fig. 3.4 a Picture of the instrumented force pedal equipped with the Look Keo platform and a retention safety strap (VélUS group, Sherbrooke University, Canada). b Exploded view of the pedal base assembly showing the different components measuring forces and angular displacement. c Representation of the tangential (F_T) and normal (F_N) components of the total force (F_{tot}) in the sagittal plane directly measured by the strain gauges. Associated with the measurement of pedal angle relative to the crank (β), both the effective force (F_{eff}) and the index of mechanical effectiveness (IE = ($F_{eff} \cdot F_{tot}^{-1}$) × 100) can be calculated and expressed as a function of the crank angle (θ measured by a second encoder). Adapted from Drouet et al. (2008); Hug et al. (2008)

been scarce until recently. However, during submaximal exercise, it was demonstrated that it increases with the level of power output (Bini et al. 2013; Patterson and Moreno 1990; Rossato et al. 2008) and decreases with increased pedaling rate (Neptune and Herzog 1999; Sanderson et al. 2000). Based on these findings, IE may be of particular interest under sprint conditions and valuable to interpret the force and power changes associated with the increase in pedaling rate during force–velocity tests. One key question is whether the force–velocity characteristics also reflect the ability of the muscle to effectively orient the total force on the pedal. In the study mentioned in the previous section (Dorel et al. 2010), which specifically investigated this in a population of active males, we observed three important insights regarding the role of IE (Fig. 3.5):

Fig. 3.5 Power output (**a**), total force (**b**, open circles), effective force (**b**, filled circles) and index of effectiveness (**c**) as a function of the pedaling rate during a force–velocity cycling test. A second-order polynomial was used for modeling the power–pedaling, total force–pedaling and index of effectiveness–pedaling rate relationships and linear regression for effective force–pedaling rate relationship. Gray lines (individual models, with the mean of individuals R^2) and black lines (mean trend curves) are shown for information and clarity purposes. Power, force and pedaling rate are normalized to their maximal value extracted from the model

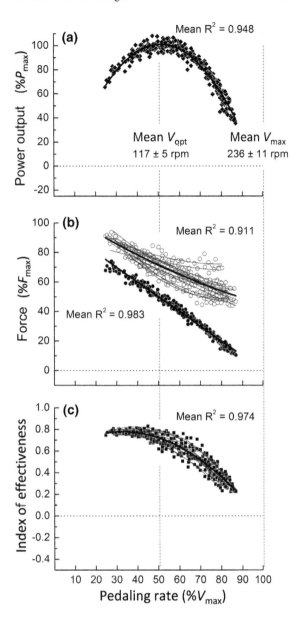

- IE values for the complete revolution (between 76 ± 4 and 73 ± 5% from 30 to 50% of V_{max}: 70–120 rpm) were very high compared with endurance cycling confirming the idea that maximization of power output is associated with optimization of the total force orientation at the pedals.
- The great reduction in IE to 35% observed at high velocities (around 190 rpm) highlights the difficulty of efficiently orienting the force on the pedals at these

cadences, especially during the upstroke, BDC and TDC phases. Hence, this decrease of IE could explain, at least partially, the descending portion of the power–velocity relationship for high pedaling rates.
- In line with these results, the IE–velocity relationship exhibited a high between-subject variability. Hence, IE appears to be a significant explanatory factor for differences in power production during the upstroke and around the TDC at low pedaling rates (i.e., below V_{opt}) and during all phases at higher cadences.

3.3 Joint-Specific Power and Interest in Inverse Dynamics

3.3.1 *Approach and Principle*

From a mechanical standpoint, the primary aim to be successful in sprint cycling is to produce a maximal torque at the crank level. However, it requires moving the lower-limb segments in such a way that the foot moves on the circular trajectory of the pedal. The work produced by the muscles (indirectly reflected by the EMG activity level) is then primarily used to move the segments (the thigh, the shank and the foot) and is ultimately transformed into mechanical work at the crank level (Driss and Vandewalle 2013; Hull and Jorge 1985; Kautz et al. 1994). Therefore, it is important to keep in mind that pedal force can be decomposed into a muscular component directly related to intersegmental net joint torques and a nonmuscular component due to gravitational and inertial effects (i.e., linked to change in mechanical energy of the segments; Fregly and Zajac 1996).

A classical approach to better estimate muscular power is to evaluate the net muscle torque or power at each joint using the well-known inverse dynamics approach. This technique entails three steps: (i) choosing a biomechanical model and the inertial parameters of the segments, (ii) recording the joint/segment kinematics and (iii) measuring the external forces applied on the lower limb of the cyclist (at the shoe–pedal interface). Since the pioneer studies (Gregor et al. 1985; Hull and Jorge 1985; van Ingen Schenau et al. 1990) important technological and methodological improvements in data recording and processing have been made, especially in the last decade, which allow assessing these parameters accurately, relatively quickly and easily. Modeling includes definition of the segments and the location of axes of rotation (joint centers). Classically, the model of the lower limb in cycling considers four rigid body segments (pelvis, thigh, shank and feet) linked by three joints (hip, knee and ankle). Because cycling movement is considered to mainly occur in the sagittal plane, it is reasonable to apply a 2D inverse dynamic procedure focusing on the efforts of joints around the transverse axis. Finally, the power produced at each joint is estimated by the dot product of net joint torque and joint angular velocity.

3.3.2 Information Regarding Force and Power Capabilities in Cycling

The main knowledge about joint-specific power distribution during maximal cycling is now well established and provides a better insight into muscle coordination (contribution of the main muscle groups), which is not apparent when only observing overall power/force at the pedal (Barratt et al. 2011; Elmer et al. 2011; Korff et al. 2009; Martin and Brown 2009; McDaniel et al. 2014). Classically, the power produced at the hip, knee and ankle is calculated. Ultimately, a last component corresponding to the effect of the power transferred across the hip joint by upper-body muscles can be estimated (by the dot product of the hip joint reaction force and linear velocity). Total muscular power is then evaluated by adding up all these power components, which allows observing the difference between the curves corresponding to pedal power and muscular power in order to appreciate the influence of nonmuscular variables (Fig. 3.6). A first practical message is that the power measured at the pedal necessarily represents an overestimation of the muscular power really produced during the downstroke and an underestimation of the muscular power produced during the upstroke. Indeed, the gravitational effect causes the weight of the limb to always act downward, providing additional power to the pedal during extension but consuming power during flexion. Moreover, depending on both the inertia of the segments and kinematics, this discrepancy may be also influenced by both the pedaling rate the anthropometric characteristics of each individual (Driss and Vandewalle 2013).

Furthermore, the main interest is to plot the profile of each joint-specific power in order the estimate the contribution of each action (hip, knee, ankle extension and flexion) to total muscular power produced and to put forward in which part of the cycle it occurs. Using this methodology on trained endurance cyclists, Martin and Brown (2009) and Elmer et al. (2011) demonstrated that maximal cycling power (at around 120 rpm) is produced mainly by the muscles involved in hip extension and to a lesser extent by those involved in knee extension, knee flexion and plantar flexion (Fig. 3.6). Two main items of information should be highlighted: (i) the main power producers are the hip extensors with more than 40% of the power produced over the entire cycle, and (ii) the total power produced during the flexion phase (i.e., upstroke) represents almost 21% of the total power, confirming this phase as a significant contributor (see Scct. 3.2.1). Interestingly, the knee flexion power produced during the upstroke is almost the same as the knee extension power produced during the extension ($\sim 22\%$). Note that a small portion of power is also produced due to power transfer at the hip joint (5–6%), suggesting that the upper body may play a small role even in the seating position.

Considering the power–velocity relationship in cycling, one main interesting question is to determine how the pedaling rate influences the contribution of the action of each joint in such a way that can explain the ascending and descending part of the power–velocity curve. In other words, we can wonder whether P_{max} and the corresponding optimal velocity (V_{opt}) may actually match the situation where all

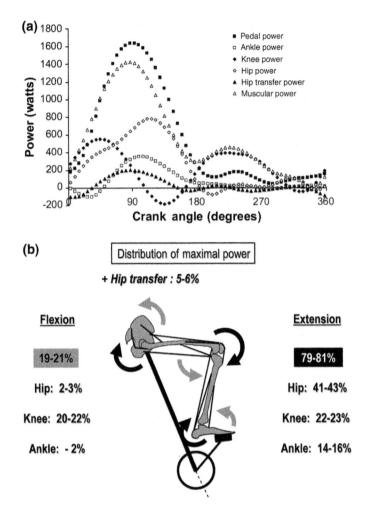

Fig. 3.6 a Mean profiles of pedal, muscular and each joint-specific power over the crank cycle during the first 3 s of isokinetic maximal cycling performed at 120 rpm. The sum of the power produced at all the joints plus the power transferred across the hip corresponds to total muscular power (adapted from Martin and Brown 2009). Note that pedal power is significantly higher than muscular power during the downstroke (0–180°) and inversely during the upstroke due to the change in mechanical energy of the lower limb (i.e., potential and kinetic energies). **b** Schematic representation of the relative contribution (in %) of each joint action (flexion and extension at the hip, knee and ankle) to total muscular power produced to generate maximal pedal power. Additional power is produced across the hip due to the involvement of upper-body musculature

the power at each joint is also at its maximal level. In fact, this question is very complex and further studies are needed to elucidate exactly the force–velocity condition of each muscle group during sprint cycling. However, it is highly unlikely that all the muscles would be in optimal condition to produce their maximal power when pedaling at V_{opt} (Driss and Vandewalle 2013; van Soest and Casius 2000).

A recent study in this area (McDaniel et al. 2014) supports this after evaluating joint-specific power at the ankle, knee and hip during a maximal force–velocity test in cycling (i.e., five isokinetic exercises performed over a range of pedaling rates from 60 to 180 rpm). The results clearly showed that the contribution of each joint action to total cycling power is dependent on pedaling rate and that ankle, knee and hip joint power–pedaling rate relationships are very different both in their amplitude and shape (Fig. 3.7c). The key findings summarized in the figure are as follows:

- as the ankle's range of motion decreases its angular velocity is almost constant with the increased pedaling rate from 90 to 180 rpm (Panel A), resulting in a decrease in plantar flexion power (Panels B and C) but a maintenance of high torque;

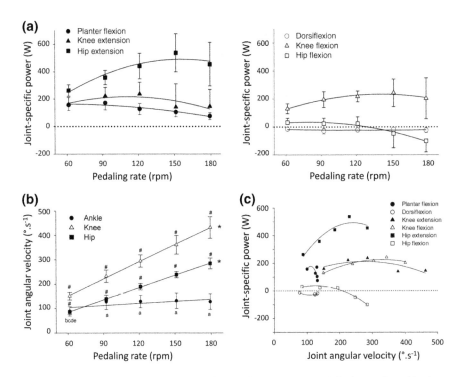

Fig. 3.7 a Mean power produced by each joint action (extension and flexion at the ankle, knee and hip) in relation to the pedaling rate during five isokinetic sprint cycling exercises (60–180 rpm). Black lines: mean trend curves (quadratic regression) are shown for information and clarity purposes. b Joint angular velocity under the five pedaling rate conditions. Knee and hip angular velocity increased with pedaling rate. (*Significant linear relationship. a, b, c, d and e Significantly different from 60, 90, 120, 150 and 180 rpm, respectively. #Significant difference from all other pedaling rates.) c Relation between the power produced by individual joint actions and joint angular velocities during the five sprints. Mean trend curves (quadratic model) are shown for information. Adapted from McDaniel et al. (2014)

- the main power producers, the hip extensors, continue to increase their power even above 120 rpm ($\sim V_{opt}$ of pedal power), reaching a maximal value at 150–180 rpm, and then significantly increases their relative contribution under these power–velocity conditions;
- the same tendency is observed for knee flexion power leading to a slight increased contribution at the higher pedaling rate whereas knee extension power decreases above 120 rpm which explains that its contribution to total power remains nearly constant.

The discrepancy in power–velocity relationships between the actions of joints is very useful to understanding how the pedaling rate induces different force–velocity and power–velocity constraints across the main muscle groups. It is particularly valuable when using sprint cycling as a muscular training or rehabilitation modality. Inversely, it is also interesting to adapt the modalities of the local strength training program (i.e., isolated single-joint exercises) when the goal is to improve cycling power in a specific part of the power–velocity relationship. Nevertheless, these joint-specific power characteristics necessarily also reflect the effect of pedaling rate on activation dynamics and muscle coordination, which are discussed in the next section.

3.4 Muscle Activity and Muscle Coordination

Muscle coordination can be defined as the distribution of activations or forces among individual muscles to produce a given combination of joint moments (Prilutsky 2000) or more generally to perform a given task (Kautz et al. 2000). While the inverse dynamics approach provides information about this coordination, it is limited as it does not give a direct measure of muscle force, which would represent the ultimate factor to understanding the muscle coordination. Indeed, the net torque estimated at a given joint can result from very different combinations of simultaneous extension and flexion torques generated by angonist and antagonist muscles. EMG activity is interesting here as it can be recorded on a specific muscle. EMG signal amplitude is mainly related to neural output from the spinal cord and, thus, to the number of activated motor units and their discharge rates. So, it is a means to have a good overview of the involvement of the main lower-limb muscles, especially how (level of activity) and when (timing of activity) they mostly contribute.

As a lower-extremity multijoint cyclic movement, cycling requires a specific muscle coordination (Hug and Dorel 2009). EMG analyses have demonstrated that most of the lower-limb muscle groups are involved during pedaling, each with their own function(s) regarding the six basic joint movements: hip extension and flexion, knee extension and flexion, and ankle plantar flexion and dorsiflexion (Fig. 3.8). The coordination strategy illustrated by the sequence of muscle activations partly underlies the joint-specific torque and pedal effective force profiles presented previously. As a consequence, recording EMG activity can provide interesting

3 Mechanical Effectiveness and Coordination: New Insights ... 47

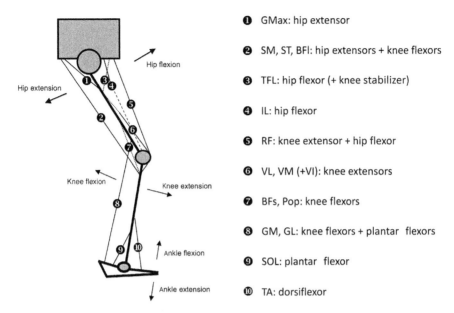

Fig. 3.8 Schematic representation of the main lower-limb muscles involved in pedaling and their functions. These include the single-joint monofunctional muscles such as the gluteus maximus (GMax), vastus medialis (VM) and vastus lateralis (VL), psoas iliacus (PI), biceps femoris short head (BFs), popliteus (Pop), soleus (SOL) and tibialis anterior (TA); as well as two-joint bifunctional muscles such as the rectus Rectus femoris (RF), semimembranosus (SM), semitendinosus (ST), biceps femoris long head (BFl), gastrocnemius medialis (GM) and lateralis (GL). Muscles that are possibly recorded with surface EMG are GMax, TFL, VM, VL, RF, SM, ST, BFl, GM, GL, SOL and TA

information about the involvement of the different muscles and, hence, interpretation of the macroscopic cycling force and power capacities.

3.4.1 The Specificity of Muscle Coordination in Sprint Cycling

Both EMG activity level and activation patterns of the different muscles over the pedaling cycle can be used to describe muscle coordination. While these data are well documented during submaximal cycling exercises (for a review see Hug and Dorel 2009), the literature regarding all-out sprint conditions is much poorer. Conceptually, contrary to endurance cycling during which muscles are involved at a submaximal level, during sprinting all the agonist muscles need to be very highly recruited in order to maximize total mechanical output. However, does this mean that all the muscles are maximally activated and produce a maximal level of force to produce a maximal pedal force? Studies performed in the last decade showed that it is not exactly the case and suggest that muscle coordination plays a key role in high power production and has some specificities in sprint cycling (Dorel et al. 2012; Wakeling et al. 2010; O'Bryan et al. 2014).

Recent studies nicely demonstrated that increases in power during cycling from very low to high levels are achieved that are not matched by similar increases in EMG amplitude, whereas they are associated with significant changes in timings of activation (Blake and Wakeling 2015). As regards sprint cycling in particular, our group investigated changes in the EMG amplitude and the timing of activity during all-out sprint pedaling by elite sprint cyclists (Dorel et al. 2012). To specifically answer the aforementioned questions, comparisons were made with both submaximal exercises and maximal voluntary contractions (MVCs) performed during single-joint tasks (Fig. 3.9). Results showed that muscle coordination during all-out

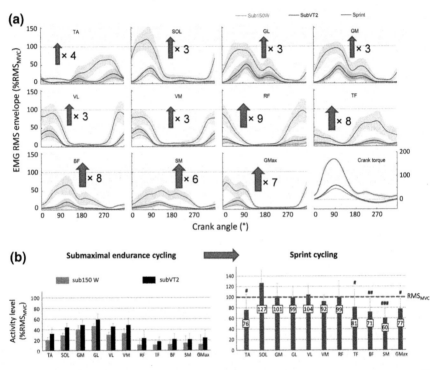

Fig. 3.9 a Mean (±SD) EMG patterns of the 11 lower-limb muscles and crank torque profile (bottom right panel) for the two lots of endurance cycling (performed at 150 W, Sub150 W in red, and at the secondary ventilator threshold, SubVT2 in black) and one lot of all-out sprint cycling (Sprint in blue). The patterns correspond to an average 30 consecutive pedaling cycles (for Sub150 W and SubVT2) and 6–7 (for Sprint) and are expressed as a function of the crank position (highest pedal position: TDC = 0°). b Activity level (mean peak value of EMG pattern) of each muscle under the three conditions, expressed as a percentage of maximal activity recorded during the MVC procedure (RMS_{MVC}). Only the statistics regarding a significant difference between values measured during Sprint and RMS_{MVC} for each muscle (on the right) are presented. ($^\#P < 0.05$, $^{\#\#}P < 0.01$, $^{\#\#\#}P < 0.001$.) See Fig. 3.8 for nomenclature of muscles. Adapted from Dorel et al. (2012) ($n = 15$, elite sprint cyclists: 5 women, 10 men) (Colour in online)

sprinting is characterized by a very large increase in EMG activity level for the hip flexors (a factor of 7–9 from 150 W submaximal condition to the sprint condition) and the knee flexors and hip extensors (a factor of 5–7), whereas plantar flexors and knee extensors demonstrate a lower increase (a factor of 2–3). Second, although all muscles greatly increase their activity during sprint cycling, the activity level failed to reach a maximal value for some muscles, especially for hamstrings and the gluteus maximus (i.e., 70–80% of peak EMG activity during MVCs), Finally, an increase in the duration of activity was reported for almost all of the muscle groups suggesting that muscle coordination represent an interesting coordination strategy during all-out sprinting to enhance the work generated by each muscle.

As a whole EMG analysis in sprint cycling gives four important insights or confirmations regarding the participation of the different muscle groups and the role of coordination and pedaling technique in maximal power capability:

- the maximal level of activity observed for the quadriceps and triceps surae suggests that the intrinsic force-generating capacities of these muscles remain likely one of the most relevant factors in performance during the extension phase;
- in addition, the greater increased activity of hip extensors and knee flexors under sprint conditions compared with the other muscles confirms the previous statement (Sects. 3.2 and 3.3), whereby maximal power in cycling also reflects the large contribution of both the most powerful hip extensors in the downstroke phase and the knee flexors in the upstroke phase;
- associated with the longer period of activation of all the muscles, these changes in coordination are overall a good explanation of the very high mechanical effectiveness reported during sprinting;
- however, non-maximal activity level of hamstrings coupled with the important interindividual variability observed while they were active (and of the tibialis anterior and gastrocnemii) are in good agreement with the interindividual variability also observed in the mechanical effectiveness during the same phases (i.e., BDC and upstroke), suggesting a link between these variables.

3.4.2 Coordination of Monoarticular and Biarticular Muscles

As aforementioned, the activation pattern of muscles during sprinting should maximize the pedal effective force, meaning that both the magnitude of the force vector and its orientation on the pedals should be optimized. It is important to understand that this is not simply achieved by increasing the activity level of the different agonist muscle groups. Some main points have to be underscored here to characterize muscle coordination in cycling and to account for the function of each muscle group. One important point is the fact that both monoarticular and biarticular muscles are involved, leading to some complexity. By definition biarticular muscles have two functions and generate a torque simultaneously on both joints they cross. If one function corresponds to an agonist powerful action at one joint (e.g., hip extension during the downstroke phase for BF and SM, in addition to GMax), the other function generally corresponds to an antagonist action producing negative power at the other joint (i.e., a knee flexion for BF and SM at the same time as knee extension occurs in this phase). A similar observation can be made for other biarticular muscles: the gastrocnemii and the rectus femoris.

Interestingly, using different simulation and modeling approaches, studies have demonstrated that these periods of coactivation of biarticular muscles (as antagonists of monoarticular muscle) are not detrimental to power production. Indeed, it is now widely accepted that they serve two main goals: (i) ensuring transfer of mechanical energy between joints and (ii) participating to effectively orient the force on the pedal (Fregly and Zajac 1996; Prilutsky 2000; Prilutsky and Gregory 2000; Raasch and Zajac 1999; van Ingen Schenau et al. 1992; Zajac 2002). As an illustration, for the first principle, the coactivation of calf muscles with the quadriceps during the downstroke phase participates in transferring a part of extension power at the knee to plantar flexion power at the ankle. For the directional constraint principle, the co-activation of hamstrings with the quadriceps in the second part of the downstroke allows limiting the forward component of the resultant force on the pedal (due to knee extension torque) and then participates in orienting it more downward and backward (after the 90° position of the pedal), representing a means of enhancing effective pedal force here (i.e., better effectiveness). Accordingly, it is often considered that monoarticular muscles have an invariant role as primary force/power producers while biarticular muscles act primarily to transfer energy between the hip, knee and ankle in critical phases of the cycle and to control the direction of force production at the shoe–pedal endpoint.

Other simulation approaches (Fregly and Zajac 1996; Kautz and Neptune 2002; Zajac 2002), estimating mechanical energy generation and transfer between the muscle, limb and crank, also highlighted the important role of the synergistic actions of agonist muscles crossing different joints, irrespective of whether they are monoarticular or biarticular muscles (i.e., coactivation of the gluteus and triceps surae during the downstroke phase). As the hip and knee are responsible for large energy transfer by accelerating the lower limb during this extension phase, the activation in synergy of the triceps surae allows this energy to be transferred from the leg to the crank. That is, much of the energy produced by the knee and hip extensors is delivered to the crank to produce cycling power. Some authors propose an interesting simplified model of muscle coordination during pedaling here, consisting of four functional muscle group controls or synergies (Hug et al. 2010; Raasch and Zajac 1999): (1) extensor group (hip, knee and ankle monoarticular extensors), (2) extension to flexion transition group (hamstrings, triceps surae), (3) flexor group (hip, knee monoarticular flexors) and (4) flexion to extension transition group (rectus femoris and tibialis anterior). Interestingly, these groups are almost chronologically activated in the same four functional sectors presented in Sect. 3.2.1 (sectors 2, 3, 4 and 1, respectively, Fig. 3.1).

3.4.3 Muscle Coordination and Torque–Velocity Relationship

Based on the maximal force–velocity model of isolated muscle, it can be theoretically assumed that the amplitude of neural excitation arriving at the muscle (represented by EMG activity) should always be maximal notwithstanding the force–velocity condition. As for the inverse dynamic, identifying potential changes in the level and the timing of activity of the main muscles at different points of the maximal torque–velocity relationship may provide insights into how their contribution may alter and/or potential adjustments of muscle coordination. One study has already pointed out a modification of muscle coordination of thigh muscles at high pedaling rates during sprint cycling (Samozino et al. 2007). Although they observed an earlier activation as a result of an increased pedaling rate which was suggested to compensate for the effect of electromechanical delay, they hypothesized that was not sufficient to maintain an appropriate effective force profile at pedaling rates beyond V_{opt}. Hence, this means that the descending part of the power–velocity relationship may be, at least partially, explained by a difficulty to maintain proper coordination.

To investigate this question more precisely, we analyzed both the level and timing of muscle activity for 11 muscles during all-out sprinting performed at different points of the maximal torque–velocity relationship (from 60 to 140% of V_{opt}; Dorel et al. 2011). The results indicated that the mean activity level decreased with increased pedaling rate for knee flexors, plantar flexors and/or hip extensors (SOL, GM, GL, BF, SM and GMax) while there was no change for knee extensors (VL, VM) or dorsiflexors (TA). There was an increase for hip flexors (RF and TFL). However, these findings were not accompanied by alteration of the peak activity level, suggesting that pedaling rate primarily influences the timing of activation of the muscles. Indeed, analysis of the EMG patterns showed the following specific adjustments (Fig. 3.10):

- some leg and thigh muscles (GL, GM, BF, SM, GMax) are recruited for longer period at lower pedaling rates than higher, reflecting a strategy to enhance mechanical output during the downstroke for the gluteus maximus, and active knee flexion torque around the bottom dead center and the upstroke for the hamstrings and the gastrocnemii;
- earlier activation of all the thigh muscles is found as the pedaling rate increases confirming previous findings (Neptune et al. 1997; Samozino et al. 2007) but no change is shown for the soleus and tibial anterior and slightly later activation is even observed for the gastrocnemii.

To further explore the limits to neuromuscular adaptations, we carried out an additional experiment to investigate whether coordination changes even more when reaching the fastest possible pedaling movement ($\sim V_{max}$). The same population of elite sprint cyclists performed an all-out sprint at their optimal pedaling rate (V_{opt}: ~ 120 to 130 rpm) and another close to their maximal pedaling rate ($\sim V_{max}$: ~ 210 to 240 rpm).

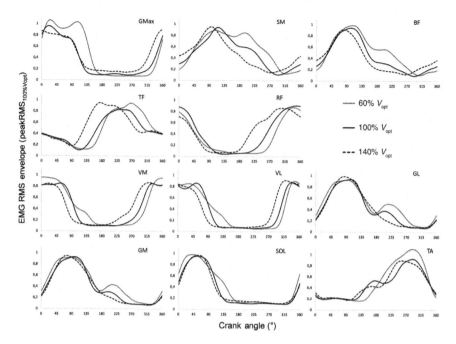

Fig. 3.10 Mean RMS EMG envelopes (averaged over four to six consecutive cycles) for 11 lower-limb muscles during cycling sprints performed at three extreme pedaling rates: low (60% $V_{opt} \sim$ 75 rpm), optimal (100% $V_{opt} \sim$ 130 rpm) and high (140% $V_{opt} \sim$ 180 rpm). For ease of comparison the pattern of each muscle is expressed as a percentage of the peak activity recorded under V_{opt} conditions (peakRMS$_{100\%Vopt}$). See Fig. 3.8 for nomenclature of muscles (n = 17, elite sprint cyclists: 6 women, 11 men)

Again, we demonstrated the ability to maintain almost the same peak activity levels under V_{max} conditions. Moreover, a large muscle excitation phase advance during V_{max} was observed for all the thigh muscles (−30 to −40° of the crank cycle) but once again not for the leg muscles. Overall, this confirms that earlier activation seems to be an interesting adjustment of coordination to address the growing effects of electromechanical delay at the very high pedaling rate. Nevertheless, opposite changes in the timing of leg muscles compared with thigh muscles can decrease their period of coactivation during the downstroke phase. Since the leg muscles transmit a part of the knee and hip extensors' power and energy of the segments to the crank thanks to these periods of coactivation (Neptune and Kautz 2001; O'Bryan et al. 2014; Raasch and Zajac 1999; Zajac 2002), this finding might be interpreted as a limiting factor for V_{max}. Interestingly, the extreme decrease of the index of effectiveness (IE = F_{eff}/F_{tot}) from 66.2 ± 10% at V_{opt} to 11 ± 4.8% at V_{max} partly reflects this difficulty at extreme velocities to maintain (i) proper muscle coordination and hence (ii) a substantial effective force in all parts of the crank cycle even though the important total force is still applied on the pedals (especially around BDC).

In fact, the cyclical nature of movement during pedaling imposes a crucial constraint on the control of muscle coordination. Indeed, the higher the pedaling rate, the lower the time for activating and deactivating the muscles. This activation–deactivation dynamics becomes an issue at extreme pedaling rates because these processes are not instantaneous. As the relative sector during which power can be produced is reduced, it is now well acknowledged that this frequency constraint is one of the principal factors in the power decrease for pedaling rates beyond V_{opt} (Martin 2007; Neptune and Kautz 2001; van Soest and Casius 2000).

3.5 Practical Implications and Perspectives for Testing and Performance

3.5.1 Pedaling Effectiveness, Muscle Coordination and Performance: What's the Link?

A key challenge is to express the influence the distribution of force throughout the pedaling cycle, the influence mechanical effectiveness, and the influence muscle coordination have on the global pedaling power capability. This is quite complex but some recent outcomes on the field suggest that coordination and technical aspects should account for a significant part of maximal cycling performance.

The complex case of submaximal cycling. During endurance cycling, although researchers and coaches commonly suggest that substantial differences exist between subjects regarding their power generation techniques, it appears that, for a given intensity–pedaling rate combination, the effective force (or torque) as well as IE profile as a function of the crank angle are relatively stereotypical (Sanderson et al. 2000; van Ingen Schenau et al. 1990) and with low interindividual variability even between novices and experts (Hug et al. 2008; Mornieux et al. 2008; Sanderson 1991). At the same time, due to the concept of redundancy (Bernstein 1967), EMG activity of the lower-limb muscles shows some between-subject variability, especially for the muscles involved in the critical phases of the cycle (TDC and upstroke phase) and those that have two functions (i.e., biarticular muscles; Hug et al. 2008). From the motor control point of view, muscle coordination is suggested to depend on different cost functions such as effort (e.g., metabolic demand, fatigue), neural activity and/or error variability (Prilutsky 2000). As a consequence, during submaximal cycling, minimization of these cost functions is very complex: (i) no single factor can be considered as the governing mechanism in muscle coordination (Kautz et al. 2000), (ii) a trade-off between these cost functions may exist and interactions with the biomechanical constraints of the task may occur (Brochner Nielsen et al. 2016) and (iii) a part of the interindividual variability observed could indicate that many good-enough solutions are available for the central nervous system to select (de Rugy et al. 2012). As a consequence, it remains very difficult though attractive to express the exact link between muscle coordination, torque/effectiveness profiles and *in fine* cycling performance during submaximal cycling. In other words, it is reasonable for an expert to advocate a specific but slightly different coordination strategy than another (expert or non-expert), associated with some subtle differences in the application of pedal force while cycling, but as yet there is practically no means of identifying clearly whether this coordination strategy (and associated pedaling technique) is better than another or whether it may be optimized. However, this does not mean that coaches and athletes should have no interest in analyzing pedaling technique and coordination during this type of exercise. What matters is to isolate some factors of optimization and to monitor whether a potential change in pedaling technique would be responsible for improving this factor (e.g., (i) a significant increase in muscular efficiency at a given level of power, (ii) lower or delayed fatigue at a given power allowing maintaining the task for a longer period; (iii) a reduction of a great bilateral imbalance to limit the risk of musculoskeletal injuries, etc.).

Valuable inputs for sprint cycling performance. During sprint cycling, as the effort is maximal most of the cost functions previously mentioned may be insignificant, particularly if the effort is short lived and performed in the absence of fatigue. Interestingly, this means the motor control "challenge" can be reduced to simply maximizing total pedal power. The question here is whether the coordination strategy and pedaling technique are related to improvement of the overall mechanical output, independently of the intrinsic mechanical properties of muscles. The first evidence of this is now emerging thanks to EMG and/or pedal force measurements in elite athletes and under field conditions.

One issue when investigating this question is that commercial portable power-meter systems (cited in the previous chapter) do not currently allow the measurement of total pedal force and, hence, neither the index of effectiveness nor the contribution of each leg in net mechanical output. In this context, our group in collaboration with the VélUS Group from the Laboratory of Mechanical Engineering at Sherbrooke University in Canada used specific instrumented pedals to quantify for the first time 2D pedal forces during sprint cycling performed under ecological conditions (on the track). Some preliminary results have been reported in the literature (Dorel et al. 2008, 2015; Drouet et al. 2008). The purpose of this work was to investigate the possible relationship between maximal force and mechanical effectiveness, and to analyze the potential link with chronometric performance. Fifteen world-class track sprint cyclists (11 men and 4 women) performed a 125 m all-out exercise on the track (without initial velocity- from a standing position). Mechanical data from instrumented pedals (i.e., forces applied on the pedals, and pedal and crank angular positions) allowed the effective force (F_{eff}) and the index of effectiveness (IE) to be calculated. A typical record is depicted in Fig. 3.11. During this exercise, we demonstrated that performance over the first 50 m was significantly related to effective force relative to body mass produced at the beginning of the effort. That is, there is a significant relationship between the 50-m time and maximal effective force produced during the first cycle (F_{max}) in $N \cdot kg^{-1}$;

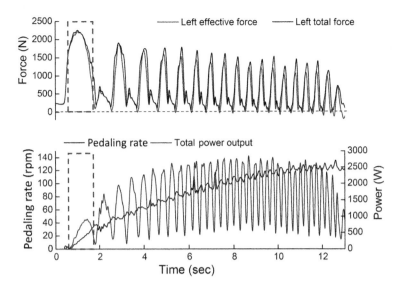

Fig. 3.11 Typical evolution of total and effective forces on the left pedal (top panel) and the pedaling rate and total power output (i.e., sum of both pedals, bottom panel) during a specific 125-m all-out effort from a standing start performed on the track by a world-class sprinter. The rectangle made up of dashed lines represents the first stroke from a starting crank position of almost 60–65° up to 180° (corresponding to the downstroke) for the left pedal and the concomitant upstroke of the right pedal (not presented here)

$P < 0.001$, $R^2 = 0.661$). Regarding the biomechanics of force application, two main results were observed. First, the positive contribution of the upstroke phase amounted to $18.9 \pm 2.1\%$ (range: 14.9–21.3) of the force produced over the entire cycle and was significantly related to F_{max} ($P < 0.001$, $R^2 = 0.686$) and to 50-m performance ($P < 0.001$, $R^2 = 0.782$). Second, the values of IE during this first cycle are extremely high ($87.0 \pm 1.6\%$; range: 83.9–89.2) and, despite a moderate interindividual variability, a significant relationship was obtained between IE and F_{max} ($P < 0.001$, $R^2 = 0.642$) and to a lower extent between IE and 50-m performance ($P < 0.05$, $R^2 = 0.348$). These findings again confirm the role played by the flexion phase in producing high force (Dorel et al. 2010; Martin and Brown 2009). They provided the first scientific evidence that IE, the ability to effectively orient the pedal force throughout the entire crank cycle (especially during the transition and upstroke phases) is an important parameter affecting maximal crank torque and, hence, could be considered an interesting factor in 50-m performance on the field, and in sprint cycling as a whole.

Although EMG activity was not recorded during this experiment, we can reasonably argue that a large part of the variability in pedaling effectiveness observed between the athletes may result from significant differences in muscle coordination. Indeed, even in the expert population tested on stationary ergometers in a previous study (Dorel et al. 2012), we already observed significant between-subject variability in both the level and the duration of EMG activity, especially for the biarticular muscles involved during the extension–flexion transition (hamstring, gastrocnemii) and also those acting during the upstroke. Moreover, we can assume that the potential difference in coordination between the athletes is certainly even more pronounced under specific track standing start conditions for which the ability to activate these muscles as high a rate as possible for as long as possible during the upstroke phase (i.e., to pull up the pedal more effectively) should be a key factor of performance.

At the same time, the difference in force-generating capacities necessarily influences the amount of force produced by these muscles and then may affect mechanical effectiveness and total pedal effective force. Nevertheless, the link between the amplitude of pedal effective force and (i) the cyclist's ability to orient the pedal force (represented by IE), (ii) muscle coordination and finally (iii) the intrinsic mechanical properties of all the activated muscles remains very complex and the relative contribution of each component to the overall performance is not fully established. In a previous study (Dorel et al. 2012), the additional MVC procedure performed on hip, knee and ankle joints in both flexion and extension came up with some preliminary insights. Indeed, we observed a significant relationship between crank torque (in $Nm \cdot kg^{-1}$) produced during a sprint performed at 80% of V_{opt} and the sum of the six maximal isometric torques (in $Nm \cdot kg^{-1}$) performed on a single-joint ergometer (Contrex^{TM}, $P < 0.01$, $R^2 = 0.458$). Interestingly, relationships were also obtained regarding more specific phases and muscle groups; for instance, between the torque produced during the upstroke phase and the sum of the three flexor maximal torques ($P < 0.01$, $R^2 = 0.504$). So, even in an elite population of sprint cyclists, force-generating capacity remains a key factor of pedal force and, hence, maximal power capability. On the other hand, the values of the coefficient of determination emphasize that almost 50% of variance in force during cycling is not related to intrinsic muscle properties. Although this link would be higher using a more specific procedure to test maximal muscle force (i.e., dynamic contraction with better specificity in terms of velocity and change in muscle length), this clearly supports muscle coordination as the other main candidate explaining much of the variance in maximal pedal force/power.

On this point, the analysis of individual results can already give us some interesting information and confirmation. For example, they demonstrated that the athlete with lower ability to generate pedal force in upstroke phase during sprint cycling ($\sim 12\%$ of total force over the full cycle) was (i) among the three athletes with the lowest torque-generating capacities of the hip, knee and ankle flexors, but also (ii) one who exhibited some "weakness" in the muscle coordination such as lower (for hamstring) or absence (for gastrocnemius) of activity at the beginning of the upstroke phase. As a whole, further studies are now necessary to determine more precisely the contribution not only of the mechanical properties of each muscle group but also to elucidate the general rules governing the optimization of muscle coordination during sprint cycling.

3.5.2 Outcomes Regarding the Meaning of Force–Velocity and Power–Velocity Relationships in Cycling and Perspectives for Testing and Training

Overall, the different sections of this chapter have described the main aspects of the neuromechanics of multijoint sprint cycling. They provide interesting new insights into interpretation of the macroscopic force–velocity and power–velocity models detailed in Chap. 2, which makes it possible to propose new practical ways of improving the different qualities (i.e., F_{max}, P_{max} and V_{max}).

Cycling F_{max} and P_{max}. The maximal power capability (P_{max}) is strongly correlated with the maximal theoretical effective force (F_{max}, see Sect. 2.3.2). First, in light of the above, this quality is largely dependent on the force-generating capacity of the lower-limb muscles and more precisely we can highlight the specific roles of:

- The main monoarticular muscles involved during the extension phase because they are the main power producers: the hip extensors which contribute heavily and the quadriceps which are maximally activated during the most important powerful extension phase. The goal of improving neuromuscular capacities and volume by specific strength-training programs remains of major importance for both.
- The plantar flexors, which are mostly activated during the extension phase, directly produce a nonnegligible part of total muscle power at the ankle ($\sim 15\%$) and, more importantly, also ensure energy transfer between joints. These muscles have to be very strong to be able to transfer the heavy work generated by the hip and knee extensors at the shoe–pedal interface and to avoid being the "weakest link of the chain."
- All the flexor muscles involved during the upstroke phase, the mechanical output of which can reach up to 25% of total output. Interesting opportunities to enhance sprint performance exist by optimizing the muscular capacity of these muscles (especially the knee flexors, hamstrings and to a lower extent the hip flexors and psoas iliacus by means of strength-training programs).

Second, it is clear that intermuscular coordination is a key factor in limiting global power output and explaining differences that are not related to the maximum power from any one muscle itself. On this point we can state that:

- All the monoarticular muscles involved have to be recruited at the highest level possible to enhance total muscular power and its transfer to the pedal.
- The skill to effectively orient the pedal force (expressed by IE) especially in critical parts of the cycle (BDC and TDC) can help enhance effective pedal force in sprint cycling.
- Pedaling effectiveness is associated with the ability to highly recruit the muscles responsible for the extension–flexion transition (gastrocnemii and hamstrings), flexion (hamstrings and psoas iliacus) and the flexion–extension transition (rectus femoris, psoas iliacus and tibialis anterior).

- Despite its inherent limitations, IE is a very useful index for sprints performed at low cadence and high level of force (from 0 rpm to V_{opt}) and, hence, will become a very interesting tool to monitor the progress in pedaling technique made by an athlete and the potential benefits on its force and power capabilities.
- The power generation technique and muscle coordination should be optimized using specific methods such as online or postsession visual feedback of force production or new training devices (e.g., independent cranks that constrain the cyclist to activate targeted muscles; Hug et al. 2013). There is no doubt that new technologies and portable systems to reliably measure the orientation of pedal force will emerge in the near future and may have the potential to assist athletes and coaches in training. Such devices would provide promising perspectives to improve mechanical effectiveness and, hence, performance, at least for sprint cycling.

Cycling V_{opt} or V_{max}. A high value of these indexes represent the capacity to maintain a high level of power even at high pedaling rates. They have been demonstrated to be related, at least partly, to fiber-type distribution (see Sect. 2.3.2) and, specifically, the percentage of fast-twitch muscle fibers in the quadriceps muscle (Hautier et al. 1996; McCartney et al. 1983). Based on the additional information described here we can propose further explanations to interpret these indexes and the shape of the power–velocity relationship:

- Although the peak activity level of the different muscles is not dramatically altered, EMG activity patterns are clearly modified for cadences higher than V_{opt} with different changes in timing of activation for thigh and leg muscles.
- While the pedaling rate is often considered a simple variable reflecting velocity capability, it is important to keep in mind the cyclic characteristic of the task. Consequently, the aforementioned adjustments in muscle coordination are largely governed by the excitation–relaxation kinetics of the muscles involved. This confounding factor (called activation dynamics) necessarily limits the capacity of the neuromuscular system to reach pedaling rates higher than 300 rpm in humans. Then, some of the challenge here may be reflected in the V_{max} or V_{opt} index.
- As pedaling rate increases, the non-muscular component due to variations in leg mechanical energy within one revolution also increases (especially, kinetic energy; Capmal and Vandewalle 2010; Driss and Vandewalle 2013). This necessarily results in a higher contribution of non-muscle forces to pedal effective force and an explanation of the alteration of its profile over the complete cycle (e.g., higher torque at the end of the downstroke and very low torque during the upstroke). One practical consequence is that IE at a high pedaling rate remains difficult to interpret and should be carried out with care regarding the optimization of coordination and, hence, performance.

- Finally, all these factors lead to the fact that the global force–velocity relationship in cycling cannot be considered simply as a perfect picture of force-velocity caracteristics of each muscle. Indeed, ankle joint velocity does not increase above V_{opt} and the associated muscles (especially, the biarticular gastrocnemii) can be fully recruited without increasing their shortening velocity (i.e., quasiisometric contraction). At the same time, the power produced by other muscles like the hip extensors continues to increase for pedaling rates ~ 150 to 180 rpm (i.e., much higher than V_{opt}), while others seem to be in better agreement with the decreased power observed in the global cycling power–velocity relationship (e.g., knee extensors).

As a whole, we can conclude that V_{opt} or V_{max} extracted from this relationship and the shape of the power–velocity curve result from the complex interaction between the (i) force–velocity properties of the different muscles, (ii) adjustments in the muscle coordination of these muscles to attempt to maintain the highest mechanical output despite the increase in pedaling rate (by optimizing the pedal force orientation), and finally (iii) growing influence of two fundamental issues with increased pedaling rate: activation dynamics and increased variation in segmental energy. Even though it is of great interest from the motor control point of view, the possibility to optimize these neuromuscular capabilities remains to be deeply explored before being able to investigate the potential effect of specific training (see Sect. 2.3.2 and the paragraph "What about the improvement possibilities by training?" in Chap. 2).

3.6 Conclusion

Besides assessment of the global force–velocity and power–velocity relationships in cycling (and associated F_{max}, V_{max} and P_{max}) presented in Chap. 2, here it was described whether these qualities can be examined in more detail by (i) measuring mechanical output at the pedal more precisely, (ii) estimating the power produced at each joint and (iii) determining the level and timing of activity of the different muscles involved (i.e., both reflecting intermuscular coordination). Depending on the accessibility of tools and level of understanding targeted, each approach can give very useful outcomes with practical implications for training and testing in explosive cycling performance. Provided that the dynamometer is capable of dissociating the force produced by each leg with sufficient temporal resolution, it is possible to analyze the pedal force profile throughout the cycle. If the dynamometer allows the orientation of force to be measured, it will add interesting information regarding the capacity of the athlete to effectively orient the total pedal force (i.e., the index of effectiveness). Despite their inherent limitations, these data greatly help in understanding the force-generating capacity of big muscle groups (extensors and flexors) as well as pedaling technique, especially for exercise performed at low and intermediate pedaling rates (below and around V_{opt}). Using inverse dynamics (including calculation of both muscular and nonmuscular components) and measuring EMG activity clearly take our interpretation to a new level by better characterizing the involvement of each muscle group. It is in practice possible to identify potential weaknesses in the contribution of specific muscle groups (extensor and flexors of the hip, knee and ankle) and to characterize muscle coordination and its key role as a limiting factor of maximal power. As a whole, the question of optimizing muscular capacities (i.e., intrinsic mechanical properties) as well as muscle coordination by training procedures still remains a challenge for the future. For now, some of useful data presented here remain not readily available for all coaches and athletes. However, given the speed of technological progress and market developments, it is important for sport scientists, coaches and strength trainers interested in cycling to take ownership of these questions now. Ultimate objective is to be ready and able to use the data provided by these devices in a near future with clear objectives in terms of performance improvement.

References

Barratt PR, Korff T, Elmer SJ, Martin JC (2011) The effect of crank length on joint-specific power during maximal cycling. Med Sci Sports Exerc 43:1689–1697. https://doi.org/10.1249/MSS. 0b013e3182125e96
Beelen A, Sargeant AJ, Wijkhuizen F (1994) Measurement of directional force and power during human submaximal and maximal isokinetic exercise. Eur J Appl Physiol Occup Physiol 68 (2):177–181
Bernstein N (1967) Coordination and regulation of movements. Pergamon, Oxford
Bini RR, Hume PA, Croft J, Kilding AE (2013) Pedal force effectiveness in cycling: a review of constraints and training effects. 2(1)
Blake OM, Wakeling JM (2015) Muscle coordination limits efficiency and power output of human limb movement under a wide range of mechanical demands. J Neurophysiol 114(6): 3283–3295. https://doi.org/10.1152/jn.00765.2015
Brochner Nielsen NP, Hug F, Guével A, Fohanno V, Lardy J, Dorel S (2016) Motor adaptations to unilateral quadriceps fatigue during a bilateral pedaling task. Scand J Med Sci Sports. https://doi.org/10.1111/sms.12811
Capmal S, Vandewalle H (2010) Interpretation of crank torque during an all-out cycling exercise at high pedal rate. Sports Eng 13(1):31–38. https://doi.org/10.1007/s12283-010-0051-2
Davis RR, Hull ML (1981) Measurement of pedal loading in bicycling: II. Analysis and results. J Biomech 14(12):857–872
de Rugy A, Loeb GE, Carroll TJ (2012) Muscle coordination is habitual rather than optimal. J Neurosci 32(21):7384–7391. https://doi.org/10.1523/jneurosci.5792-11.2012
Dorel S, Boucher M, Fohanno V, Rao G (2015) Neuromechanics of cycling: opportunities for optimizing performance. In: Conference proceedings archive (33 international conference of biomechanics in sports), pp 1368–1373
Dorel S, Couturier A, Lacour JR, Vandewalle H, Hautier C, Hug F (2010) Force-velocity relationship in cycling revisited: benefit of two-dimensional pedal forces analysis. Med Sci Sports Exerc 42(6):1174–1183. https://doi.org/10.1249/MSS.0b013e3181c91f35
Dorel S, Drouet JM, Hug F, Lepretre PM, Champoux Y (2008) New instrumented pedals to quantify 2D forces at the shoe-pedal interface in ecological conditions: preliminary study in elite track cyclists. Comput Methods Biomech Biomed Eng 11(S1):89–90. https://doi.org/10.1080/10255840802297275

Dorel S, Guilhem G, Couturier A, Hug F (2012) Adjustment of muscle coordination during an all-out sprint cycling task. Med Sci Sports Exerc 44(11):2154–2164. https://doi.org/10.1249/MSS.0b013e3182625423

Dorel S, Hug F, Couturier A, Guilhem G (2011) Are EMG patterns and activity levels modified at different points of the cycling torque-velocity relationship? In: XXIII congress of the international society of biomechanics, Brussels

Driss T, Vandewalle H (2013) The measurement of maximal (anaerobic) power output on a cycle ergometer: a critical review. Biomed Res Int 2013:589361. https://doi.org/10.1155/2013/589361

Drouet JM, Champoux Y, Dorel S (2008) Development of multi-platform instrumented force pedals for track cycling. In: Estivalet M, Brisson P (eds) The Engineering of Sport 7, vol 1. Springer, Paris, pp 263–271

Elmer SJ, Barratt PR, Korff T, Martin JC (2011) Joint-specific power production during submaximal and maximal cycling. Med Sci Sports Exerc 43(10):1940–1947. https://doi.org/10.1249/MSS.0b013e31821b00c5

Ericson MO, Nisell R (1988) Efficiency of pedal forces during ergometer cycling. Int J Sports Med 9(2):118–122. https://doi.org/10.1055/s-2007-1024991

Fregly BJ, Zajac FE (1996) A state-space analysis of mechanical energy generation, absorption, and transfer during pedaling. J Biomech 29(1):81–90

Gregor RJ, Cavanagh PR, LaFortune M (1985) Knee flexor moments during propulsion in cycling —a creative solution to Lombard's Paradox. J Biomech 18(5):307–316

Hautier CA, Linossier MT, Belli A, Lacour JR, Arsac LM (1996) Optimal velocity for maximal power production in non-isokinetic cycling is related to muscle fibre type composition. Eur J Appl Physiol Occup Physiol 74(1–2):114–118

Hug F, Boumier F, Dorel S (2013) Altered muscle coordination when pedaling with independent cranks. Front Physiol 4:232. https://doi.org/10.3389/fphys.2013.00232

Hug F, Dorel S (2009) Electromyographic analysis of pedaling: a review. J Electromyogr Kinesiol 19 (2):182–198. doi: https://doi.org/10.1016/j.jelekin.2007.10.010 [pii] S1050-6411(07)00181-2

Hug F, Drouet JM, Champoux Y, Couturier A, Dorel S (2008) Interindividual variability of electromyographic patterns and pedal force profiles in trained cyclists. Eur J Appl Physiol 104 (4):667–678. https://doi.org/10.1007/s00421-008-0810-y

Hug F, Turpin NA, Guével A, Dorel S (2010) Is interindividual variability of EMG patterns in trained cyclists related to different muscle synergies? J Appl Physiol 108(6):1727–1736. https://doi.org/10.1152/japplphysiol.01305.2009 01305.2009 [pii]

Hull ML, Jorge M (1985) A method for biomechanical analysis of bicycle pedalling. J Biomech 18 (9):631–644

Kautz SA, Hull ML, Neptune RR (1994) A comparison of muscular mechanical energy expenditure and internal work in cycling. J Biomech 27(12):1459–1467

Kautz SA, Neptune RR (2002) Biomechanical determinants of pedaling energetics: internal and external work are not independent. Exerc Sport Sci Rev 30(4):159–165

Kautz SA, Neptune RR, Zajac FE (2000) General coordination principles elucidated by forward dynamics: minimum fatigue does not explain muscle excitation in dynamic tasks. Motor Control 4(1):75–80; discussion 97-116

Korff T, Hunter EL, Martin JC (2009) Muscular and non-muscular contributions to maximum power cycling in children and adults: implications for developmental motor control. J Exp Biol 212(Pt 5):599–603

Korff T, Romer LM, Mayhew I, Martin JC (2007) Effect of pedaling technique on mechanical effectiveness and efficiency in cyclists. Med Sci Sports Exerc 39(6):991–995. https://doi.org/10.1249/mss.0b013e318043a235

LaFortune M, Cavanagh PR (1983) Effectiveness and efficiency during bicycle riding. In: Matsui H, Kobayashi K (eds) Biomechanics VIIB: international series on biomechanics 4B. Human Kinetics, Champaign, IL, pp 928–936

Martin JC (2007) Muscle power: the interaction of cycle frequency and shortening velocity. Exerc Sport Sci Rev 35(2):74–81. https://doi.org/10.1097/jes.0b013e31803eb0a0

Martin JC, Brown NA (2009) Joint-specific power production and fatigue during maximal cycling. J Biomech 42(4):474–479. doi: https://doi.org/10.1016/j.jbiomech.2008.11.015 [pii] S0021-9290(08)00601-5

McCartney N, Heigenhauser GJ, Jones NL (1983) Power output and fatigue of human muscle in maximal cycling exercise. J Appl Physiol Respir Environ Exerc Physiol 55(1 Pt 1):218–224

McDaniel J, Behjani NS, Elmer SJ, Brown NA, Martin JC (2014) Joint-specific power-pedaling rate relationships during maximal cycling. J Appl Biomech. https://doi.org/10.1123/jab.2013-0246

Mornieux G, Stapelfeldt B, Gollhofer A, Belli A (2008) Effects of pedal type and pull-up action during cycling. Int J Sports Med 29(10):817–822

Neptune RR, Herzog W (1999) The association between negative muscle work and pedaling rate. J Biomech 32(10):1021–1026

Neptune RR, Kautz SA (2001) Muscle activation and deactivation dynamics: the governing properties in fast cyclical human movement performance? Exerc Sport Sci Rev 29(2):76–80

Neptune RR, Kautz SA, Hull ML (1997) The effect of pedaling rate on coordination in cycling. J Biomech 30(10):1051–1058

O'Bryan SJ, Brown NA, Billaut F, Rouffet DM (2014) Changes in muscle coordination and power output during sprint cycling. Neurosci Lett 576:11–16. https://doi.org/10.1016/j.neulet.2014.05.023

Patterson RP, Moreno MI (1990) Bicycle pedalling forces as a function of pedalling rate and power output. Med Sci Sports Exerc 22(4):512–516

Prilutsky BI (2000) Coordination of two- and one-joint muscles: functional consequences and implications for motor control. Mot Control 4(1):1–44

Prilutsky BI, Gregory RJ (2000) Analysis of muscle coordination strategies in cycling. IEEE Trans Rehabil Eng 8(3):362–370

Raasch CC, Zajac FE (1999) Locomotor strategy for pedaling: muscle groups and biomechanical functions. J Neurophysiol 82(2):515–525

Rossato M, Bini RR, Carpes FP, Diefenthaeler F, Moro AR (2008) Cadence and workload effects on pedaling technique of well-trained cyclists. Int J Sports Med 29(9):746–752

Samozino P, Horvais N, Hintzy F (2007) Why does power output decrease at high pedaling rates during sprint cycling? Med Sci Sports Exerc 39(4):680–687. https://doi.org/10.1249/MSS.0b013e3180315246

Sanderson DJ (1991) The influence of cadence and power output on the biomechanics of force application during steady-rate cycling in competitive and recreational cyclists. J Sports Sci 9(2):191–203

Sanderson DJ, Hennig EM, Black AH (2000) The influence of cadence and power output on force application and in-shoe pressure distribution during cycling by competitive and recreational cyclists. J Sports Sci 18(3):173–181

van Ingen Schenau GJ, Boots PJ, de Groot G, Snackers RJ, van Woensel WW (1992) The constrained control of force and position in multi-joint movements. Neuroscience 46(1):197–207

van Ingen Schenau GJ, van Woensel WW, Boots PJ, Snackers RW, de Groot G (1990) Determination and interpretation of mechanical power in human movement: application to ergometer cycling. Eur J Appl Physiol Occup Physiol 61(1–2):11–19

van Soest O, Casius LJ (2000) Which factors determine the optimal pedaling rate in sprint cycling? Med Sci Sports Exerc 32(11):1927–1934

Wakeling JM, Blake OM, Chan HK (2010) Muscle coordination is key to the power output and mechanical efficiency of limb movements. J Exp Biol 213(3):487–492

Zajac FE (2002) Understanding muscle coordination of the human leg with dynamical simulations. J Biomech 35(8):1011–1018

Part II
Ballistic Movements of Upper and Lower Limbs

Chapter 4
A Simple Method for Measuring Lower Limb Force, Velocity and Power Capabilities During Jumping

Pierre Samozino

Abstract Lower limb ballistic movement performance (jump, sprint start, change of direction) is considered to be a key factor in many sport activities and depends directly on the mechanical capabilities of the neuromuscular system. This chapter presents the force-velocity (F-v) and power-velocity (P-v) relationships, and their associated variables of interest, as an interesting tool to evaluate the different lower limb muscle mechanical capabilities during ballistic push-off: maximal power and F-v profile. In this chapter, we will present the different laboratory and field methods to directly or indirectly assess these muscle properties, as well as their respective limits. We will also present an accurate and reliable simple field method to determine these muscle capabilities with a precision similar to that obtained with specific laboratory ergometers, while being convenient for field use because the computations only require loaded jumps (accurately standardized and performed) and three parameters rather easily measurable out of laboratory: body mass, jump height and push-off distance. The use of this simple method as routine test gives interesting information to coaches or physiotherapists: individual maximal power output and F-v profile. Validation studies, practical testing considerations and limits of this simple method will be presented. This method makes possible the follow-up of athlete muscle capabilities during a season or over several years, but also the comparison between athletes, which can help to optimize training and individualize loads and exercise modalities in strength training.

P. Samozino (✉)
Laboratoire Inter-universitaire de Biologie de la Motricité, Université de Savoie Mont Blanc, Campus Scientifique, 73000 Le Bourget du Lac, Chambéry, France
e-mail: pierre.samozino@univ-smb.fr

© Springer International Publishing AG 2018
J.-B. Morin and P. Samozino (eds.), *Biomechanics of Training and Testing*,
https://doi.org/10.1007/978-3-319-05633-3_4

4.1 Introduction

Accelerating its own body mass is involved in most of animal locomotor behaviors, notably in survival movements, whether to escape from predators or to capture preys. Natural selection acts within populations to further the phenotypic traits that improve such kind of abilities. Examining the factors that determine performance in these ballistic movements could allow to better understand the morphological and physiological adaptations characterizing a large variety of animals. In contrast, for humans (who no longer require ballistic movements to survive), ballistic performances are key factor in numerous sport activities (e.g., jumping, starting phase in sprint running, changing of direction, throwing). The understanding and determination of the physical characteristics underlying ballistic movement optimization is highly valued by performance professionals, such as strength and conditioning coaches. Beyond sport situations, some of movements occurred during daily-life activities can be considered as "ballistic" movements, such as standing up from a sitting position for the elderly or individuals with diseases (Corcos et al. 1996; Mak et al. 2003; Janssen et al. 2002). For these individuals, a sit-to-stand task represents a dynamic maximum effort for which reduced strength or inability to perform rapid muscle contractions can lead to impaired mobility and eventually to institutionalization (Mak et al. 2003; Corcos et al. 1996). So, one of the main questions interesting sport practitioners and physiotherapists is to know which muscle qualities to train or rehabilitate in order to increase (or to get back to a given level of) ballistic performances.

Ballistic performances can be defined as the ability to accelerate a mass as much as possible to reach the highest velocity in the shortest time, be it its one's own mass (e.g., sprints or jumps) or an external mass (e.g., throws or shots). Accelerating a mass can be performed during one unique movement (acyclic movements), often a ballistic push-off when it consists in a lower or upper limb extension (e.g. jumps, first steps of a sprint, throws), or during an acceleration phase over many consecutive movements or steps (cyclic movements, as sprint running acceleration, Chaps. 10–13). Ballistic movements, notably jumping or sprint running, have often been investigated to better understand the mechanical limits of skeletal muscle function in vivo, be it in animals (James et al. 2007; Lutz and Rome 1994) or in humans (Cormie et al. 2007d, 2011a; Jaric and Markovic 2009; Rabita et al. 2015; Samozino et al. 2012).

From Newton's law of motion, accelerating a mass require to apply external force and to produce mechanical work on its center of mass. Being very schematic, this requires "physical" capabilities, to produce the mechanical energy, and "technical" abilities to effectively transfer this mechanical energy from active muscles to center of mass of the body or of other moving system. These "technical" abilities have to respect the specific constraints and demands of the targeted sport activity via optimal motor behavior, segmental configuration or coordination. The present chapter, as well as the next ones, will only focus on "physical" capabilities to produce the maximum mechanical work possible during one or more (see Chap. 6) ballistic movement(s). This can give new insights for strength and conditioning training, while keeping in mind that producing high mechanical work is interesting only when associated to good technical skills specific to the task.

The ability to quickly accelerate the body from a resting position is considered to be particularly important for successful performance in many sport activities, be it in track and field events or in most of team, racket or combat sports. This ability in ballistic movements is directly related to mechanical properties of the neuromuscular system, and notably to power capabilities. Jumping movements have often been investigated to better understand and evaluate the lower limb mechanical properties. Vertical jump is the most widely used movement to assess "explosive" or "ballistic" qualities of lower limbs due to its simplicity, its very short duration (<0.5 s) and so very high intensity (it is considered as one of the most "explosive" movements) and its high correlation with peak power output (Davies and Young 1984).

Even if jumping tests will be briefly presented in this chapter, lower limb mechanical capabilities during dynamic extension cannot be only sum-up by maximal power output, and even less by a single performance. Since power is the product of force and velocity, muscle mechanical properties also include force and velocity capabilities which represent two different properties: the ability to develop very high levels of force and the ability to contract muscles—or extend limbs—at very high velocities. Even if both affect maximal power, these two muscle capabilities are independent. A good way to well understand this independency between force and velocity qualities is the analogy between muscle mechanical properties and vehicle engine capabilities. The latter are well described by the relationship between the torque the engine can produce and the velocity at which it turns. And when we compare this relationship between different vehicles (tractors and city or sport cars), we can observe big differences resulting in different properties (Fig. 4.1, Gülch 1994). The city or sport cars engine can go at high velocities, but cannot keep moving when they have to tow a big and heavy trailer. At the opposite, the tractor can move forward towing a trailer weighting several tons. And when the trailer is removed, it does not go very faster, while we can expect that it intensively accelerates once the heavy load is off, since it was able to tow very high loads. This is quite the same thing when the force generator is our neuromuscular system. The athlete's ability to move against very high loads is not associated to the capability to move fast when the mechanical constraints are low. Force and velocity capabilities are independent.

In this way, these different muscle capabilities cannot be distinguished from mechanical outputs obtained from only one single test (Jaric 2016b). The maximum vertical jump height is one of the best example since it has been indiscriminately interpreted as either an index of force (Kawamori et al. 2006), velocity (Yamauchi and Ishii 2007) or power (Cormie et al. 2011b) capabilities. During lower limb dynamic extension, the muscle mechanical capabilities are well and entirely described only by the force-velocity and power-velocity relationships (as presented and detailed for pedaling movement in Chap. 2).

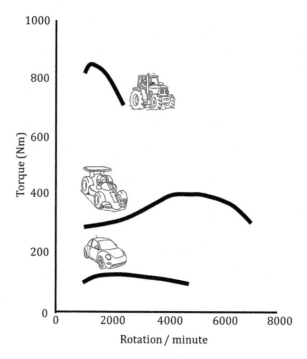

Fig. 4.1 Mechanical properties of the engine of three typical vehicles: a tractor, a sport car and a city car (from Gülch 1994).

In this chapter, we will define the force- and power-velocity relationships during ballistic lower limb push-off, the typical methods commonly used to determine these muscle capabilities, and a simple method we developed to compute them accurately out of labs. Practical applications will then be presented and discussed as well as different technologies that can be used to measure the mechanical inputs of this simple method.

4.2 Force, Velocity, Power Mechanical Profile

4.2.1 Force-Velocity and Power-Velocity Relationships in Jumping

When we focus on all-out or ballistic performances, i.e. movements performed with the highest intensity (force, velocity or power) possible, the performance is mainly limited by the mechanical capabilities of the neuromuscular system during concentric contractions. The overall lower limb dynamic mechanical capabilities have been well described by inverse force-velocity (F-v) and parabolic power-velocity (P-v) relationships (Fig. 4.2). While the F-v relationships obtained on isolated

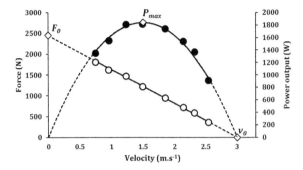

Fig. 4.2 Schematic force-velocity (open circles) and power-velocity (black filled circles) relationships for lower limb extensions during a typical leg press movement. Each point represent a lower limb extension performed against a given load. Black lines represent linear and polynomial regressions for force- and power-velocity relationships, respectively, dashed lines being their respective extrapolations. The mechanical outputs F_0, v_0, P_{max} are represented by the open diamonds, S_{fv} being the slope of the force-velocity relationship

muscles or mono-articular movements are described by a hyperbolic equation (Hill 1938; Thorstensson et al. 1976), linear relationships were consistently obtained for multi-joint functional tasks such as pedaling, squat, leg press or sprint running movements (e.g. Rahmani et al. 2001; Vandewalle et al. 1987a; Yamauchi and Ishii 2007; Bosco et al. 1995; Samozino et al. 2007; Bobbert 2012; Dorel et al. 2010; Morin et al. 2010; Rabita et al. 2015; Jaric 2015).

During ballistic push-offs (e.g. leg press, squat, Samozino et al. 2012; Yamauchi and Ishii 2007; Rahmani et al. 2001; Samozino et al. 2014), these relationships describe the changes in the maximal capability of lower limbs to produce external force and power output with increasing movement velocity over one extension. They characterize the external mechanical limits of the entire neuromuscular system. All the area under the F-v curve (Fig. 4.2) represents combinations of force and velocity outputs than can be produced during submaximal movements, and the F-v curve corresponds to outputs that can be produced during all-out push-offs. The other force and velocity combinations beyond the F-v line are not possible for muscles. These relationships put forward different typical parameters representing the different muscle mechanical capabilities (for details, see Morin and Samozino 2016). The force-axis intercept of the F-v relationships (F_0) represents the maximal external force lower limbs could produce during a theoretical extension movement at null velocity. The velocity-axis intercept (v_0) corresponds to the maximal velocity at which lower limbs could extend during a theoretical extension under zero load. It should be interpreted as the maximal extension velocity until which lower limb muscles can produce force. In other words, this is the athlete's ability to produce force at very high extension velocities. F_0 and v_0 are two purely theoretical values which cannot be measured experimentally, but they have to be considered as targeted values towards which maximal muscle capabilities tend when velocity decreases or increases, respectively. Hence, F_0 and v_0 have to be understood as the "force" and "velocity" maximal capabilities of the entire lower limbs. The apex of the P-v relationships (P_{max}) is the maximal power output lower limbs can produce over one extension and refers to the power capabilities. Under these conditions, the relationship among these three parameters can be described by the following mathematical equation: $P_{max} = F_0 \cdot v_0/4$ (details in Chap. 2, Vandewalle et al. 1987b; Samozino et al. 2012).

These mechanical properties obtained from multi-joint F-v and P-v relationships characterize the mechanical limits of the entire neuromuscular function and so are a complex integration of the different mechanisms involved in the total external force produced during one limb extension. They encompass individual muscle mechanical properties (e.g. intrinsic F-v and length-tension relationships, rate of force development), some morphological factors (e.g. cross sectional area, fascicle length, pennation angle, tendon properties), neural mechanisms (e.g. motor unit recruitment, firing frequency, motor unit synchronization, inter-muscular coordination) and segmental dynamics (Cormie et al. 2011a; Bobbert 2012; Cormie et al. 2010a, b).

The inverse F-v relationship is often misunderstood since we have in mind that if we increase the force applied to an object (or to the ground), we increase the velocity of the object (or of our center of mass). And the F-v relationship says the opposite: force and velocity change in opposite ways. In fact, there is no opposition between these two observations. They just do not refer to the same mechanical constraints. The first one (velocity increases when force applied increases or when resistance decreases) is the expression of the fundamental principles of dynamics well known through the Newton's laws of motion. They are the mechanical constraints imposed by Earth physical laws on human (or all other objects) movements. The second one (force decreases when velocity increases) corresponds to the mechanical properties of the neuromuscular system, and so to the mechanical constraints imposed by the biology on human movements performed with maximal effort (all-out). When physics says that velocity depends on force (2nd Newton's law of motion), physiology says that force depends on velocity (F-v relationship). During sport activities, physical laws are the same for everybody, muscle mechanical properties are not. And ballistic push-off performance is the best solution of both mechanical constraints (this will be detailed in Chap. 5).

4.2.2 Force-Velocity Mechanical Profile in Jumping

Beyond maximal force, velocity and power capabilities, the F-v relationships bring out another interesting information for scientists and coaches: the lower limb force-velocity mechanical profile. It is the slope of the F-v relationship (S_{FV}) and represents the individual ratio between force and velocity qualities ($S_{FV} = -F_0/v_0$). When the F-v relationship is graphically represented with a vertical force-axis, the steeper the slope, the more negative its value, the more "force-oriented" the F-V profile, and vice versa. The F-v profile is very interesting for the following reasons.

Independent from P_{max}. The athlete's F-v profile put forward insights about muscle mechanical qualities independently from power capabilities. Two athletes

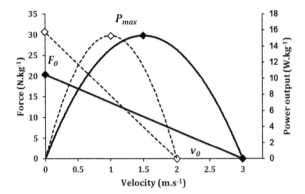

Fig. 4.3 Schematic force-velocity and power-velocity relationships for two hypothetical athletes presenting the same maximal power output (P_{MAX}) and two opposite force-velocity profiles: an athlete with a "force" profile characterized by a high maximal force (F_0) and low maximal velocity (v_0) (white diamonds and dashed lines), an athlete with a "velocity" profile characterized by a low F_0 and a high v_0 (black filled circles and continuous lines)

can present the same P_{max} values with different F-v profiles (Fig. 4.3). The athlete with a "force profile" (i.e. a F-v profile more oriented toward force capabilities) develops his P_{max} at a higher force and a lower velocity than an athlete with a "velocity profile", while both produce the same P_{max}. Since there is no direct relationship with P_{max}, the F-v profile brings another valuable information about the individual mechanical muscle properties.

Sensible to strength training. The F-v profile is sensible to training intervention, which is of great interest for strength and conditioning coaches. Changes in the F-v relationship, notably in its slope, can be achieved by specific strength training (Cormie et al. 2007c, 2010a, b, 2011b; Kaneko et al. 1983; McBride et al. 2002). The maximal force capabilities can be improved through strength training with heavy loads (>75–80% of one repetition maximum) while velocity capabilities can be increased by training with maximal efforts and light (e.g. <30% of one repetition maximum) or negative loading, which is often referred to as "ballistic" or "power" training (McBride et al. 2002; Cormie et al. 2010a, b; Cronin et al. 2001; Markovic et al. 2011; Argus et al. 2011). Note that when training focus on maximal force improvements, the velocity capabilities does not change, and vice versa (Cormie et al. 2010a). This underlined the independency between force and velocity capabilities, both ballistic and heavy strength training being associated to different physiological and neural adaptations (Cormie et al. 2010a, 2011a). This is quite the same history as for vehicle engines presented in the introduction section of this chapter.

Large inter-individual differences. Besides to be sensible to strength training, the other important point which confers great interest in F-v profile for training purposes is the fact that it presents very big differences between individuals. The initial works on F-v relationships obtained during pedaling exercises of Vandewalle and colleagues at the end of eighties well showed that force, velocity and power capabilities were very different across individuals of different ages, expertise levels or sport activities (Vandewalle et al. 1987a, b). During lower limb push-off, same large inter-individual differences can be observed between men and women or between different sport activities (Bosco et al. 1995; Giroux et al. 2016). Giroux et al. (2016) recently reported F- and P-v relationships of world class men and women athletes across different sport activities (Giroux et al. 2016). Between track cyclists, sprinters, taekwondo and fencing athletes, P_{max}, F_0 and v_0 values lots of differed even if all these sports can be considered as explosive activities (Fig. 4.4). These differences can be due to the specific mechanical demands of each activities and/or to their respective strength training habits. Even if this was not the aim of these study, data of Giroux et al. (2016) and Bosco et al. (1995) showed that differences in P_{max} between men and women were mainly explained by differences in velocity capabilities (Fig. 4.4). We can see that for some world class athletes in some sports (e.g. track cyclist, sprinters or taekwondo athletes), muscle mechanical capabilities differed only in v_0 between men and women, resulting in higher P_{max}, for men. Finally, within a same team sport activity, important differences can be observed between players according to their position, their training history and/or their intrinsic properties. And sometimes, F-v profiles do not follow what we can expect. The typical example presented in Fig. 4.5 shows the F-v relationships of two professional rugby players: a forward (body mass: 102 kg, stature: 1.76 m) and a back (body mass: 86 kg, stature: 1.84 m) player. We could expect (notably from his morphology) that the forward player would present a more "force-oriented"

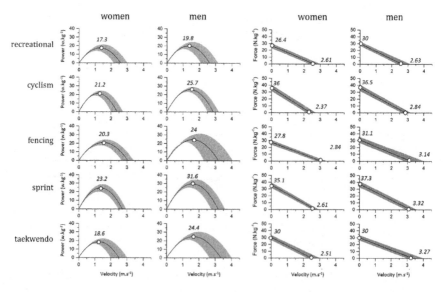

Fig. 4.4 Power- and Force-velocity relationships of men and women recreational individuals and world class sprinting, cycling, fencing and taekwondo athletes. Maximal power and maximal theoretical force and velocity values are presented (filled circles) (Modified from Giroux et al. 2016)

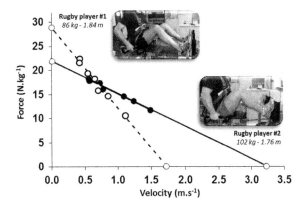

Fig. 4.5 Force-velocity relationships of two 1st-Italian league rugby players: a back (player #1, filled circles, dash line) and a forward player (player #2, black points and continuous line). Each point represents a lower limb extension performed against a given load during leg press exercises

profile (as a tractor) than the back player. This is not actually the case, mainly due to the very large and impressive velocity capabilities of the forward player. His absolute maximal force is also very high, but when expressed relatively to his body mass (which is important for ballistic performances), force capabilities are even lower than backward player's ones. During the leg press movement experimentation, the forward player's ability to fast extend his legs against low resistance was more impressive than his capacity to move against heavy loads.

The complementarity of these different variables to well characterize muscle mechanical capabilities (F_0 and v_0, or P_{max} and F-v profile), their large inter-individual differences, their sensibility to strength training and their high importance in ballistic performances (see Chap. 5 for more details) support the great interest to accurately evaluate them for sport scientists and sport practitioners.

4.3 Reference Testing Methods

4.3.1 Methodological Considerations

Muscle contraction modalities. The determination of lower limb F-v and P-v relationships require several all-out limb extensions performed in different mechanical conditions, each one corresponding to different force, velocity and power values. The different experimental methods require:

- either to control the force developed in order to maintain it constant all over the movement and to measure the movement velocity: isotonic condition. Tests are performed at different force values (e.g. Yamauchi and Ishii 2007; Yamauchi et al. 2005).
- either to control movement velocity to maintain it constant all over the movement and to measure the force produced: isokinetic condition. Tests are performed at different velocity values (e.g. Sargeant et al. 1981; Wilson et al. 1997).

- either to control and maintain constant the resistance acting against the movement and to measure both force produced and movement velocity: isoinertial or isoload condition. The resistance can be inertia (e.g. additional mass) and/or resistive force (e.g. weight of additional loads, elastic band or pneumatic force). Tests are performed at a given resistance, resulting in different force and velocity values which are both measured (e.g. Rahmani et al. 2004; Bosco et al. 1995; Samozino et al. 2014; Cuk et al. 2014; Markovic et al. 2011).

Isokinetic or isotonic testing conditions present many advantages, notably a best control of the mechanical modalities in which the task is realized and a higher safety for subjects. These two methodological concerns, very important in medical, scientific or sport fields, explain the widely use of isokinetic movements in muscle capacity testing, isotonic modality being less commonly used due to the technical difficulty to maintain constant the force produced. However, isokinetic conditions have often been challenged in sport field because some methodological requirements are not always respected (Gülch 1994), and especially because the movement is not "natural" due to constant velocities which are very rare, if not inexistent, in daily life and sport activities (Gülch 1994; Kannus 1994). Indeed, human movements are characterized by acceleration and deceleration of its whole body mass, the mass of a body segment or an external mass. Most of time in actual human performances, the mass of the moving system is constant all over the movement, but the force produced and the movement velocity change. Consequently, lower limb mechanical capabilities testing is well advised to be performed in isoload or isoinertial conditions, notably as part of exploring or training ballistic performances. That is why we will focus on isoinertial conditions in the following sections.

Peak *versus* averaged variables. Testing muscle properties, as for all kind of physical tests, requires to identify the variable(s) which better characterize(s) the muscular effort expended. When we explore the muscle maximal capabilities, the measured parameters are commonly the force produced, the movement velocity and the resulting power output. These three mechanical output can be quantified and presented in two different forms: instantaneous (often peak) values (e.g. Yamauchi et al. 2007; McCartney et al. 1983) or averaged values over one leg extension (e.g. Arsac et al. 1996; Rahmani et al. 2001; Bosco et al. 1995; Zamparo et al. 1997). The choice between instantaneous and averaged values depends on what we have to assess. In 1983, Andrews (Andrews 1983) suggested that:

- **Instantaneous values** are adapted to describe the value of a variable at a specific time of the movement (e.g. take-off during jumping or heel-ground contact during running) or to characterize extreme values of a parameter over a movement (e.g. extreme joint angles to compute range of motion, maximal running speed, minimum heart rate).
- **Averaged values** (or more generally values representing a time interval) are adapted to characterize an effort or a movement in its entirety, notably when the parameter significantly changes over the effort or the movement.

It is worth noting that the two types of values are strongly related during ballistic movements, with for instance averaged values of power output between 40 and 60% of maximal instantaneous values (Marsh 1994; Martin et al. 1997; Driss et al. 2001). So, the general shape of Force- and Power-velocity relationships are quasi the same, only value magnitudes change (Martin et al. 1997).

When we aim to evaluate muscle mechanical capabilities, we want to characterize the lower limb maximal capabilities to produce force or power over one extension. However, force production capabilities change all over lower limb extension: besides to depend on movement velocity, they are affected by the torque-angle (force-length) relationship of muscle groups involved at each joint (e.g. Thorstensson et al. 1976), by the time required for muscles to reach their maximum active state (e.g. van Soest and Casius 2000), or by muscle coordination patterns (e.g. Suzuki et al. 1982; Van Soest et al. 1994). So, only focusing on the instantaneous peak values measured during a functional movement does not make lots of sense since this values would correspond to a very specific anatomical and neuromuscular configuration and does not represent the whole dynamic lower limb capabilities. Consequently, even it is still source of debate (Dugan et al. 2004; Vandewalle et al. 1987b), we think that using force, velocity and power values averaged over the entire extension movement seems to be adapted to characterize these mechanical capabilities (e.g. Arsac et al. 1996; Bassey and Short 1990; Samozino et al. 2007, 2012; Rahmani et al. 2001). Moreover, from a purely mechanical point of view, dynamic principles show that the change in momentum of a system depend directly to the net mechanical impulse applied on it over the entire movement. So, ballistic performances do not depend on the maximum force or power output lower limb muscles are able to produce at a given (very short) time during their extension, but depend especially to force or power output muscles are able to produce over the entire extension phase allowing maximisation of the net mechanical impulse. And this is better described by averaged than by peak instantaneous values.

4.3.2 Laboratory Methods

As presented in Chaps. 2 and 3, lower limb mechanical capabilities were first described through F-v and P-v relationships during pedaling movement in eighties (Seck et al. 1995; Arsac et al. 1996; Martin et al. 1997; Hintzy et al. 1999; McCartney et al. 1983, 1985; Sargeant et al. 1981). Even if pedaling is very pertinent and convenient to assess lower limb extensor muscle capabilities, mechanical outputs produced during this cyclic movement depend directly to very specific muscle coordination (Samozino et al. 2007; van Soest and Casius 2000; Dorel et al. 2012, more details in Chap. 3) and so are quite different from what lower limbs can produce during one ballistic push-off as a jump or a sprint start.

The primary acyclic movement used to determine F-v and P-v relationships are squat or squat jump (when a take-off occurs) movements during which the body mass (and external mass) is moved up in the vertical direction. Even if the vertical jump is widely used to assess lower limb ballistic capabilities, the first studies (and the only one until the last few years) exploring F-v and P-v relationships during this kind of lower limb movements are the works of both Bosco's and Rahmani's teams (Bosco et al. 1995; Rahmani et al. 2001; Bosco and Komi 1979; Rahmani et al. 2004). They described, during squats and squat jumps, these two relationships as linear and parabolic, respectively. In contrast to pedaling, these relationships describe here the mechanical capabilities of both lower limbs acting together (bilateral movement) and require several distinct movements performed against several loads or external resistances: each load/resistance is associated to a movement velocity, a force and a power produced, and so correspond to one point of the F-v and P-v relationships (Fig. 4.2).

Different devices exist to measure the force and power output produced by lower limbs, as well as their extension velocity, during a vertical push-off phase. In the late XIXth century, Etienne-Jules Marey and George Demeny studied vertical jumps combining kinematic and dynamic analyses, notably using chronophotography and dynamograph, ancestors of optoelectronic and force plate systems, respectively (Marey and Demeny 1885). Since these first experimentations, force plates have remained the lab device the most used to measure forces applied onto the ground, but also to determine continuously the body center of mass acceleration, velocity and displacement during the push-off phase of a vertical jump (Fig. 4.6, Bosco and Komi 1979; Harman et al. 1991; Driss et al. 2001; Ferretti et al. 1987; Rahmani et al. 2001; Davies and Rennie 1968) or a contact phase during walking or running (Cavagna 1975). Note that the center of mass velocity during vertical jump has been often used to estimate lower limb extension velocity since feet do not move during push-off. The product of force and velocity data at each instant gives lower limb power output. Since a value of force, power and velocity can be obtained at each push-off phase, and so at each jump, F-v and P-v relationships can be determined from several jumps with different loads. In contrast to pedaling or sprint running movements, the force (vertical component) measured by force plate during vertical jumping corresponds to the quasi-total external force developed by lower limbs (horizontal force components are very low and negligible).

Other devices, less expensive than force plates and easier to use with strength training systems in weights room, were then developed to measure force, velocity and power during squat or jumps. In 1995, Bosco proposed to measure these parameters from only the displacement of the moving mass (body mass and additional mass) obtained from an optical sensor. This kinematic method, validated few years later by Rahmani in comparison to force plate measurements (Rahmani et al. 1998, 2000) consists in deriving twice over time the displacement signal to obtain center of mass velocity and acceleration, and then net external force developed. Based on this methodology, several systems have then been proposed using linear position transducers or accelerometers to measure directly the acceleration (Cormie et al. 2007a, b; Harris et al. 2007; Giroux et al. 2015; McMaster et al. 2014; Comstock et al. 2011; Cronin et al. 2004). Other lower limb extension movements have been also used to evaluate mechanical muscle capabilities during

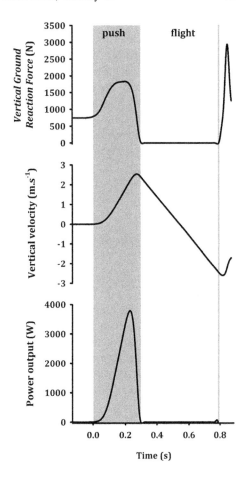

Fig. 4.6 Vertical force, velocity and power output signals obtained from force plate measurements during a squat jump (push-off and flight phases)

inclined or horizontal push-offs with different kind of ergometers measuring force, velocity and power output during push-off phase (Zamparo et al. 1997; Yamauchi et al. 2007; Avis et al. 1985; Pearson et al. 2004; Bassey and Short 1990; Macaluso and De Vito 2003; Zamparo et al. 2000; Samozino et al. 2012). The non-vertical orientation makes possible to change the gravity magnitude against which the movement is performed and to vary the body configuration (notably hip angle and range of motion (Padulo et al. 2017). Many of these devices have led to F-v and P-v relationship determination (e.g. Yamauchi and Ishii 2007; Samozino et al. 2012).

The different methodologies used in laboratory to determine force, velocity and power muscles capabilities present very high accuracy and high reliability. The standard error of measurement ranges from ~ 3 to $\sim 10\%$ and intraclass correlation coefficient from ~ 0.85 to ~ 0.99 according to the device, the protocol or the movement tested (Cuk et al. 2014; Giroux et al. 2015; Rahmani et al. 2001). Be it during vertical or horizontal movements, P_{max}, v_0 and F_0 values are very different according to subject levels and vary between 700 and 3500 W (from ~ 15 to

45 W kg^{-1}), between 1000 and 3000 N (from $\sim$20 to 50 N kg^{-1}) et between 2 and 8 m s^{-1}, respectively. However, the main limit of such lab methodologies is that they are not compatible with the daily constraints of the main users. Indeed, evaluating muscle mechanical capabilities is in the center of the training process of many sport activities. So this is of great interest for coaches and strength and conditioning coaches who look for simple, cheap and quick methods which can be easily set up in field conditions. This is not the case for such lab methods requiring very specific and expensive tools, as well as advanced skills in data acquisition and analysis. So, the above-mentioned laboratory methods on jumping movements have existed for more than 20 years, but very little used by sport practitioners. The later prefer performed tests convenient in field conditions, even if less accurate.

4.3.3 Field Methods

Before the few last years, no method have made possible the determination of P_{max} and F-v relationship out of laboratory and without skilled persons. However, maximal power, force and velocity capabilities have been assessed directly or indirectly by coaches using different simple methodologies.

Maximal strength. Out of laboratory, the maximum force the lower limb extensor muscles are able to produce is usually evaluated through the maximum load the athlete can move one time over an entire extension movement (one repetition maximum, 1RM). Different protocols exist to determine it on lower limbs during squat or a leg press exercises, all of them using increasing loads. Considering the inverse F-v relationship, the maximum muscle force can be produced at null velocity, which corresponds to isometric contraction. Even if the movement is realized at a non-negligible velocity, the one maximum repetition load is a good index of maximal dynamic strength. This test is convenient for field using and present a high reliability (McMaster et al. 2014; Seo et al. 2012; Verdijk et al. 2009). Moreover, the maximum isometric force depends directly on the joint (hip, knee and ankle for lower limb) angles at which the effort is performed and requires dynamometers to be measured. Very recently, the 1RM strength index was compared to F_0 by situating the 1RM point along the F-v relationship (Riviere et al. 2017). On the velocity axis, the 1RM point was shown to be situated at $\sim$30% from F_0 and $\sim$70% from the point corresponding to the SJ performed with the highest load. On the force axis, the 1RM point was $\sim$11% below F_0 and $\sim$16% above the highest force obtained during loaded SJs. This suggests that 1RM performance is affected partly (even slightly) by velocity qualities, and so does not represent only pure force capacities, even if it still represents a good practical index of dynamic maximal strength (Riviere et al. 2017).

Maximal extension velocity. When sport practitioners want to assess lower limb velocity capabilities of their athletes, they often measure a performance associated to high movement velocities, such a vertical jump, a sprint time or an agility test performance. However, beside to depend on muscular velocity qualities, each of these movements also require force production, and the velocity reached is far to be maximal. So, assessing the maximal extension velocity of lower limb cannot be performed measuring the performance of a movement depending on both force and velocity, and so on power output.

Maximal power output. Different field tests have been proposed to simply assess lower limb maximal power output, one of the most famous (but not commonly used) is the Margaria stair test (Margaria et al. 1966). Vertical jumps have remained the most widely used tests to assess power capabilities due to its simplicity, its very short duration and very high intensity. In 1921, Dudley Allen Sargent proposed the first field method to estimate the vertical displacement of the center of mass during a vertical jump by the difference between the maximal height reached by the hand during the jump and in standing position, both with the upper limb extended to the top. Some years later, in 1924, the jump height associated to this test was proposed as a measure of muscle power by a namesake, Sargent (1924). Other simple methods were then proposed to estimate the center of mass displacement during a jump though the roll-out of a ribbon attached to the athlete's waist (Abalakov's test) or the flight time of the jump (Bosco's test). Flight time measurement was made possible by the use of "timer" mat (e.g. Ergojump^{TM}, Bosco 1992) or photocells placed at some millimeters above the ground (e.g. Optojump$^®$), both easy to bring and use in field conditions. Jump height can be then computed using Newton's law of motion, as firstly proposed by Asmussen and Bonde-Petersen (1974). Several jumping modalities exist, each of them being associated to specific muscle mechanical properties: squat jump (starting from a crouching position), countermovement jump (starting from a standing upright position followed by a downward countermovement before jumping), drop jump (starting upright from an elevated place) and rebound jumps (hopping during a given time, see Chap. 6). Although maximal jump height was shown to be highly correlated to P_{max}, this does not give a power values, nor force or velocity ones. Consequently, the jump height as an index of power capabilities can be biased by the subject body mass and the lower limb range of motion. Two athletes of different mass and jumping at the same height do not develop the same power output, the heavier one being more powerful. In the same way, two athletes with the same body mass but different vertical push-off distances (lower limb range of motion over which the push-off is performed) and jumping at the same height do not produce the same power, the athlete using the shorter push-off distance being more powerful. So, jump height alone cannot be an accurate index of maximal power output.

For these reasons, different formulae have been proposed (still recently, see review McMaster et al. 2014) to estimate the power produced during a vertical jump from jump height and body mass (Table 4.1). Some of them were developed from fundamental principles of dynamics (Gray et al. 1962; Lewis formulae cited in Fox and Mathews 1974), but the biomechanical models from which they are based have been challenged (Harman et al. 1991). All the other formulae were statistically determined from regression equations obtained from experimental measurements (Table 4.1, Johnson and Bahamonde 1996; Canavan and Vescovi 2004; Harman et al. 1991; Sayers et al. 1999; Lara et al. 2006a, b; Shetty 2002; Bahamonde 2005;

Table 4.1 Equations previously proposed to indirectly estimate lower limb power output (from Lara et al 2006a)

Authors	Equations
Lewis	Power = $\sqrt{4.9}$ × 9.8 × body mass (kg) × $\sqrt{}$(jump height (m))
Harman et al. (1991)	Power = 61.9 × jump height (cm) + 36 × body mass (kg) − 1822
Bahamonde (2005)	Power = 78.5 × jump height (cm) + 60.6 × body mass (kg) − 15.3 × height (cm) − 1308
Sayers et al. (1999)	Power = 60.7 × jump height (cm) + 45.3 × body mass (kg) − 2055
Shetty (2002)	Power = 1925.7 × jump height (cm) + 14.7 × body mass (kg) − 66.3
Canavan and Vescovi (2004)	Power = 65.1 × jump height (cm) + 25.8 × body mass (kg) − 1413.1
Lara et al. (2006a)	Power = 62.5 × jump height (cm) + 50.3 × body mass (kg) − 2184.7

Wright et al. 2012). Besides the lack of theoretical background, the main limit of this kind of formulae is their dependence to the population from which they were obtained, which can lead to important error in the power estimation (from 3 to 40%, Lara et al. 2006a, b; McMaster et al. 2014).

In 1983, Carmelo Bosco proposed a simple test to measure lower limb power output during series of rebound jumps over 15–60 s (Bosco et al. 1983). The mathematical formulae based on Newton laws of motion give validated power values. However, this power is the mean power developed over series of jumps, and does not correspond to the maximal power output lower limbs can produce over one extension.

4.3.4 Limitations of the Reference Methods

On one side, laboratory methodologies present very accurate and reliable measurement of the different muscle mechanical qualities (force, velocity and power capabilities) but are not widely used by sport practitioners due to the need of expensive tools and very specific skills in data analysis. On the other side, simple tests used in routine in field conditions are not very accurate and valid and do not allow coaches to evaluate the entire spectrum of muscle capabilities, notably the athlete's F-v profile. It is worth noting that the different tests and formulae frequently proposed, notably to estimate maximal power output, well show the great interest, past or present (from 1962 to nowadays), to evaluate muscle mechanical capabilities with simple methodologies easily usable out of laboratories.

In order to answer to this need while considering the limitations of previous tests, we proposed a simple field method to compute accurately force, velocity and power lower limb capabilities from few data inputs that are easy to obtain in typical training practice.

4.4 A Simple Method for Measuring Force, Velocity and Power During Jumping

4.4.1 Theoretical Bases and Equations

The simple method is based on the fundamental principles of dynamics applied to the body center of mass during a vertical jump and on the analyses of its mechanical energy (kinetic and potential) at different specific instants of the movement (Fig. 4.7). This gives the following equations (for detailed computations, see Samozino et al. 2008) to compute the mean force (F) and power output (P) produced by lower limbs during one vertical jump and their mean extension velocity (v):

$$F = m.g.\left(\frac{h}{h_{PO}} + 1\right) \quad (4.1)$$

$$v = \sqrt{\frac{g.h}{2}} \quad (4.2)$$

$$P = m.g.\left(\frac{h}{h_{PO}} + 1\right) \cdot \sqrt{\frac{g.h}{2}} \quad (4.3)$$

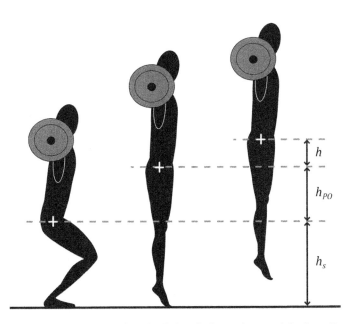

Fig. 4.7 The three key positions during a loaded vertical squat jump and the three distances used in the computations of the simple method

These 3 equations required only three input variables to estimate force, velocity and power during one jump: m the athlete body mass + additional mass (if any, in kg), h the jump height (in m) and h_{PO} the push-off distance which is the vertical distance over which the force is produced and thus corresponds to the lower limb length change between the starting position (h_s) and the moment of takeoff (Fig. 4.7). These three input variables are easy to obtain out of laboratory (see "technologies and inputs measurements" section).

Using these three equations for vertical jumps performed with different additional loads gives different points of the F-v relationship: the higher the moving mass (body mass + additional mass), the higher the force and the lower the velocity. Plotting F versus v and modeling this relationship by a linear equation gives the F-v relationship which can be extrapolated to obtain the maximal force (F_0) and velocity (v_0) values as the intercept with the force- and velocity-axis, respectively (Fig. 4.2). The maximal power output (P_{max}) can be computed as the apex of the 2nd order polynomial P-v relationship (if the number and range of the computed F-v points are sufficient), or easier by the following validated equation (Samozino et al. 2012; Vandewalle et al. 1987a):

$$P_{max} = \frac{F_0 \cdot v_0}{4} \qquad (4.4)$$

4.4.2 Limits of the Method

The theoretical computations on which were developed the above-mentioned equations are based on fundamental principles of dynamics. However, some mathematical and physical simplifying assumptions have been required to express F, v and P from only three simple parameters.

- The Newton's laws were applied to a whole body considered as a system and represented by its center of mass (e.g. Bosco and Komi 1979; Harman et al. 1990). So, only the mechanical energy applied to the center of mass was considered.
- The air resistance was neglected since it was shown to only affect very small animals jump height (body mass below 0.5 g, Scholz et al. 2006).
- It was supposed that the center of mass vertical displacement during push-off corresponds to the lower limb length change (i.e. h_{PO}), whereas the actual relative position of the center of mass within the body shifts slightly downwards due to the lower limb extension. This simplification was necessary to make h_{PO} measurement possible during field testing (see "technologies and inputs measurements" section).
- The mean power output over the push-off phase was computed as the product of the averaged values of force and velocity, which was mathematically wrong per se. Mean power is the average of instantaneous power values computed as the product of instantaneous force and velocity values. However, during an explosive movement during which the movement is only accelerated in one direction, the induced error is minor.

Knowing these different simplifying assumptions, it was important to quantify the errors they induced on mechanical variables, as well as the errors induced by the measurements of the three input variables with devices usable out of laboratories. So, the three equations and the simple method were validated in comparison to the gold standard force plate measurements.

4.4.3 Validation of the Method

The simple method has been the object of different validation protocols done by our research group, but also by others.

The first protocol aimed at validating the three equations to compute F, v and P during unloaded and loaded vertical jumps (Samozino et al. 2008). Eleven male physically active men performed two maximal squat jumps without and with 3 different additional loads (25, 50 and 75% of body weight). During each trial, F, v and P were calculated during push-off from both force plate measurements (Kistler type 9861A, Winterthur, Switzerland, 2000 Hz) and the proposed computations. For the latter, jump height was obtained from flight time and h_{PO} measured a priori as the difference between the great trochanter height in the crouch starting position (h_s, Fig. 4.7) and the extended lower limb length with maximal foot plantar flexion (great trochanter to tiptoe distance, see "technologies and inputs measurements"). The systematic bias between the two methods were not significant and lower than 2% for force, and lower than 8% for velocity and power whatever the load. The random errors ranged between 1 and 9% (Table 4.2). The concurrent validity of the computation method was also highlighted by high correlation and linear regressions close to the identity line (for full details and statistics, see Samozino et al. 2008). The low changes in the mean, low standard error of measurements and high intraclass coefficient between the two trials demonstrated the good reliability of the simple method whatever the additional load (Table 4.3). This reliability is in line with (and even better than) the reliability obtained with the reference method, which shows that the difference between both trials is due to the intra-individual variability in a jumping task (biological error, Hopkins et al. 2001). These results supported that the errors induced by the simplifying assumptions on which equations were developed are very low. The proposed method, solely based on three simple parameters easily obtained in field conditions (body mass, jump height and push-off distance), is valid and reliable to compute force, velocity and power developed by lower limb extensor muscles during loaded and unloaded squat jumps.

In 2014, an Italian research group tested the validity of the simple method to quantify power output during jumping comparing it to a method based on a multibody model that simulates the jumps processing the data obtained by a three-dimensional (3D) motion capture system and the dynamometric measurements obtained by the force platforms (Palmieri et al. 2015). The comparison gave

Table 4.2 Mean ± SD of force, velocity and power output obtained with force plate and simple methods, systematic bias and random error between the two methods and characteristics of correlations between the two methods

	Load (% body mass)	Force plate	Simple method	Systematic Bias (%)	Random error (%)	Pearson correlation coefficient (r)	
Force (N)	0	1294 ± 132	1282 ± 133	−0.88	1.94	0.98	***
	25	1451 ± 143	1433 ± 141	−1.20	1.70	0.99	***
	50	1554 ± 151	1557 ± 153	0.21	2.21	0.97	***
	75	1675 ± 175	1654 ± 170	1.20	1.43	0.99	***
	All conditions			−0.77	1.90	0.99	***
Velocity (m s^{-1})	0	1.08 ± 0.12	1.10 ± 0.12	1.62	3.15	0.96	***
	25	0.96 ± 0.11	0.95 ± 0.10	−1.30	4.63	0.92	***
	50	0.81 ± 0.11	0.80 ± 0.10	−0.18	6.17	0.87	***
	75	0.72 ± 0.07	0.66 ± 0.09	−7.93	8.84	0.72	***
	All conditions			0.05	4.89	0.96	***
Power output (W)	0	1412 ± 221	1411 ± 224	−0.14	3.04	0.98	***
	25	1389 ± 198	1357 ± 186	−2.13	4.49	0.95	***
	50	1253 ± 198	1246 ± 187	−0.30	6.64	0.88	***
	75	1182 ± 142	1085 ± 152	−8.07	8.41	0.76	***
	All conditions			−0.86	4.96	0.92	***

***$P < 0.001$

Table 4.3 Inter-trial reliability parameters for the simple method

	Load (% body mass)	Change in the mean (%)		Standard error of measurement (%)		Intraclass coefficient	
		Force plate	Simple method	Force plate	Simple method	Force plate	Simple method
Force (N)	0	0.32	0.64	2.52	2.56	0.94	0.94
	25	0.68	−0.30	1.47	1.57	0.98	0.98
	50	−1.37	0.30	2.24	1.46	0.95	0.98
	75	0.18	0.07	1.70	1.82	0.97	0.97
Velocity (m s^{-1})	0	2.22	2.32	6.23	3.84	0.70	0.89
	25	0.06	−0.60	5.70	3.27	0.78	0.91
	50	−3.79	0.21	7.93	3.60	0.66	0.93
	75	−0.15	−1.28	7.47	5.91	0.50	0.82
Power (W)	0	2.55	2.96	7.24	6.35	0.79	0.85
	25	0.79	−0.89	5.72	4.66	0.85	0.89
	50	−4.36	0.51	9.22	5.13	0.67	0.89
	75	−0.19	−1.21	8.55	7.89	0.52	0.70

errors lower than 6% on mean power output and mechanical work over the push-off, that *"can be accepted for the benefit of a very simple procedure which requires very few data as input"*. The same year, Caroline Giroux and the research team of the French National Institute of Sport determined the concurrent validity and reliability of force, velocity and power capabilities measurements provided by the simple method (jump height obtained from flight time using an optical measurement system, OptojumpNext®, Mircogate, Bolzano-Bozen, Italy), force plate (Kistler, Wintertur, Switzerland), accelerometer (Myotest Pro®, Myotest SA, Sion, Switzerland) and linear position transducer (GymAware®, Kinetic Performance, Mitchell, Australia) during loaded squats jumps (Giroux et al. 2015). Values obtained by the simple method showed very good agreement with force plate measurements, similar to accelerometer or linear position transducer (low bias between 2.6 and 3.2%, narrow confidence intervals ranging from 6.2 to 12.7%). All methods were shown to be reliable, especially the simple method (ICC = 0.97–0.99; SEM = 2.7–8.6%). The authors concluded from their results and the ease of use of the simple method that *"this method is suitable for monitoring power training sessions under field conditions"*.

Very recently, we validated the simple method computations and their validity to determine F-v relationships during countermovement jumps (CMJ) representing more commonly used movements in all sport activities (Jimenez-Reyes et al. 2016). The mechanical capabilities derived from F-v relationships (F_0, v_0, P_{max} and F-v profile) were obtained with both the simple method and force plate measurements (Bertec, Type 4060-15, Bertec Corporation, Columbus, OH, USA). Results showed high agreement between values obtained by both methods (mean absolute bias values between 0.9 and 3.7%) and high reliability of the simple method when applied during CMJ (ICC > 0.98). This supports the accuracy of the simple method, not only to compute F, v and P during one vertical jump, but also to determine F-v relationships from several loaded SJ or CMJ.

4.5 Technologies and Input Measurements

The accuracy and reliability of the simple method depends on both validity of equations (see previous section for details) and accuracy of the devices and technologies used to obtain the mechanical inputs of the models (m, h and h_{PO}), notably jump height and push-off distance.

4.5.1 Jump Height

Jump height (h) can be easily measured out of laboratories using different methodologies and devices (see Sect. 4.3.3). Most of them are based on the flight time and laws of motion to determine the center of mass elevation after take-off

(Bosco and Komi 1979). Even if the flight time method was very accurate and reliable, it requires field equipment detecting contact and no contact with the ground, and forces subjects to land in the same leg position as they take off. Such devices detect ground contact from optical measurement system just above the ground (e.g., OptojumpNext^{TM}, Glatthorn et al. 2011), contact pressure on the ground (e.g., Ergojump™), accelerometry (e.g. Myotest^{TM}, Casartelli et al. 2010) or video analysis (e.g. MyJump, Balsalobre-Fernandez et al. 2015). The main advantage of flight time methods is that the upper limb are free, which allows them to hold additional loads on shoulders. This is not the case for the well-known Sargent test (or similar) during which the athlete has to touch a wall (or other devices, e.g. Vertec^{TM}) the highest as possible with the hand. Video analysis can also be used to measure directly the vertical displacement of one point of the body which does not move relatively to the center of mass after take-off (e.g. a point at the hip or shoulders).

Whatever the methodology, the accuracy and the resolution of the jump height measurement directly affect the accuracy of force, velocity and power computations. For instance, for a device measuring flight time with a resolution of 30 data per second (as a standard video camera at 30 fps), the potential error on power values is between 7 and 19%. Increasing the resolution leads to lower errors: from 4 to 10% for 60 fps, from 2 to 6% for 120 fps, and from 1.3 to 3.5% for 180 fps. Recently, smartphones present cameras that allow for slow motion modes with up to 240 fps (e.g. Apple iPhone 6 and subsequent models). The flight time determination can be made with a resolution of $\sim$4 ms, which gives errors lower than 3% on jump height and power. This is the basis of the app named MyJump which allow users to measure accurately jumping performance in a few seconds by filming the feet on the ground during a jump and then by just clicking on the screen to select the frames corresponding to foot ground contact and take-off (Fig. 4.8). Besides to be validated to measure jump height (validation against force plate measurements, Balsalobre-Fernandez et al. 2015), this app also compute force, velocity and power variables to determine F-v relationship on the basis of the simple method presented here.

4.5.2 Push-off Distance

The distance covered by the center of mass during the push off (h_{PO}) can be easily estimated (with the slight overestimation discussed in the method limit section) in field conditions through the hip vertical elevation corresponding to the leg length change. The later can be determined before testing by the difference between the hip (great trochanter or superior iliac crest) height in the crouch starting position (Fig. 4.9) and the extended lower limb length with maximal foot plantar flexion (great trochanter/superior iliac crest to tiptoe distance, measured lying down on the back, Fig. 4.10) which is the configuration of lower limbs at take-off (Fig. 4.7). This is a very convenient and simple method requiring only a measuring tape. The

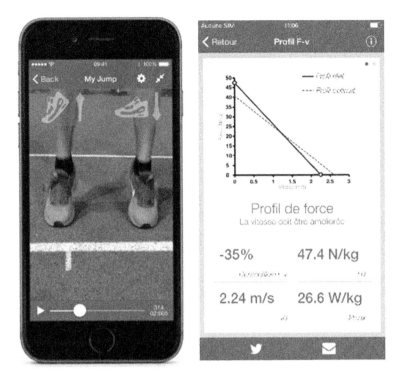

Fig. 4.8 The iPhone app "Myjump" uses the high-speed video slow motion mode to measure the aerial time of the jump, and thus compute jump height and all Force-velocity profile variables (some of whom will be presented in the next chapter)

crouch starting position can be set at a given knee angle for all athletes (e.g. 90°, as done during the validation protocol, Samozino et al. 2008) or letting athletes free to choose their preferred starting position in order to maximize jump height. Then, the starting position as to be standardized for all jumps to ensure that h_{PO} corresponds to the values determined a priori (Fig. 4.9). Note that the h_{PO} was measured in this way in the first validation protocol of the simple method, and results showed very low bias. Other techniques may be found to increase the precision of this parameter and/or the easiness of use, as what we proposed for bench press in Chap. 7 which could be adapted for jumping.

4.6 Practical Applications

The interest of such a simple method to determine F-v relationship is to be used directly in typical training practice as a relatively simple routine test of the force, velocity and power generating capacities of lower limb muscles. After a warm-up, a

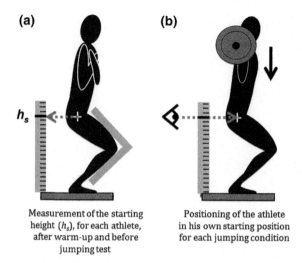

Measurement of the starting height (h_s), for each athlete, after warm-up and before jumping test

Positioning of the athlete in his own starting position for each jumping condition

Fig. 4.9 Measurement of the starting height (h_s) after warm-up and before jumping test for each athlete (**a**) and using of this value to position each athlete in his own starting position before each jumping trial (**b**)

good processing of the simple method makes possible to test 3 or 4 athletes in 30–40 min. Here we will detail the different practical points of a typical testing session using the simple method that contributes to the accuracy and the relevance of the obtained data.

Warm-up. After 5–10 min of a typical general warm-up (e.g. running, cycling or rowing), a specific warm-up has to include vertical jumps with a progressive increase in the intensity and loaded jumps. Besides to complete warm-up, the latter also aim at avoiding any apprehension of this kind of exercise and taking advantage of a potential potentiation effect from the beginning of the tests. Note that if athletes are not accustomed to loaded SJ and CMJ and/or to loaded squat exercises, an accustomed session should be done some days before. The specific warm-up should be used to check that athletes (i) perform vertical jumps without countermovement if the tests focus on SJ and (ii) contact the ground at landing with an ankle plantar flexion if jump height is measured from flight time.

Push-off distance measurement. Before, during or after warm-up, a mark has to be put on great trochanter or superior iliac crest. In a practical point of view, a mark on the superior iliac crest is easier to see during the testing session to standardize the starting position. The extended lower limb length with maximal foot plantar flexion can be measured lying down on the back (Fig. 4.10). At the end of the warm-up, the vertical height of the mark is measured when athlete is in the crouch starting position for SJ or the crouch position corresponding to the beginning of the ascendant phase for CMJ (Fig. 4.9). As previously mentioned, this position can be fixed or let free to the athletes (recommended). In the latter case, the choice of the

Fig. 4.10 Measurement of the lower limb length with maximal foot plantar flexion, which corresponds to the height of the hip at the instant of take-off during a vertical jump

starting position can be done through different trials with and without loads. Note that this choice is more relevant when performed after the warm-up than before.

Number and choice of loads. To determine a F-v relationship, several jumps with different loads are required. A minimum of two different loads is necessary (the two-load model recently proposed by Jaric 2016a), the maximum of loads depending on fatigue occurrence. The lower the number of loads, the higher the sensibility of the mechanical outputs to the potential measurement errors. So, we recommended, notably during field testing, to test 5 or 6 different loads including the condition without load. The loads can be determined at given values relative to body mass (Samozino et al. 2008), at given values relative to the one maximum repetition in squat (Giroux et al. 2015) or at absolute values (Samozino et al. 2014). Note that if several athletes are tested together in the same session, absolute values of loads are easier to set up. The exact values of the load is not so important, the idea is that the several loads cover the highest range of values as possible, the lowest load being without additional load and the highest load being the highest load with which the athlete can jump. For security and quality of movement execution, we advise to not test loads with which the athlete cannot jump higher than ~ 8 to 10 cm. In practice, the highest load is from $\sim 75\%$ of body mass for athlete non-accustomed to heavy strength training to $\sim 100\%$ of body mass for others. The number and choice of loads can be changed during testing considering the actual jumping performance realized, for instance by adding a supplementary loading condition if the athlete jump largely higher than 10 cm with the highest load or slightly decreasing the highest load if an athlete does not feel to jump with it. The different loads can be randomized.

Jumping tests. Each athlete has to perform at least two trials at each load, only the best performance being considered for analysis. If performances are too much different between the two trials, athlete should do another one to confirm or infirm the best jump previously performed. An overestimation of the performance can be due to a non-respect of the instructions: incorrect countermovement, feet too much dorsiflexed at landing (if using flight time) or important trunk extension during push-off. All these criteria have to be carefully checked at each jump to be validated. The different trials at each load can be performed successively with 20–30 s of rest between them. The different loading conditions can be separated by a 5-min rest, period during which other athletes can be tested. For each trial, athlete set up for the squat jump in a standing position while holding a barbell across their shoulders for additional loads conditions or with arms crossed on torso for the no-load jump. For SJ, athlete are asked to bend their legs and reach the a priori defined starting height carefully checked thanks to a ruler and the mark on athlete hip (Fig. 4.9). After having maintained this crouch position for about 2 s, they were asked to apply force as fast as possible and to jump for maximum height. For CMJ, athletes initiate a downward movement until the a priori defined squatting position followed immediately by a jump to maximum height. To control if the degree of crouching squatting achieved is good, an elastic band can be extended just under the athlete buttock at the height corresponding to the a priori defined position. If the athlete is too far from the elastic or if he/she touches too much it, the trial cannot be validated.

Live feedback and adjustment quality of the F-v relationship. The accuracy and reliability of the mechanical outputs (F_0, v_0, P_{max} and F-v profile) depend directly on the adjustment quality of the F-v linear regression. If the different points corresponding to the different loading conditions (best trials) are not well aligned as expected, the obtained outputs make less sense. A good objective index of the adjustment quality of the F-v relationship is the determination coefficient (r^2) which should be higher than ~ 0.95 when 5 or 6 loads (i.e. points on the F-v curve) are used. In practice, it is interesting to have a feedback of the F-v relationship and its adjustment quality in live during testing in order to ask athlete to do again a condition in case of doubts (a point strangely above or below the linear F-v curve) (Fig. 4.2). A simple spreadsheet including the different computations and F-v relationship diagram allows athletes or coaches, after entering the individual inputs data (h_{PO}, and h and m for each condition), to have this kind of feedback and to know the individual muscle capabilities directly at the end of the test.[1] The MyJump App makes also possible this kind of direct feedback.

4.7 Conclusion

This chapter present the F-v relationship as an interesting tool to evaluate the different lower limb muscle mechanical capabilities during ballistic push-off: maximal power and F-v profile. This chapter also present an accurate and reliable simple field method to determine these muscle capabilities with a precision similar to that obtained with specific laboratory ergometers, while being convenient for field use because the computations only require loaded jumps (accurately standardized and performed) and three parameters rather easily measurable out of laboratory: body mass, jump height and push-off distance.

The use of this simple method as routine test gives interesting information to coaches or physiotherapists: individual maximal power output and F-v profile. This

---

[1] A typical spreadsheet and a tutorial to use it (home-made by Morin and Samozino) can be downloaded/viewed here: https://www.researchgate.net/publication/320146284_JUMP_FVP_profile_spreadsheet

makes possible the follow-up of athlete muscle capabilities during a season or over several years, but also the comparison between athletes, which can help to optimize training and individualize loads and exercise modalities in strength training.

As for all kind of tests, determining the strengths and the weaknesses of an athlete based on the comparison with other athletes or with indexes thresholds computed from other athlete data is limited, yet interesting if no other way to interpret them. In order to optimize and individualize strength training, the idea would be rather to compare each athlete data to the data he/she should present to reach the best ballistic performance as possible. This will be addressed in the next chapter.

References

Andrews GC (1983) Biomechanical measures of muscular effort. Med Sci Sports Exerc 15(3): 199–207

Argus CK, Gill ND, Keogh JW, Blazevich AJ, Hopkins WG (2011) Kinetic and training comparisons between assisted, resisted, and free countermovement jumps. J Strength Conditioning Res 25(8):2219–2227. https://doi.org/10.1519/JSC.0b013e3181f6b0f4

Arsac LM, Belli A, Lacour JR (1996) Muscle function during brief maximal exercise: accurate measurements on a friction-loaded cycle ergometer. Eur J Appl Physiol 74(1–2):100–106

Asmussen E, Bonde-Petersen F (1974) Storage of elastic energy in skeletal muscles in man. Acta Physiol Scand 91(3):385–392

Avis FJ, Hoving A, Toussaint HM (1985) A dynamometer for the measurement of force, velocity, work and power during an explosive leg extension. Eur J Appl Physiol Occup Physiol 54(2): 210–215

Bahamonde RE (2005) Power prediction equations. Med Sci Sports Exerc 37(3):521; discussion 521-522

Balsalobre-Fernandez C, Glaister M, Lockey RA (2015) The validity and reliability of an iPhone app for measuring vertical jump performance. J Sports Sci 33(15):1574–1579. https://doi.org/10.1080/02640414.2014.996184

Bassey EJ, Short AH (1990) A new method for measuring power output in a single leg extension: feasibility, reliability and validity. Eur J Appl Physiol Occup Physiol 60(5):385–390

Bobbert MF (2012) Why is the force-velocity relationship in leg press tasks quasi-linear rather than hyperbolic? J Appl Physiol 112(12):1975–1983. https://doi.org/10.1152/japplphysiol.00787.2011

Bosco C (1992) La valutazione della forza con il test di Bosco [Strength assessment with the Bosco's Test]. In: Societa Sampa Sportiva, Roma, pp 56–57

Bosco C, Belli A, Astrua M, Tihanyi J, Pozzo R, Kellis S, Tsarpela O, Foti C, Manno R, Tranquilli C (1995) A dynamometer for evaluation of dynamic muscle work. Eur J Appl Physiol Occup Physiol 70(5):379–386

Bosco C, Komi PV (1979) Potentiation of the mechanical behavior of the human skeletal muscle through prestretching. Acta Physiol Scand 106(4):467–472

Bosco C, Luhtanen P, Komi PV (1983) A simple method for measurement of mechanical power in jumping. Eur J Appl Physiol Occup Physiol 50(2):273–282

Canavan PK, Vescovi JD (2004) Evaluation of power prediction equations: peak vertical jumping power in women. Med Sci Sports Exerc 36(9):1589–1593

Casartelli N, Muller R, Maffiuletti NA (2010) Validity and reliability of the Myotest accelerometric system for the assessment of vertical jump height. J Strength Conditioning Res 24(11):3186–3193. https://doi.org/10.1519/JSC.0b013e3181d8595c

Cavagna GA (1975) Force platforms as ergometers. J Appl Physiol 39(1):174–179
Comstock BA, Solomon-Hill G, Flanagan SD, Earp JE, Luk HY, Dobbins KA, Dunn-Lewis C, Fragala MS, Ho JY, Hatfield DL, Vingren JL, Denegar CR, Volek JS, Kupchak BR, Maresh CM, Kraemer WJ (2011) Validity of the Myotest® in measuring force and power production in the squat and bench press. J Strength Conditioning Res 25(8):2293–2297. https://doi.org/10.1519/JSC.0b013e318200b78c
Corcos DM, Chen CM, Quinn NP, McAuley J, Rothwell JC (1996) Strength in Parkinson's disease: relationship to rate of force generation and clinical status. Ann Neurol 39(1):79–88. https://doi.org/10.1002/ana.410390112
Cormie P, Deane R, McBride JM (2007a) Methodological concerns for determining power output in the jump squat. J Strength Conditioning Res 21(2):424–430
Cormie P, McBride JM, McCaulley GO (2007b) Validation of power measurement techniques in dynamic lower body resistance exercises. J Appl Biomech 23(2):103–118
Cormie P, McCaulley GO, McBride JM (2007c) Power versus strength-power jump squat training: influence on the load-power relationship. Med Sci Sports Exerc 39(6):996–1003
Cormie P, McCaulley GO, Triplett NT, McBride JM (2007d) Optimal loading for maximal power output during lower-body resistance exercises. Med Sci Sports Exerc 39(2):340–349
Cormie P, McGuigan MR, Newton RU (2010a) Adaptations in athletic performance after ballistic power versus strength training. Med Sci Sports Exerc 42(8):1582–1598
Cormie P, McGuigan MR, Newton RU (2010b) Influence of strength on magnitude and mechanisms of adaptation to power training. Med Sci Sports Exerc 42(8):1566–1581
Cormie P, McGuigan MR, Newton RU (2011a) Developing maximal neuromuscular power: part 1—biological basis of maximal power production. Sports Med 41(1):17–38
Cormie P, McGuigan MR, Newton RU (2011b) Developing maximal neuromuscular power: part 2—training considerations for improving maximal power production. Sports Med 41(2):125–146
Cronin J, McNair PJ, Marshall RN (2001) Velocity specificity, combination training and sport specific tasks. J Sci Med Sport 4(2):168–178
Cronin JB, Hing RD, McNair PJ (2004) Reliability and validity of a linear position transducer for measuring jump performance. J Strength Conditioning Res 18(3):590–593. https://doi.org/10.1519/1533-4287(2004)18<590:RAVOAL>2.0.CO;2
Cuk I, Markovic M, Nedeljkovic A, Ugarkovic D, Kukolj M, Jaric S (2014) Force-velocity relationship of leg extensors obtained from loaded and unloaded vertical jumps. Eur J Appl Physiol 114(8):1703–1714. https://doi.org/10.1007/s00421-014-2901-2
Davies CT, Rennie R (1968) Human power output. Nature 217(5130):770–771
Davies CT, Young K (1984) Effects of external loading on short term power output in children and young male adults. Eur J Appl Physiol Occup Physiol 52(3):351–354
Dorel S, Couturier A, Lacour JR, Vandewalle H, Hautier C, Hug F (2010) Force-velocity relationship in cycling revisited: benefit of two-dimensional pedal forces analysis. Med Sci Sports Exerc 42(6):1174–1183. https://doi.org/10.1249/MSS.0b013e3181c91f35
Dorel S, Guilhem G, Couturier A, Hug F (2012) Adjustment of muscle coordination during an all-out sprint cycling task. Med Sci Sports Exerc 44(11):2154–2164. https://doi.org/10.1249/MSS.0b013e3182625423
Driss T, Vandewalle H, Quièvre J, Miller C, Monod H (2001) Effects of external loading on power output in a squat jump on a force platform: a comparison between strength and power athletes and sedentary individuals. J Sports Sci 19(2):99–105
Dugan EL, Doyle TLA, Humphries B, Hasson CJ, Newton RU (2004) Determining the optimal load for jump squats: a review of methods and calculations. J Strength Conditioning Res 18(3):668–674
Ferretti G, Gussoni M, Di Prampero PE, Cerretelli P (1987) Effects of exercise on maximal instantaneous muscular power of humans. J Appl Physiol 62(6):2288–2294
Fox EL, Mathews DK (1974) Interval training: conditioning for sports and general fitness. In: Saunders WB (eds) Philadelphia, pp 257–258

Giroux C, Rabita G, Chollet D, Guilhem G (2015) What is the best method for assessing lower limb force-velocity relationship? Int J Sports Med 36(2):143–149. https://doi.org/10.1055/s-0034-1385886

Giroux C, Rabita G, Chollet D, Guilhem G (2016) Optimal balance between force and velocity differs among world-class athletes. J Appl Biomech 32(1):59–68. https://doi.org/10.1123/jab.2015-0070

Glatthorn JF, Gouge S, Nussbaumer S, Stauffacher S, Impellizzeri FM, Maffiuletti NA (2011) Validity and reliability of Optojump photoelectric cells for estimating vertical jump height. J Strength Conditioning Res 25(2):556–560. https://doi.org/10.1519/JSC.0b013e3181ccb18d

Gray R, Start K, Glenross D (1962) A test of leg power. Res Q 33:44–50

Gülch RW (1994) Force-velocity relations in human skeletal muscle. Int J Sports Med 15(Suppl 1):S2–S10

Harman EA, Rosenstein MT, Frykman PN, Rosenstein RM (1990) The effects of arms and countermovement on vertical jumping. Med Sci Sports Exerc 22(6):825–833

Harman EA, Rosenstein MT, Frykman PN, Rosenstein RM, Kraemer WJ (1991) Estimation of human power output from vertical jump. J Appl Sport Sci Res 5(3):116–120

Harris NK, Cronin JB, Hopkins WG (2007) Power outputs of a machine squat-jump across a spectrum of loads. J Strength Conditioning Res 21(4):1260–1264

Hill AV (1938) The heat of shortening and the dynamic constants of muscle. Proc R Soc Lond B Biol Sci 126B:136–195

Hintzy F, Belli A, Grappe F, Rouillon JD (1999) Optimal pedalling velocity characteristics during maximal and submaximal cycling in humans. Eur J Appl Physiol Occup Physiol 79(5): 426–432

Hopkins WG, Schabort EJ, Hawley JA (2001) Reliability of power in physical performance tests. Sports Med 31(3):211–234

James RS, Navas CA, Herrel A (2007) How important are skeletal muscle mechanics in setting limits on jumping performance? J Exp Biol 210(Pt 6):923–933

Janssen WG, Bussmann HB, Stam HJ (2002) Determinants of the sit-to-stand movement: a review. Phys Ther 82(9):866–879

Jaric S (2015) Force-velocity relationship of muscles performing multi-joint maximum performance tasks. Int J Sports Med 36(9):699–704. https://doi.org/10.1055/s-0035-1547283

Jaric S (2016a) Two-load method for distinguishing between muscle force, velocity, and power-producing capacities. Sports Med. https://doi.org/10.1007/s40279-016-0531-z

Jaric S (2016b) Two-load method for distinguishing between muscle force, velocity, and power-producing capacities. Sports Med 46(11):1585–1589. https://doi.org/10.1007/s40279-016-0531-z

Jaric S, Markovic G (2009) Leg muscles design: the maximum dynamic output hypothesis. Med Sci Sports Exerc 41(4):780–787

Jimenez-Reyes P, Samozino P, Pareja-Blanco F, Conceicao F, Cuadrado-Penafiel V, Gonzalez-Badillo JJ, Morin JB (2016) Validity of a simple method for measuring force-velocity-power profile in countermovement jump. Int J Sports Physiol Perform. https://doi.org/10.1123/ijspp.2015-0484

Johnson DL, Bahamonde R (1996) Power output estimate in university athletes. J Strength Conditioning Res 10(3):161–166

Kaneko M, Fuchimoto T, Toji H, Suei K (1983) Training effect of different loads on the force-velocity relationship and mechanical power output in human muscle. Scand J Sports Sci 5(2):50–55

Kannus P (1994) Isokinetic evaluation of muscular performance: implications for muscle testing and rehabilitation. Int J Sports Med 15(Suppl 1):S11–S18

Kawamori N, Rossi SJ, Justice BD, Haff EE, Pistilli EE, O'Bryant HS, Stone MH, Haff GG (2006) Peak force and rate of force development during isometric and dynamic mid-thigh clean pulls performed at various intensities. J Strength Conditioning Res 20(3):483–491. https://doi.org/10.1519/18025.1

Lara AJ, Abian J, Alegre LM, Jimenez L, Aguado X (2006a) Assessment of power output in jump tests for applicants to a sports sciences degree. J Sports Med Phys Fitness 46(3):419–424

Lara AJ, Alegre LM, Abian J, Jimenez L, Urena A, Aguado X (2006b) The selection of a method for estimating power output from jump performance. J Hum Mov Stud 50:399–410

Lutz GJ, Rome LC (1994) Built for jumping: the design of the frog muscular system. Science 263(5145):370–372

Macaluso A, De Vito G (2003) Comparison between young and older women in explosive power output and its determinants during a single leg-press action after optimisation of load. Eur J Appl Physiol 90(5–6):458–463

Mak MK, Levin O, Mizrahi J, Hui-Chan CW (2003) Joint torques during sit-to-stand in healthy subjects and people with Parkinson's disease. Clin Biomech (Bristol, Avon) 18(3):197–206. doi:S0268003302001912 [pii]

Marey JE, Demeny G (1885) Locomotion humaine, mécanisme du saut. Comptes rendus des séances de l'Académies des Sciences; séance du 24 août 1885, vol 101. Institut de France, Académies des Sciences, Paris

Margaria R, Aghemo P, Rovelli E (1966) Measurement of muscular power (anaerobic) in man. J Appl Physiol 21(5):1662–1664

Markovic G, Vuk S, Jaric S (2011) Effects of jump training with negative versus positive loading on jumping mechanics. Int J Sports Med 32:1–8

Marsh RL (1994) Jumping ability of anuran amphibians. Adv Vet Sci Comp Med 38B:51–111

Martin JC, Wagner BM, Coyle EF (1997) Inertial-load method determines maximal cycling power in a single exercise bout. Med Sci Sports Exerc 29(11):1505–1512

McBride JM, Triplett-McBride T, Davie A, Newton RU (2002) The effect of heavy- vs. light-load jump squats on the development of strength, power, and speed. J Strength Conditioning Res 16(1):75–82

McCartney N, Heigenhauser GJ, Sargeant AJ, Jones NL (1983) A constant-velocity cycle ergometer for the study of dynamic muscle function. J Appl Physiol 55(1 Pt 1):212–217

McCartney N, Obminski G, Heigenhauser GJ (1985) Torque-velocity relationship in isokinetic cycling exercise. J Appl Physiol 58(5):1459–1462

McMaster DT, Gill N, Cronin J, McGuigan M (2014) A brief review of strength and ballistic assessment methodologies in sport. Sports Med 44(5):603–623. https://doi.org/10.1007/s40279-014-0145-2

Morin JB, Samozino P, Bonnefoy R, Edouard P, Belli A (2010) Direct measurement of power during one single sprint on treadmill. J Biomech 43(10):1970–1975

Morin JB, Samozino P (2016) Interpreting power-force-velocity profiles for individualized and specific training. Int J Sports Physiol Perform 11(2):267–272

Padulo J, Migliaccio GM, Ardigo LP, Leban B, Cosso M, Samozino P (2017) Lower limb force, velocity, power capabilities during leg press and squat movements. Int J Sports Med. http://doi.org/10.1055/s-0043-118341

Palmieri G, Callegari M, Fioretti S (2015) Analytical and multibody modeling for the power analysis of standing jumps. Comput Methods Biomech Biomed Engin 18(14):1564–1573. https://doi.org/10.1080/10255842.2014.930135

Pearson SJ, Cobbold M, Harridge SD (2004) Power output of the lower limb during variable inertial loading: a comparison between methods using single and repeated contractions. Eur J Appl Physiol 92(1–2):176–181

Rabita G, Dorel S, Slawinski J, Saez de villarreal E, Couturier A, Samozino P, Morin JB (2015) Sprint mechanics in world-class athletes: a new insight into the limits of human locomotion. Scand J Med Sci Sports. doi:https://doi.org/10.1111/sms.12389

Rahmani A, Dalleau G, Viale F, Belli A, Lacour JR (1998) Mesure de la force dynamique par un test ballistique. Science et Sports 13:241–242

Rahmani A, Dalleau G, Viale F, Hautier CA, Lacour JR (2000) Validity and reliability of a kinematic device for measuring the force developed during squatting. J Appl Biomech 16:26–35

Rahmani A, Locatelli E, Lacour JR (2004) Differences in morphology and force/velocity relationship between Senegalese and Italian sprinters. Eur J Appl Physiol 91(4):399–405

Rahmani A, Viale F, Dalleau G, Lacour JR (2001) Force/velocity and power/velocity relationships in squat exercise. Eur J Appl Physiol 84(3):227–232

Riviere JR, Rossi J, Jimenez-Reyes P, Morin JB, Samozino P (2017) Where does the one-repetition maximum exist on the force-velocity relationship in squat? Int J Sports Med. http://doi.org/10.1055/s-0043-116670

Samozino P, Edouard P, Sangnier S, Brughelli M, Gimenez P, Morin JB (2014) Force-velocity profile: imbalance determination and effect on lower limb ballistic performance. Int J Sports Med 35(6):505–510. https://doi.org/10.1055/s-0033-1354382

Samozino P, Horvais N, Hintzy F (2007) Why does power output decrease at high pedaling rates during sprint cycling? Med Sci Sports Exerc 39(4):680–687

Samozino P, Morin JB, Hintzy F, Belli A (2008) A simple method for measuring force, velocity and power output during squat jump. J Biomech 41(14):2940–2945

Samozino P, Rejc E, Di Prampero PE, Belli A, Morin JB (2012) Optimal force-velocity profile in ballistic movements. Altius: citius or fortius? Med Sci Sports Exerc 44(2):313–322

Sargeant AJ, Hoinville E, Young A (1981) Maximum leg force and power output during short-term dynamic exercise. J Appl Physiol 51(5):1175–1182

Sargent LW (1924) Some observations on the Sargent test of neuromuscular efficiency. Am Phys Educ Rev 29:47–56

Sayers SP, Harackiewicz DV, Harman EA, Frykman PN, Rosenstein MT (1999) Cross-validation of three jump power equations. Med Sci Sports Exerc 31(4):572–577

Scholz MN, Bobbert MF, Knoek van Soest AJ (2006) Scaling and jumping: gravity loses grip on small jumpers. J Theor Biol 240(4):554–561

Seck D, Vandewalle H, Decrops N, Monod H (1995) Maximal power and torque-velocity relationship on a cycle ergometer during the acceleration phase of a single all-out exercise. Eur J Appl Physiol Occup Physiol 70(2):161–168

Seo DI, Kim E, Fahs CA, Rossow L, Young K, Ferguson SL, Thiebaud R, Sherk VD, Loenneke JP, Kim D, Lee MK, Choi KH, Bemben DA, Bemben MG, So WY (2012) Reliability of the one-repetition maximum test based on muscle group and gender. J Sports Sci Med 11(2):221–225

Shetty AB (2002) Estimation of leg power: a two-variable model. Sports Biomech 1(2):147–155

Suzuki S, Watanabe S, Homma S (1982) EMG activity and kinematics of human cycling movements at different constant velocities. Brain Res 240(2):245–258

Thorstensson A, Grimby G, Karlsson J (1976) Force-velocity relations and fiber composition in human knee extensor muscles. J Appl Physiol 40(1):12–16

Van Soest AJ, Bobbert MF, Van Ingen Schenau GJ (1994) A control strategy for the execution of explosive movements from varying starting positions. J Neurophysiol 71(4):1390–1402

van Soest O, Casius LJ (2000) Which factors determine the optimal pedaling rate in sprint cycling? Med Sci Sports Exerc 32(11):1927–1934

Vandewalle H, Peres G, Heller J, Panel J, Monod H (1987a) Force-velocity relationship and maximal power on a cycle ergometer. Correlation with the height of a vertical jump. Eur J Appl Physiol Occup Physiol 56(6):650–656

Vandewalle H, Peres G, Monod H (1987b) Standard anaerobic exercise tests. Sports Med 4(4):268–289

Verdijk LB, van Loon L, Meijer K, Savelberg HH (2009) One-repetition maximum strength test represents a valid means to assess leg strength in vivo in humans. J Sports Sci 27(1):59–68. https://doi.org/10.1080/02640410802428089

Wilson GJ, Walshe AD, Fisher MR (1997) The development of an isokinetic squat device: reliability and relationship to functional performance. Eur J Appl Physiol Occup Physiol 75(5):455–461

Wright GA, Pustina AA, Mikat RP, Kernozek TW (2012) Predicting lower body power from vertical jump prediction equations for loaded jump squats at different intensities in men and women. J Strength Conditioning Res 26(3):648–655. https://doi.org/10.1519/JSC.0b013e3182443125

Yamauchi J, Ishii N (2007) Relations between force-velocity characteristics of the knee-hip extension movement and vertical jump performance. J Strength Conditioning Res 21(3):703–709

Yamauchi J, Mishima C, Fujiwara M, Nakayama S, Ishii N (2007) Steady-state force-velocity relation in human multi-joint movement determined with force clamp analysis. J Biomech 40(7):1433–1442

Yamauchi J, Mishima C, Nakayama S, Ishii N (2005) Torque-velocity relation of pedaling movement against stepwise increase in load. Int J Sport Health Sci 3:110–115

Zamparo P, Antonutto G, Capelli C, di Prampero PE (2000) Effects of different after-loads and knee angles on maximal explosive power of the lower limbs in humans. Eur J Appl Physiol 82(5–6):381–390

Zamparo P, Antonutto G, Capelli C, Girardis M, Sepulcri L, di Prampero PE (1997) Effects of elastic recoil on maximal explosive power of the lower limbs. Eur J Appl Physiol Occup Physiol 75(4):289–297

Chapter 5
Optimal Force-Velocity Profile in Ballistic Push-off: Measurement and Relationship with Performance

Pierre Samozino

Abstract Training or rehabilitation programs have to induce changes in force(F)-velocity(v)-power(P) capabilities according to both mechanical demands of the targeted task and actual athlete's muscle capabilities. To determine individual strengths and weaknesses and then individualize strength training modalities, it is essential to know which mechanical capabilities lower limb muscles have to present to maximize ballistic push-off performances. In this chapter, we explore the relationship between the different lower limb muscle mechanical capabilities and ballistic push-off performances. A biomechanical model is presented to bring new insights on the effect of F-v profile on ballistic performances, notably on the existence of an optimal F-v profile. The latter can be accurately determined for each athlete using equations given in this chapter and usual squat jump FvP profile evaluations, including testing using the simple field method presented in Chap. 4. This makes possible the determination of F-v imbalance (towards force or velocity capabilities) and the quantification of the magnitude of the associated force or velocity deficits. These indexes constitute interesting tools to individualize athlete's training programs aiming to improve athletes' ballistic performance. These individual programs should focus on increasing lower limb maximal power and/or decreasing force-velocity imbalance. The effectiveness of such an individualized "optimized" training was shown to be greater than a traditional strength training similar for all athletes. This supports the great interest for strength and conditioning coaches, who aim to improve athlete's ballistic performance, to evaluate FvP profile on each of their athlete and to consider F-v imbalance to design individually training regimen.

P. Samozino (✉)
Laboratoire Inter-universitaire de Biologie de la Motricité, Université de Savoie Mont Blanc, Campus Scientifique, 73000 Le Bourget du Lac, Chambéry, France
e-mail: pierre.samozino@univ-smb.fr

© Springer International Publishing AG 2018
J.-B. Morin and P. Samozino (eds.), *Biomechanics of Training and Testing*,
https://doi.org/10.1007/978-3-319-05633-3_5

5.1 Introduction

The previous chapter presents a simple method to evaluate power muscle capabilities and F-v mechanical profile of lower limbs during ballistic push-off in real practice field conditions (see Chap. 4 for more details on these concepts and method). Training or rehabilitation programs have to induce changes in these capabilities according to both mechanical demands of the targeted task and actual athlete's muscle capabilities. To use the simple jumping method to determine individual strengths and weaknesses and then individualize strength training modalities, it is essential to know which mechanical capabilities lower limb muscles have to present to maximize ballistic push-off performance. Notably, one of the main questions scientists, coaches, or athletes ask when exploring factors for optimizing ballistic performance is which mechanical quality of the neuromuscular system is more important: "force" or "velocity" mechanical capability?

Acyclic ballistic performances are determined by the capability to accelerate a mass (its own body mass or an external mass) as much as possible to reach the highest velocity at the end of a push-off, such a common limb extension. From Newton's second law of motion, the velocity reached by the body center of mass (CM) at the end of a push-off (or take-off velocity) directly depends on the mechanical impulse developed in the movement direction, i.e. the integration of the force produced over the time (Winter 2005; McBride et al. 2010; Knudson 2009). However, the ability to develop a high impulse cannot be considered as a mechanical capabilities of the neuromuscular system: we cannot say that an athlete presents a high mechanical impulse ability. The mechanical impulse is directly associated to the movement/task constraints and not only to the individual properties. It is important to differentiate mechanical outputs characterizing the movement/task (e.g. external force, movement velocity, power output, impulse, mechanical work) from mechanical capabilities of the neuromuscular system which represent the maximal limit of what the athlete muscles can produce. So, the issue is to identify which mechanical muscle capability(ies) determine(s) the ability to produce a high mechanical impulse.

Developing a high impulse during a lower or upper limb push-off, and in turn accelerating a mass as much as possible, has often been assumed to depend on muscle power capabilities (Vandewalle et al. 1987; James et al. 2007; Yamauchi and Ishii 2007; Newton and Kraemer 1994; Frost et al. 2010; Samozino et al. 2008; McBride et al. 2010). It is why many sports performance practitioners, interested in ballistic performances, focus in improving muscular power (Cormie et al. 2011b; Frost et al. 2010; Cronin and Sleivert 2005; McBride et al. 2002; Cormie et al. 2011a). However, maximal power output may be improved by increasing the ability to develop high levels of force at low velocities (henceforth called "force capabilities" or "maximal strength") and/or lower levels of force at high velocities (henceforth called "velocity capabilities") (McBride et al. 2002; Cormie et al. 2011b; Cronin et al. 2001). One of the main issues for strength and conditioning coaches is to determine where to place the training "cursor" within the continuum between these two extreme training modalities: maximal force and maximal velocity strength training. Which are the muscle mechanical capabilities to train to improve power capabilities and in turn ballistic performance at best?

In this chapter, we will explore the relationship between the different lower limb muscle mechanical capabilities and ballistic push-off performances. First, this will be studied from previous experimental data. Then, a biomechanical model will be presented to bring new insights on the effect of F-v profile on ballistic performances, notably on the existence of an optimal F-v profile. Experimental data and practical applications and examples will be put forward to support this biomechanical model and its associated findings.

5.2 Force, Velocity, Power Capabilities & Performance

5.2.1 Performance and Maximal Power Output

As previously mentioned, ballistic push-off performances are often been explored through maximal jumping performances. Jump height has mostly been related to lower limb maximal power output (P_{max}, Newton and Kraemer 1994), the latter being measured during pedaling (Driss et al. 1998; Vandewalle et al. 1987; Hautier et al. 1996), jumping (Samozino et al. 2014; McBride et al. 2010) or horizontal leg press movement (Yamauchi and Ishii 2007). The high magnitudes and occurrences of correlation between jump height and P_{max} (from 0.65 to 0.87 when P_{max} is expressed relatively to body weight) well support the important influence of P_{max} on ballistic push-off performances (Fig. 5.1). Interestingly, the understanding of the muscle properties determining jumping performance has also been extensively studied in animals, as frogs or bushbabies, with a similar high relationship with muscle power capabilities (James et al. 2007; Aerts 1998; Marsh 1994). This explains the wide interest of sports performance practitioners in improving muscular power (Cormie et al. 2011b; Frost et al. 2010; Cronin and Sleivert 2005; McBride et al. 2002; Cormie et al. 2011a).

However, although high and significant, the correlations between P_{max} and jumping performances are not perfect and shows (interpreting the r^2 values) that only ∼40–80% of the differences in jumping performances between athletes can be explained by differences in P_{max} (Fig. 5.1). Even if a part of this non-explained variability in jump height can be attributed to the specific experimental conditions and to the inevitable measurement noise and variability, the potential effects of others muscle qualities, as F-v profile, cannot be ruled out. This was one of the conclusions of the review of Cronin and Sleivert about the understanding of the influence of maximal power training on improving athletic performance (Cronin and Sleivert 2005): "*power is only one aspect that affects performance and it is quite likely that other strength measures may be equally if not more important for determining the success of certain tasks*".

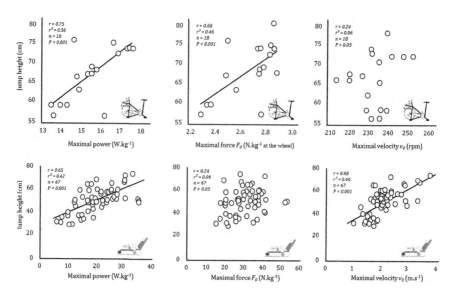

Fig. 5.1 Correlations between jumping performance and maximal power, force and velocity capabilities, the latter being obtained either during pedalling (from Driss et al 1998, on the top) or leg press movement (from Yamauchi and Ishii 2007, on the bottom). Correlation and determination coefficients, number of subjects and correlation significance are presented for each relationship

5.2.2 Performance and Force-velocity Mechanical Profile

The effect of force and velocity lower limb capabilities on jumping performance is controverted. Driss and colleagues showed a significant correlation between jump height and the maximal force (F_0) normalized to athlete body mass (r = 0.67; P < 0.01, Driss et al. 1998), this correlation being shown as non-significant by Yamauchi & Ishii (r = 0.24; P > 0.05, Yamauchi and Ishii 2007) (Fig. 5.1). These contrasted results about the association between jumping performance and maximal force capabilities (often evaluated through one maximum repetition tests) have been previously reported and discussed (e.g. Ugrinowitsch et al. 2007; Young et al. 1999; Davis et al. 2003; Cormie et al. 2011b; Stone et al. 2002). Some studies show the effect of force capabilities on jump height only when force is assessed at high movement velocities and not maximal force values (Eckert 1968; Young et al. 1999; Genuario and Dolgener 1980). So, velocity capability also seems to affect jumping performance, even if correlations with the maximal lower limb extension velocity (v_0) were shown to be significant (r = 0.68; P < 0.001, Yamauchi and Ishii 2007) or not (r = 0.24; P > 0.05, Driss et al. 1998) (Fig. 5.1).

All these results do not give a clear view of the effects of lower limb force and velocity capabilities on jumping performance, and even less of the effect of F-v profile (ratio between force and velocity capabilities, details in previous chapter).

The previously mentioned relationships were tested through correlation analyses or comparisons between athletes with different training backgrounds. This inevitably presents potential interactions with covariant parameters, making thus uncertain the causal link between each muscle capability and jumping performance. For instance, a significant association between maximal strength and jump height (Driss et al. 1998; Ugrinowitsch et al. 2007) may be due (or partly due) to the fact that, in the athlete sample tested, the strongest athletes were also the most powerful, which is likely possible since power is a combination of force and velocity capabilities. In this case, the lower limb maximal force indirectly may affect jumping performance only because it positively influences power capabilities. This is also true in studies testing different training interventions on ballistic performances in which the various athletes F-v profiles were also associated with various P_{max} values among subjects, making it impossible to identify the sole effect of the F-v profile (Cronin et al. 2001; Harris et al. 2008; Cormie et al. 2007a, 2009, 2010, 2011b; McBride et al. 2002).

So, such experimental analyses have often failed to inform us about the effect of force and velocity capabilities on ballistic performances independently from the well-known effect of P_{max}. Otherwise, it is very difficult to experimentally consider or control all the parameters which can affect, together or independently, jumping ability. This partly explains the previous contradictory results. For these reasons, biomechanical models have been proposed to better understand the mechanical factors affecting jumping performance.

5.2.3 Biomechanical Models Applied to Ballistic Push-off

The most famous biomechanical models applied to jumping performance are forward simulation models that integrate a large amount of morphological, physiological or neuromuscular variables (Alexander 1995; Bobbert and Van Soest 1994; Pandy and Zajac 1991). These models have brought new insights about optimal muscular coordination strategies or effects of different jumping techniques, what could not have been shown experimentally (e.g. Pandy and Zajac 1991; Alexander 1995; Bobbert and Casius 2005). However, the numerous inputs used do not represent the overall mechanical capabilities of the neuromuscular system, but specific biological features (e.g. isolated muscle force and contraction velocity, muscle activation dynamics or architecture) that affect these mechanical properties, several of them acting on the same quality. So, they did not isolate the basic lower limb mechanical capabilities determining jumping ability, which makes more difficult the direct transfer to training applications. Other theoretical frameworks have explored these overall mechanical characteristics affecting jumping ability through the investigation of the mechanical requirements of maximal jumps. Indeed, dynamics analyses of jumping allow one to identify the mechanical outputs of push-off determining jumping performance at best. For instance, jumping performance has been mathematically shown to be proportional to the mechanical work produced during push-off (Emerson 1985; Alexander 2003) or to the mean power developed before take-off (Marsh 1994; Minetti 2002). These models, based on dynamics principles, showed that maximal jumping performances are determined by the capability of lower limbs to produce mechanical work or power, but did not identify which muscle capabilities, notably which F-v profile, are associated with such outputs.

In order to have a better understanding of the implication of the different lower limb mechanical capabilities, and notably the F-v profile, on ballistic push-off performances, we used a biomechanical model only based on simple physical and physiological principles applied to the athlete's body CM during jumping.

5.3 An Optimal Force-Velocity Mechanical Profile During Jumping

5.3.1 Theoretical Bases and Equations of the Biomechanical Model

During a ballistic movement, the mechanical outputs are the results of two different mechanical constraints: the fundamental laws of dynamics and the mechanical properties of the neuromuscular system. These two mechanical constraints affect jumping performance through the interdependence of force production and movement velocity.

Previous biomechanical models (e.g. Minetti 2002) have well shown the relationship between the external force applied to the body CM and other mechanical outputs during a jump, notably movement velocity. For a given mass and lower limb extension range, the higher the force produced onto the ground, the higher the CM acceleration and in turn the CM velocity. This is the mechanical constraint imposed by the movement dynamics, well synthesized by the second Newton's law of motion. So, to jump high, or to reach the highest velocity at the end of a push-off, the athlete has to present high movement velocity. To do that, the athlete has thus to develop the highest force as possible onto the ground. But, the higher the movement velocity, the lower the athlete's force production capability. This is the mechanical constraint associated to the properties of the neuromuscular system well represented through the F-v relationship: the higher the lower limb extension velocity, the lower the maximal force that can be developed. Consequently, a maximal ballistic performance is a "circular" interaction between these two mechanical constraints which describe opposite relationships between external force and movement velocity: increase in velocity when force applied increases (movement dynamics constraints) and decrease in force produced when velocity increases (neuromuscular constraints) (Samozino et al. 2010; Fig. 5.2). The highest ballistic push-off performance an athlete can reach is thus achieved in the push-off condition respecting these two mechanical constraints, and so corresponding to both muscle maximal

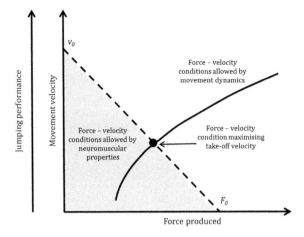

Fig. 5.2 Theoretical representation of the mechanical constraints imposed by movement dynamics (solid line) and lower limb neuromuscular properties (dashed line) during a typical vertical jump. The solid line represents the take-off velocity according to the force produced over push-off. The force–velocity conditions (mean force produced and mean velocity over push-off) allowed by lower limb mechanical capabilities are represented for sub maximal efforts (grey area) and for all-out efforts (dashed line). Note that here, force and velocity axes are switched compared to usual figures of force-velocity relationships presented elsewhere in this book, which does not modify the curve interpretation. (From Samozino et al. 2010)

capabilities and movement dynamics. The idea of the present biomechanical model is to put this "circular" interaction into equations to express the athlete maximal performance as a mathematical function of his/her muscle mechanical capabilities.

The biomechanical model is developed from an analysis of the dynamics of the body CM during ballistic performance through maximal jumps at different push-off angles. The entire lower limb neuromuscular system is considered as a force generator characterized by an inverse linear F-v relationship (as described in Chap. 4) and a given range of motion (h_{PO}). The maximal ballistic performance (be it during jumping or start sprinting) can be well represented by the maximal CM velocity reached at the end of the push-off, i.e. at take-off (v_{TOmax}), and can be expressed as (for detailed computations, see Samozino et al. 2010, 2012):

$$v_{TOmax} = h_{PO}\left(\sqrt{\frac{S_{Fv}^2}{4} + \frac{2}{h_{PO}} \cdot \left(2\sqrt{-P_{max} \cdot S_{Fv}} - g \cdot \sin\alpha\right)} + \frac{S_{Fv}}{2}\right) \quad (5.1)$$

when g is the gravitational acceleration (9.81 m s^{-2}), α the push-off angle with respect to the horizontal (in deg), h_{PO} the distance covered by the CM during push-off corresponding to the extension range of lower limbs (in m), and S_{Fv} the F-v mechanical profile of lower limbs corresponding to the slope of the linear F-v relationship ($S_{Fv} = -F_0/v_0$).

This equation does not aim to predict the performance of a given athlete knowing his power capabilities and his F-v profile (you have just to make him jump to have this information). This equation help to better understand how the different lower limb mechanical capabilities are involved in ballistic performance. First of all, it shows that the maximal velocity an athlete can reach at the end of a ballistic push-off depends directly on the inclination of the push-off and on lower limb maximal power, F-v profile and range of extension. Then, quantification of the effect magnitudes of each properties to performance can be obtained simulating this equation, i.e. "playing" with the different variables one after the other. Before that (Sect. 5.3.3), it is important to support the validity and the relevance of the proposed biomechanical model comparing the equation predictions to experimental data.

5.3.2 Validation of the Model

The validity of the biomechanical model was tested by comparing performances predicted by Eq. 5.1 based on mechanical properties measured on different athletes to actual performances performed by the same athletes (Samozino et al. 2012).

Fourteen male athletes practicing physical activities including explosive efforts (e.g. basket-ball, rugby, soccer) performed two series of maximal lower limb push-offs: (i) horizontal extensions with different resistive forces allowing us to determine lower limb F-v relationships; and (ii) inclined jumps used to compare experimental performances to theoretical predictions. Tests were realized on the Explosive Ergometer of the Department of Biomedical Sciences and Technologies of the University of Udine which consists of a metal frame supporting one rail on which a seat, fixed on a carriage, was free to move (for more details, see Rejc et al. 2010; Samozino et al. 2012, Fig. 5.3). For each test, athletes were asked to push on two force plates (LAUMAS PA 300, Parma, Italy) positioned perpendicular to the rail to accelerate themselves and the carriage seat backward as fast as possible. The velocity of the carriage seat along the rail was continuously recorded by a wire tachometer (LIKA SGI, Vicenza, Italy). During the first series of tests, athletes performed horizontal maximal push-offs against 7 different resistive forces (from 0 to 240% of body weight) to determine individual F-v and P-v relationships and associated P_{max}, S_{Fv} and h_{PO} values. For each athlete, the latter values were used as input in the Eq. 5.1 to compute the maximal velocities (v_{TOmax}) he could reach at the end of push-offs inclined at different angles from the horizontal (10, 20 and 30°). These theoretical predicted values were compared to the actual take-off velocities measured during the second series of tests consisting of inclined (10, 20 and 30°) jumps performed as fast as possible. Results showed no significant difference between actual and predicted values with low absolute errors from 4 to 6.6% between both (Table 5.1). This supports the validity of the biomechanical model and the associated Eq. 5.1, which allowed us to use it for simulations to explore how the different mechanical variables, and notably F-v profile, affects ballistic performance.

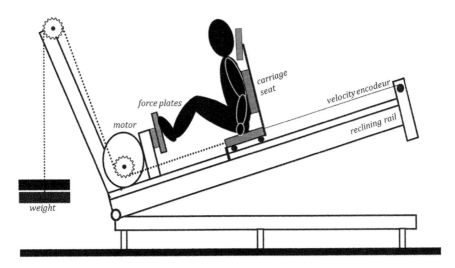

Fig. 5.3 Schematic view of the Explosive Ergometer (EXER). The rail can be tilted upward. To allow the motor to act on the seat without delays, a constant tension (weight = 196 N) is applied to the steel chain (dotted line). (From Samozino et al. 2012)

Table 5.1 Mean ± SD of take-off velocity obtained with experimental and theoretical approaches, absolute bias between these two approaches, and t-test comparison results (from Samozino et al 2012)

Push-off angle (°)	Experimental values (m s^{-1})	Theoretical values (m s^{-1})	t-test	Absolute Bias (%)
10	2.45 ± 0.22	2.43 ± 0.18	ns	4.40 ± 4.94
20	2.32 ± 0.25	2.25 ± 0.16	ns	6.56 ± 5.46
30	2.14 ± 0.23	2.07 ± 0.15	ns	5.73 ± 3.89

ns Non significative difference between experimental and theoretical values

5.3.3 Muscular Capabilities Determining Jumping Performance

The interest of such a model and equation is to isolate the effect of each factor from the others, which is very difficult, if not impossible, to do experimentally (Samozino et al. 2012). The lower limb extension range (h_{PO}) corresponds to the optimal or usual one for each athlete, and so depends on individual intrinsic morphological or biological features. Its effect on performance is not of great interest for training purposes, it was kept constant here for the different model simulations, but the following results were the same whatever the h_{PO} values.

The first result was the large effect of P_{max} on ballistic performance with the greatest weight among all the other muscular capabilities. This confirms all the previous experimental data showing a significant (but not perfect) correlation

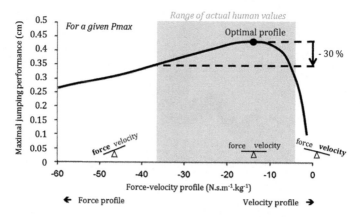

Fig. 5.4 Changes in maximal jumping performance that can be reached by an athlete as a function of the changes in the force-velocity profile (slope of the F-v relationship) for a given maximal power values. The black point represents the highest performance that can be reached for an optimal F-v profile. Considering actual human values of F-v profile, an unfavourable profile can lead to loss of performance until 30%, while maximal power is similar. (From Samozino et al. 2012)

between lower limb P_{max} and jumping ability (see Sect. 5.2.1). This supports the fact that the ability to develop a high impulse against the ground, and in turn the ability to reach maximal CM velocity at the end of a push-off, is highly related to the maximal power the lower limbs can produce (over a given extension range).

However, the main finding was not here the well-known positive effect of P_{max}, but was rather to show that power capabilities is not the only muscular properties involved in jumping performance. Two athletes with the same P_{max} (and the same h_{PO}) can achieve different velocities at the end of a vertical or horizontal push-offs. These differences are due to their respective F-v profiles (S_{Fv}), i.e. to their respective ratios between maximal force (F_0) and velocity (v_0) capabilities. Indeed, for a given P_{max}, v_{TOmax} changes as a function of the F-v profile in a curvilinear shape, with a maximum value reached for an optimal F-v profile ($S_{Fv}opt$, Fig. 5.4), the latter corresponding to the best balance between maximal force and velocity capabilities. The more this F-v profile differs from the optimal one, the lower the performance in comparison to the one that could be reached with the same power capabilities. Consequently, the jumping performances of the two above-mentioned athletes presenting the same power capabilities can be 30% different only due to a pronounced unfavorable F-v profile for one of them (Fig. 5.4). The optimal F-v profile is individual since it depends on P_{max} and h_{PO} individual values. It can be accurately determined for each athlete using the following equation representing the real solution cancelling out the first mathematical derivative of Eq. 5.1 with respect to S_{Fv} (more details in (Samozino et al. 2012)):

$$S_{Fv}opt = -\frac{(g \cdot sin\alpha)^2}{3P_{max}}$$
$$-\frac{\left(-\left((g \cdot sin\alpha)^4\right) \cdot h_{PO}^4 - 12g \cdot sin\alpha \cdot h_{PO}^3 \cdot P_{max}^2\right)}{3h_{PO}^2 \cdot P_{max} \cdot Z(P_{max}, h_{PO})} + \frac{Z(P_{max}, h_{PO})}{3h_{PO}^2 \cdot P_{max}} \quad (5.2)$$

with

$$Z(P_{max}, h_{PO}) = \left(-\left((g \cdot sin\alpha)^6\right) \cdot h_{PO}^6 - 18(g \cdot sin\alpha)^3 \cdot h_{PO}^5 \cdot P_{max}^2 - 54h_{PO}^4 \cdot P_{max}^4 \right.$$
$$\left. + 6\sqrt{3} \cdot \sqrt{2(g \cdot sin\alpha)^3 \cdot h_{PO}^9 \cdot P_{max}^6 + 27h_{PO}^8 \cdot P_{max}^8}\right)^{1/3} \quad (5.3)$$

These formulae can appear a little bit complicated, but remain easy to use in a spreadsheet, a typical spreadsheet and a tutorial to use it (home-made by Morin and Samozino) can bedownloaded/viewed here: https://www.researchgate.net/publication/320146284_JUMP_FVP_profile_spreadsheet. Note that complete computations of the individual optimal profile, and subsequent indexes (see below), are proposed in the smartphone application MyJump concomitantly to the determination of the maximal power output and the F-v profile (see Sect. 4.5.1 in Chap. 4).

The optimal F-v profile also depends on push-off angle (α in Eqs. 5.1–5.3), and more generally on the magnitude of the gravity component opposing motion: the lower the push-off angle, the more the optimal F-v profile is oriented towards velocity capabilities. If you are a strength and conditioning coach of volley ball players and you want to improve their vertical jump height ($\alpha = 90°$), the optimal F-v profile will be more oriented towards force qualities than if you aim to increase the change in velocity during the first steps of a track and field sprinter or a team sport player ($\alpha \approx 30–40°$), and still more than if you focus on the diving push off of a swimmer ($\alpha \approx 0°$ or negative) (Fig. 5.5).

Whatever the kind of push-off, the individual F-v profile can then be expressed relatively to the optimal one ($S_{Fv\%}$, in percentage) to underline the unfavorable balance between force and velocity capabilities:

$$S_{Fv\%} = \frac{S_{Fv}}{S_{Fv}opt} \cdot 100 \quad (5.4)$$

A $S_{Fv\%}$ value higher than 100% represents a F-v profile too much oriented towards force capabilities, and in turn correspond to a velocity deficit. In contrast, a $S_{Fv\%}$ value lower than 100% is associated to a F-v profile rather oriented to velocity qualities with a force deficit.

These finding support that the F-v profile, and its comparison to the optimal one, represents a muscular quality that has to be considered attentively by scientists working on muscle function during maximal efforts, but also by coaches for training purposes.

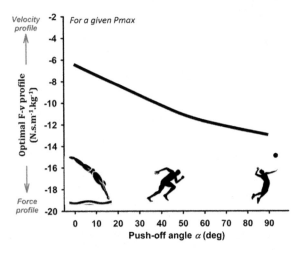

Fig. 5.5 Changes in optimal force-velocity profile as a function of the push-off angle for a given maximal power output and push-off distance. In each sport activity, athletes mainly perform push-offs at specific push-off angles, as here in swimming diving, horizontal sprint starting (track and field, team sport) or volley ball. (From Samozino et al. 2012)

5.3.4 FV Imbalance & Performance

The above-mentioned results are "only" based on theoretical simulations of a biomechanical model, it was in turn important to bring experimental evidence of the effect of F-v profile on ballistic performance. Does the magnitude of F-v imbalance actually affect negatively ballistic performances independently from the effect of power capabilities?

To answer to this question, F-v relationships and P_{max} were determined on forty eight high-level athletes (soccer players, sprinters, rugbymen) who performed maximal squat jumps with additional loads from 0 to 100% of body mass and using the simple field method presented in the previous section (complete details on this study in Samozino et al. 2014). For each athlete, F-v imbalance (Fv_{IMB}) was computed as the normalized difference between actual and optimal F-v profile as:

$$Fv_{IMB} = 100 \cdot \left| 1 - \frac{S_{Fv}}{S_{Fv}opt} \right| \qquad (5.5)$$

A multiple regression analysis showed that, when considered together, each of the three predictor variables P_{max}, Fv_{IMB} and h_{PO} accounted for a significant amount of jumping performance variability ($P < 0.001$). The negative effect of Fv_{IMB} on jumping performance put forward that, even if P_{max} remains the main determinant in ballistic performance, a F-v imbalance is associated to a lower performance. This is an experimental support for (i) the influence of the normalized F-v profile (characterized by the slope of the F-v relationship, S_{Fv}) on jumping performance independently from the large effect of P_{max}, and (ii) the existence of an optimal F-v profile maximizing performance for each individual.

The negative effect of the F-v imbalance on performance was quantified for each athlete through the difference between their actual jump height reached during unloaded condition and the maximal height they would have reached if they presented the same P_{max} with an optimal F-v profile. This hypothetic maximal jump each athlete could reach with an optimal F-v profile (h_{max}, in m) was computed from his actual P_{max} and h_{PO} values and using the following equation derived from the previously validated biomechanical model (Samozino et al. 2010, 2012):

$$h_{max} = \frac{h_{PO}^2}{2g}\left(\sqrt{\frac{S_{Fv}opt^2}{4} + \frac{2}{h_{PO}}(2\sqrt{-\bar{P}_{max}S_{Fv}opt} - g)} + \frac{S_{Fv}opt}{2}\right)^2 \quad (5.6)$$

Figure 5.6 presents the jump height actually reached by each athlete relatively to their respective h_{max} according to the individual F-v profile normalized to $S_{Fv}opt$ ($S_{Fv}\%$). This shows the loss of performance due to the F-v imbalance illustrated here by the grey area. Athletes presenting a F-v profile close to the optimal one can jump to the maximal performance associated to their power capabilities, the others could not. The higher Fv_{IMB}, the higher the loss of performance. The individual loss of performance due to the F-v imbalance ranged from 0% for subjects presenting optimal profiles to $\sim$30% for one rugby player who exceptionally presented a Fv_{IMB} of $\sim$250% (Fig. 5.6). This is in line with the above-mentioned theoretical simulation. In fact, an athlete presenting a F-v imbalance (as computed here from $S_{Fv}opt$ maximizing vertical jumps) means that he does not develop his P_{max} against his body mass during a vertical jump, and so he does not make the most of his power capabilities during ballistic movement performed at body weight (Samozino et al. 2012). This is associated to the question of the optimal load maximizing

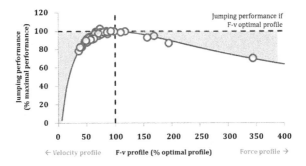

Fig. 5.6 Actual jump height reached in unloaded condition (jumping performance expressed relatively to maximal performance athlete could reach if he presented an optimal F-v profile) according to F-v profile (expressed relatively to their personal F-v optimal profile). Each circle represents a subject. The solid line represents theoretical changes predicted by the biomechanical model. The grey area represents the individual loss of performance due to the F-v imbalance and ranging from 0 for subjects presenting optimal profile to $\sim$30% for the rugby player represented by the circle on the extreme right of the figure. (From Samozino et al. 2014)

power production during ballistic movement (Jaric and Markovic 2009). The optimal load for an athlete presenting an optimal F-v profile for vertical jumps is his own body mass. However, an athlete with a "force" profile (i.e. with a "velocity" deficit), as most of the rugby players tested in this study, developed their P_{max} during vertical jumps against loads higher than their body mass (i.e. for additional loads from 25 to 100% of their body mass). Note that this is not necessarily an issue if ballistic performance at body weight is not the primary quality to optimize for such players to who it may have been asked (for some of them) to be efficient against resistance (e.g., for forward players). At the opposite, the optimal load for an athlete with a "velocity" profile is lower than his body mass, and so he needs unloaded conditions to develop his P_{max}. The large range of F-v imbalance observed in this study support the recently proposed and debated influence of training history on the optimal load-maximizing power during vertical jumps (Jaric and Markovic 2009; Markovic and Jaric 2007; Nuzzo et al. 2011; Pazin et al. 2011, 2013).

These data well supported the theoretical findings brought out by the biomechanical model on squat jump movements. However, squat jumps, involving just a sole concentric action, does not occur very frequently in many sport activities in which movements with successive eccentric and concentric muscle actions (e.g. sprint running, jumps, changes of direction) are more usual. This is why counter movement jumps (CMJ) remain the most commonly used task in sports training and testing. Even if nothing could challenge the transfer to CMJ of the theoretical and experimental results showed in SJ, the same protocol as described above was conducted on CMJ (Jimenez-Reyes et al. 2014). As in SJ, CMJ performance was shown to depend on F-v imbalance, independently from the effect of P_{max}, with the existence of an individual optimal F-v profile, the negative effect of Fv_{IMB} being even larger in CMJ. Moreover, this study clearly showed that the linear F-v relationship in CMJ was shifted to the right in comparison to SJ, with a larger shift in F_0 than in v_0, and in turn a higher P_{max} which explained the increase in CMJ maximal height compared to SJ.

5.4 Practical Applications

5.4.1 F-v Profile & F-v Imbalance Indices

Testing F-v profiles when aiming to identify the optimal balance between force and velocity capabilities may be of interest to set training loads and programs, as previously proposed using power-load relationships (Cormie et al. 2011b; Sheppard et al. 2008; Jaric and Markovic 2009; McBride et al. 2002). F-v profile individual values allow comparisons among athletes independently from their power capabilities, which is not possible from only F_0 and v_0 values. This makes possible to compare different athletes and to know whether an individual is characterized by a

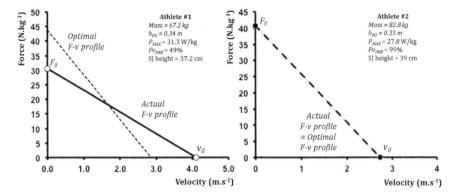

Fig. 5.7 Force-velocity profiles of 2 track and field athletes obtained from maximal loaded squat jumps (SJ). Despite a higher maximal power output (P_{max}) value, athlete #1's SJ performance is lower because his Fv_{IMB} (magnitude of the relative difference between the slope of the linear force-velocity relationship and the optimal slope) is greater than for athlete #2. For athlete #1, the black line indicates the actual profile, and the dashed line, the optimal profile. Note that athlete #2's profile if almost optimal, and therefore the actual and optimal relationships are confounded in the right panel (gray line and black dashed line). (From Morin and Samozino 2016)

"force" or a "velocity" profile compared to another one (Figs. 4.3 and 5.7). Only Carmelo Bosco proposed in the nineties an index to compare athletes' F-v profiles through the ratio between jump height reached with an additional load (100% of body mass) and unloaded jump height (Bosco 1992). The higher this index, the higher the force capabilities compared to velocity ones. However, Bosco's index does not allow the orientation of training loads for a given athlete according to his/her own strengths and weaknesses and to movement specificities. This is why the individual value of S_{Fv}, expressed relatively to $S_{Fv}opt$ ($S_{Fv\%}$), is a good and practical index to characterize F-v profile and to design appropriate training programs. The difference between $S_{Fv\%}$ and 100% (i.e. Fv_{IMB}) allows the quantification of the magnitude of force or velocity deficits. Consequently, $S_{Fv\%}$ and Fv_{IMB} values should be individually determined when testing Force-velocity-Power (FvP) profile, using the simple field method proposed in the previous chapter or other methods. This can be easily done using a simple homemade spreadsheet or program writing, or even easier using MyJump application which automatically computes these indexes in addition to measure FvP profile variables (see previous chapter).

Beyond to bring new insights into the understanding of the relationship between mechanical function of the lower limbs neuromuscular system and maximal human performances, the interest of the FvP profile approach, including the optimal profile, is that it allows for a more individualized athlete's evaluation, monitoring and training practices. When a training program is designed to improve athletes' ballistic push-off performance (e.g. jumps, single maximal push-offs, change of direction), the focus should be placed on increasing P_{max} and/or decreasing Fv_{IMB}. With regards to athletes displaying significant imbalance in mechanical capacities, training programs should prioritize training the lacking mechanical capability, in order to shift S_{Fv} towards $S_{Fv}opt$. Here we will present different data supporting the great interest to consider F-v profile, notably F-v imbalance, to design training program focused on ballistic performances.

5.4.2 FV Imbalance & Case Reports

The importance of F-v profile in ballistic performance can be well illustrated with data from two track and field athletes (#1 and #2) with similar lower limb extension ranges but different P_{max} and S_{Fv} (Fig. 5.7, Morin and Samozino 2016). Athlete #1 has a higher P_{max}, but he presents a F-v imbalance (about 50% towards a force deficit) while athlete #2 has an optimal F-v profile (only a negligible 1% imbalance). This results to a higher squat jump performance for athlete #2. The present approach would thus suggest, for athlete #1, a training program focusing in priority the development of maximal force capabilities in order to both correct his imbalance and increase P_{max}. Once this goal is achieved, he may transition into similar training to athlete #2, in order to improve his P_{max}, while maintaining his corrected (i.e. optimal) profile.

Another obvious illustration of the significance of F-v profile is the comparison of muscle mechanical capabilities and performance of two young players (#3 and #4, Fig. 5.8) from the same soccer team (French first league professional club

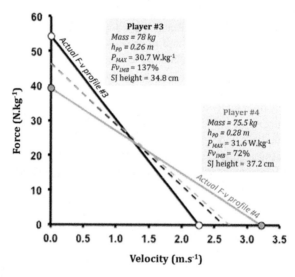

Fig. 5.8 Force-velocity profiles of 2 elite young (under-19) soccer players obtained from maximal loaded squat jumps (SJ). Player#4 (a goalkeeper) has a force deficit (magnitude of the relative difference between the slope of the linear force-velocity relationship and the optimal slope—Fv_{IMB}—of 72%), whereas Player#3 (a central defender) has a velocity deficit (Fv_{IMB} of 137%). (From Morin and Samozino 2016)

academy U19). They have similar P_{max} and $S_{Fv}opt$ values, but present opposing Fv_{IMB} characteristics. Player #3 displays a velocity deficit, while player#4 shows a force deficit, but with an absolute difference with his own $S_{Fv}opt$ lower than for player#3 (28% versus 37%). This relatively smaller Fv_{IMB} and the slightly higher P_{max} in player#4 explain his higher squat jump performance. Knowing these discrepancies in F-v profile between these two players, the most efficient way to train and improve ballistic push-off performance in both of them would be an individualized program (indexed on each player's Fv_{IMB}) that targets the development of totally different capabilities.

5.4.3 FV Profile and Training

As briefly mentioned in the Chap. 4 (Sect. 4.2.2), the F-v profile is sensible to the kind of strength training performed (Cormie et al. 2010), which is of great interest for sport practitioners, but most importantly a prerequisite to individually correct F-v imbalance in order to maximize ballistic performances. Whatever the initial athlete's F-v profile, training should be ideally designed to both increase P_{max} and decreasing Fv_{IMB}.

For athletes presenting a force deficit, training should aim to increase force capabilities (F_0) as a priority (Samozino et al. 2012). The effectiveness of heavy strength training, involving the use of high loads (>70% RM) in order to achieve the maximal neuromuscular adaptations in periods ranging from 6 to 12 weeks, to specifically increase maximal force capabilities has been clearly shown (Cormie et al. 2007a, 2010; McBride et al. 2002). They result in improvements in maximal strength parameters, as the well-known one repetition maximum (1RM) load.

Contrastingly, for athletes presenting a velocity deficit, training should focus in increasing P_{max} by improving maximal velocity capabilities, i.e. the capacity to produce force at very high contraction velocities. This can be achieved using maximal velocity efforts during high accelerated movements with minimal or null braking phase. Different training modalities have been proposed to increase movement velocity: maximal effort removing deceleration phase at the end of the movement (with bar throw or jump, Frost et al. 2010) using low loads (<30% of the repetition maximum, Cormie et al. 2010), negative loads (Argus et al. 2011; Markovic et al. 2011) or only inertia (Djuric et al. 2016). Plyometric contraction mode can also been used to increase the contraction velocity during the concentric phase of the movement. This type of training -commonly referred to as ballistic— may result in a training-induced shift in force-time curves and force-velocity relationships towards more velocity-related capabilities (Djuric et al. 2016).

Finally, for athletes presenting low or no deficit, i.e. with an actual F-v profile close to the optimal one, the training program should target a balanced combination of force, velocity and power in order to shift the entire F-v relationship to the top and right. This would increase P_{max} as a priority (e.g. Cormie et al. 2007b; Kotzamanidis et al. 2005; McBride et al. 2002; Cormie et al. 2011b) while maintaining the F-v profile close to the optimal value (and thus Fv_{IMB} close to 0%). The effects of studies aiming to both increase maximal power and shift the entire F-v curve show how combining a wide range of loads (heavy, optimal and low loads) is an appropriate stimulus (Cormie et al. 2007b; Kotzamanidis et al. 2005; McBride et al. 2002; Cormie et al. 2011b; Kaneko et al. 1983).

5.4.4 FV Imbalance, "Optimized" Training & Performance

The previous section shows that the sensitivity of the F-v profile to specific training programs can result in either maximal force or velocity capabilities improvements. Most often, these specific training prescriptions are proposed similarly for all athletes, sometimes considering the mechanical requirements of the tasks involved in a given sport activity, but rarely (if not never) considering the initial F-v profile, and so the needs, of each individual. This results in contrasting findings, as to the effects on jumping performance (e.g. Chelly et al. 2009; Cormie et al. 2007a, 2010; McBride et al. 2002; Kotzamanidis et al. 2005), likely because of the various levels and F-v characteristics of the populations tested. Indeed, a training program leading to improve P_{max} while increasing Fv_{IMB} could result in a lack of change, or even a decrease in jumping performance.

In light of the above-mentioned points, we can thus reasonably assume that specific strength training aiming at improving ballistic performance should be designed on an individual basis to both reduce Fv_{IMB} (i.e. to increase preferably the F_0 or v_0 component of an individual's F-v profile and shift it towards his optimal profile) and increase P_{max}. This can been termed "optimized training" or "individualized training based on Fv_{IMB}". A recent study aimed at experimentally testing and quantifying the effectiveness of such a training on ballistic performance compared to a traditional resistance training common to all subjects and designed without taking account of individual Fv_{IMB} (Jimenez-Reyes et al. 2016). Eighty four subjects were assigned to three training intervention groups based on their initial Fv_{IMB}: an optimized group divided into velocity-deficit, force-deficit and well-balanced sub-groups based on subjects' Fv_{IMB}, a "non-optimized" group for which the training program was not specifically based on Fv_{IMB} and a control group (without training). All subjects underwent a 9-week specific resistance training program before and after which F-v profile, Fv_{IMB}, P_{max} and maximal vertical jump height (used here as the index of performance since it represents the archetype of ballistic movements) were tested. The programs were designed to reduce Fv_{IMB} for the optimized groups (with specific strength training regimen for sub-groups formed depending on individual Fv_{IMB} values), while the non-optimized group followed a classical program exactly similar for all subjects targeting a balanced combination of force, velocity and power qualities. During training, the force-deficit sub-group performed mainly force-oriented training (using very high loads >80% of 1RM), while the velocity-deficit sub-group performed velocity-oriented (ballistic, unloaded and band assisted SJ or horizontal assisted roller push-off) training. The well-balanced sub-group followed a training program covering the entire force-velocity spectrum in equal proportions: heavy loads, power and ballistic training. All the subjects of the "non-optimized" training group followed the latter kind of training, independently from each subject's Fv_{IMB}.

All subjects in the three optimized training sub-groups (velocity-deficit, force-deficit and well-balanced) increased their jumping performance (very likely to most likely very large effects of +7.2 to +14.2% on average) with jump height improvement for all subjects, whereas the results were much more variable and unclear in the non-optimized and control groups (trivial change of +2.3% on average, 10 subjects out of 18 improved) (Fig. 5.9). This greater change in jump height was associated with a markedly reduced Fv_{IMB} for both force-deficit (57.9 ± 34.7% decrease in Fv_{IMB}) and velocity-deficit (20.1 ± 4.3% decrease in Fv_{IMB}) subjects, and unclear or small changes in P_{max} (−0.40 ± 8.4% and +10.5 ± 5.2%, respectively). Both non-optimized and control groups presented trivial change in Fv_{IMB}, P_{max} and jumping performance.

These results clearly showed that an optimized and individualized training program specifically addressing the F-v imbalance is more efficient for improving jumping performance than a traditional resistance training common to all subjects regardless of their force-velocity imbalance and optimal force-velocity profile (Fig. 5.9). Most of the jumping performance improvements were associated to reduce F-v imbalance with little or no change in P_{max}. Normalized F-v profile and Fv_{IMB} could therefore be considered as a potentially useful variables for prescribing optimal resistance training to improve ballistic performances. These experimental results confirmed the theoretical principles of the optimized training approach (Morin and Samozino 2016; Samozino et al. 2012, 2014) that ballistic performance depends not only on maximal power output, but also on an optimal force-velocity profile (Jimenez-Reyes et al. 2016).

5.5 Conclusion

This chapter presents the implication of lower limb muscle mechanical capabilities in ballistic push-off performances. Biomechanical model simulations and experimental data showed that ballistic performances (e.g. jump, sprint start, change of direction) depend on both lower limb maximal power capabilities and F-v profile, with the existence of an individual optimal F-v profile. The latter can be accurately determined for each athlete using equations given in this chapter and usual squat jump power-force-velocity profile evaluations, including testing using the simple field method presented in the previous chapter. This makes possible the determination of F-v imbalance (towards force or velocity capabilities) and the quantification of the magnitude of the associated force or velocity deficits. These indexes constitute interesting tools to individualize athlete's training program aiming to improve athletes' ballistic performance. These individual programs should focus on increasing lower limb maximal power and/or decreasing force-velocity imbalance.

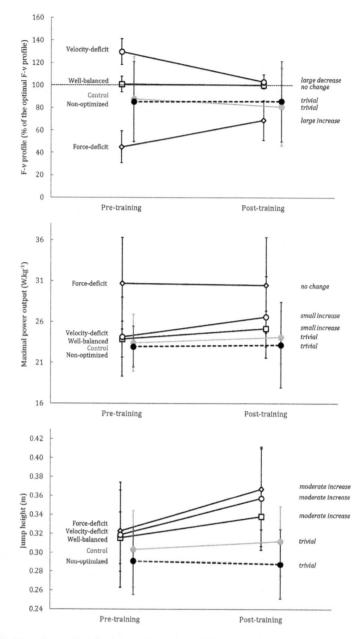

Fig. 5.9 Mean ± standard deviation of pre-post training changes in Force-velocity (F-v) profile, maximal power output and jumping performance for the intervention (filled symbols and black continuous lines), non-optimized (black points and dashed lines) and control (grey points and lines) groups. The intervention group is divided in three subgroups: the force-deficit group (filled diamonds), the velocity-deficit group (filled circles) and the well-balanced group (filled squares). The magnitude-based inference changes are presented on the right. The short-dashed line on the top panel represents the optimal force-velocity profile. (From Jimenez-Reyes et al. 2016)

The effectiveness of such an individualized "optimized" training was shown to be higher than a traditional strength training similar for all athletes. This support the great interest for strength and conditioning coaches, who aim to improve athlete's ballistic performance, to evaluate power-force-velocity profile on each of their athlete and to consider force-velocity imbalance to design individually training regimen.

References

Aerts P (1998) Vertical jumping in galago senegalensis: the quest for an obligate mechanical power amplifier. Philos Trans R Soc Lond B Biol Sci 353:1607–1620

Alexander RM (1995) Leg design and jumping technique for humans, other vertebrates and insects. Philos Trans R Soc Lond B Biol Sci 347(1321):235–248

Alexander RM (2003) Climbing and jumping. In: Principles of animal locomotion. Princeton University Press, pp 146–165

Argus CK, Gill ND, Keogh JW, Blazevich AJ, Hopkins WG (2011) Kinetic and training comparisons between assisted, resisted, and free countermovement jumps. J Strength Cond Res 25(8):2219–2227. https://doi.org/10.1519/JSC.0b013e3181f6b0f4

Bobbert MF, Casius LJ (2005) Is the effect of a countermovement on jump height due to active state development? Med Sci Sports Exerc 37(3):440–446

Bobbert MF, Van Soest AJ (1994) Effects of muscle strengthening on vertical jump height: a simulation study. Med Sci Sports Exerc 26(8):1012–1020

Bosco C (1992) La valutazione della forza con il test di Bosco [Strength assessment with the Bosco's Test]. In: Societa Sampa Sportiva, Roma, pp 56–57

Chelly MS, Fathloun M, Cherif N, Ben Amar M, Tabka Z, Van Praagh E (2009) Effects of a back squat training program on leg power, jump, and sprint performances in junior soccer players. J Strength Cond Res 23(8):2241–2249. https://doi.org/10.1519/JSC.0b013e3181b86c40

Cormie P, McBride JM, McCaulley GO (2009) Power-time, force-time, and velocity-time curve analysis of the countermovement jump: impact of training. J Strength Cond Res 23(1):177–186

Cormie P, McCaulley GO, McBride JM (2007a) Power versus strength-power jump squat training: influence on the load-power relationship. Med Sci Sports Exerc 39(6):996–1003

Cormie P, McCaulley GO, Triplett NT, McBride JM (2007b) Optimal loading for maximal power output during lower-body resistance exercises. Med Sci Sports Exerc 39(2):340–349

Cormie P, McGuigan MR, Newton RU (2010) Adaptations in athletic performance after ballistic power versus strength training. Med Sci Sports Exerc 42(8):1582–1598

Cormie P, McGuigan MR, Newton RU (2011a) Developing maximal neuromuscular power: part 1—biological basis of maximal power production. Sports Med 41(1):17–38

Cormie P, McGuigan MR, Newton RU (2011b) Developing maximal neuromuscular power: part 2—training considerations for improving maximal power production. Sports Med 41(2):125–146

Cronin J, McNair PJ, Marshall RN (2001) Velocity specificity, combination training and sport specific tasks. J Sci Med Sport 4(2):168–178

Cronin J, Sleivert G (2005) Challenges in understanding the influence of maximal power training on improving athletic performance. Sports Med 35(3):213–234

Davis DS, Briscoe DA, Markowski CT, Saville SE, Taylor CJ (2003) Physical characteristics that predict vertical jump performance in recreational male athletes. Phys Ther Sport 4(4):167–174

Djuric S, Cuk I, Sreckovic S, Mirkov D, Nedeljkovic A, Jaric S (2016) Selective effects of training against weight and inertia on muscle mechanical properties. Int J Sports Physiol Perform. https://doi.org/10.1123/ijspp.2015-0527

Driss T, Vandewalle H, Monod H (1998) Maximal power and force-velocity relationships during cycling and cranking exercises in volleyball players. Correlation with the vertical jump test. J Sports Med Phys Fitness 38(4):286–293

Eckert HM (1968) Angular velocity and range of motion in the vertical and standing broad jumps. Res Q 39(4):937–942

Emerson SB (1985) Jumping and leaping. In: Hildebrand ME, Bramble DM, Liem KF, Wake DB (eds) Functional vertebrate morphology. Harvard university Press, Cambridge, pp 58–72

Frost DM, Cronin J, Newton RU (2010) A biomechanical evaluation of resistance: fundamental concepts for training and sports performance. Sports Med 40(4):303–326

Genuario SE, Dolgener FA (1980) The relationship of isokinetic torque at two speeds to the vertical jump. Res Q Exerc Sport 51(4):593–598

Harris NK, Cronin JB, Hopkins WG, Hansen KT (2008) Squat jump training at maximal power loads vs. heavy loads: effect on sprint ability. J Strength Cond Res 22(6):1742–1749

Hautier CA, Linossier MT, Belli A, Lacour JR, Arsac LM (1996) Optimal velocity for maximal power production in non-isokinetic cycling is related to muscle fibre type composition. Eur J Appl Physiol Occup Physiol 74(1–2):114–118

James RS, Navas CA, Herrel A (2007) How important are skeletal muscle mechanics in setting limits on jumping performance? J Exp Biol 210(Pt 6):923–933

Jaric S, Markovic G (2009) Leg muscles design: the maximum dynamic output hypothesis. Med Sci Sports Exerc 41(4):780–787

Jimenez-Reyes P, Samozino P, Brughelli M, Morin JB (2016) Effectiveness of an individualized training based on force-velocity profiling during jumping. Front Physiol 7:677. https://doi.org/10.3389/fphys.2016.00677

Jimenez-Reyes P, Samozino P, Cuadrado-Penafiel V, Conceicao F, Gonzalez-Badillo JJ, Morin JB (2014) Effect of countermovement on power-force-velocity profile. Eur J Appl Physiol. https://doi.org/10.1007/s00421-014-2947-1

Kaneko M, Fuchimoto T, Toji H, Suei K (1983) Training effect of different loads on the force-velocity relationship and mechanical power output in human muscle. Scand J Sports Sci 5(2):50–55

Knudson DV (2009) Correcting the use of the term "power" in the strength and conditioning literature. J Strength Cond Res 23(6):1902–1908

Kotzamanidis C, Chatzopoulos D, Michailidis C, Papaiakovou G, Patikas D (2005) The effect of a combined high-intensity strength and speed training program on the running and jumping ability of soccer players. J Strength Cond Res 19(2):369–375. https://doi.org/10.1519/R-14944.1

Markovic G, Jaric S (2007) Positive and negative loading and mechanical output in maximum vertical jumping. Med Sci Sports Exerc 39(10):1757–1764

Markovic G, Vuk S, Jaric S (2011) Effects of jump training with negative versus positive loading on jumping mechanics. Int J Sports Med 32:1–8

Marsh RL (1994) Jumping ability of anuran amphibians. Adv Vet Sci Comp Med 38B:51–111

McBride JM, Kirby TJ, Haines TL, Skinner J (2010) Relationship between relative net vertical impulse and jump height in jump squats performed to various squat depths and with various loads. Int J Sports Physiol Perform 5(4):484–496

McBride JM, Triplett-McBride T, Davie A, Newton RU (2002) The effect of heavy- vs. light-load jump squats on the development of strength, power, and speed. J Strength Cond Res 16(1):75–82

Minetti AE (2002) On the mechanical power of joint extensions as affected by the change in muscle force (or cross-sectional area), ceteris paribus. Eur J Appl Physiol 86(4):363–369

Morin JB, Samozino P (2016) Interpreting power-force-velocity profiles for individualized and specific training. Int J Sports Physiol Perform 11(2):267–272. https://doi.org/10.1123/ijspp.2015-0638

Newton RU, Kraemer WJ (1994) Developing explosive muscular power: implications for a mixed methods training strategy. Strength Conditioning 16(5):20–31

Nuzzo JL, McBride JM, Dayne AM, Israetel MA, Dumke CL, Triplett NT (2011) Testing of the maximal dynamic output hypothesis in trained and untrained subjects. J Strength Cond Res 24 (5):1269–1276

Pandy MG, Zajac FE (1991) Optimal muscular coordination strategies for jumping. J Biomech 24 (1):1–10

Pazin N, Berjan B, Nedeljkovic A, Markovic G, Jaric S (2013) Power output in vertical jumps: does optimum loading depend on activity profiles? Eur J Appl Physiol 113(3):577–589. https://doi.org/10.1007/s00421-012-2464-z

Pazin N, Bozic P, Bobana B, Nedeljkovic A, Jaric S (2011) Optimum loading for maximizing muscle power output: the effect of training history. Eur J Appl Physiol 111(9):2123–2130. https://doi.org/10.1007/s00421-011-1840-4

Rejc E, Lazzer S, Antonutto G, Isola M, di Prampero PE (2010) Bilateral deficit and EMG activity during explosive lower limb contractions against different overloads. Eur J Appl Physiol 108 (1):157–165

Samozino P, Edouard P, Sangnier S, Brughelli M, Gimenez P, Morin JB (2014) Force-velocity profile: imbalance determination and effect on lower limb ballistic performance. Int J Sports Med 35(6):505–510. https://doi.org/10.1055/s-0033-1354382

Samozino P, Morin JB, Hintzy F, Belli A (2008) A simple method for measuring force, velocity and power output during squat jump. J Biomech 41(14):2940–2945

Samozino P, Morin JB, Hintzy F, Belli A (2010) Jumping ability: a theoretical integrative approach. J Theor Biol 264(1):11–18

Samozino P, Rejc E, Di Prampero PE, Belli A, Morin JB (2012) Optimal force-velocity profile in ballistic movements. Altius: citius or fortius? Med Sci Sports Exerc 44(2):313–322

Sheppard J, Cormack S, Taylor K, McGuigan M, Newton R (2008) Assessing the force-velocity characteristics of the leg extensors in well-trained athletes: the incremental load power profile. J Strength Cond Res 22(4):1320–1326

Stone MH, Moir G, Glaister M, Sanders R (2002) How much strength is necessary? Phys Ther Sport 3:88–96

Ugrinowitsch C, Tricoli V, Rodacki AL, Batista M, Ricard MD (2007) Influence of training background on jumping height. J Strength Cond Res 21(3):848–852

Vandewalle H, Peres G, Heller J, Panel J, Monod H (1987) Force-velocity relationship and maximal power on a cycle ergometer. Correlation with the height of a vertical jump. Eur J Appl Physiol Occup Physiol 56(6):650–656

Winter EM (2005) Jumping: power or impulse. Med Sci Sports Exerc 37(3):523–524

Yamauchi J, Ishii N (2007) Relations between force-velocity characteristics of the knee-hip extension movement and vertical jump performance. J Strength Cond Res 21(3):703–709

Young W, Wilson G, Byrne C (1999) Relationship between strength qualities and performance in standing and run-up vertical jumps. J Sports Med Phys Fitness 39(4):285–293

Chapter 6
A Simple Method for Measuring Lower Limb Stiffness in Hopping

Teddy Caderby and Georges Dalleau

Abstract Lower limb stiffness is of great of interest to the scientific and sporting communities, given its implication in sporting performance and in musculoskeletal injury risk. In the literature, lower limb stiffness has been extensively studied during hopping, as it constitutes a simple bouncing gait. Characterization of lower limb stiffness in hopping is commonly based on a biomechanical model called the "spring-mass model". This model assimilates the whole-body to an oscillating system consisting of a mass supported by a single spring, which represents the mechanical behaviour of the lower limbs during the ground contact phase of hopping. The stiffness of the spring, referred to as "leg spring", represents an overall stiffness of the musculoskeletal system of the lower limbs. In this chapter, we will describe the biomechanical aspects related to this concept of leg stiffness in hopping and we will present a simple method for measuring it. This method enables the calculation of leg stiffness from just the body mass of the individual and the contact and flight times during hopping, both of which may be obtained with simple technical equipment. This simple method may be particularly advantageous for assessing leg stiffness in a field environment, as well as in laboratory conditions.

6.1 Introduction

Jump tests are commonly used in the sporting field for assessing athletes' physical qualities and in particular for determining the mechanical characteristics of their lower extremities (Joyce and Lewindon 2016). The most popular jump tests are probably vertical jumps - such as the counter movement jump, squat jump or drop jump—which consist of athletes raising their centre of mass as high as possible above the ground. Since it requires the production of a high force level over a very

---

T. Caderby · G. Dalleau (✉)
Department of Physical Activity and Sports Science - IRISSE EA4075,
University of La Réunion, 117 Rue Du Général Ailleret,
97430 Le Tampon, La Réunion, France
e-mail: georges.dalleau@univ-reunion.fr

© Springer International Publishing AG 2018
J.-B. Morin and P. Samozino (eds.), *Biomechanics of Training and Testing*,
https://doi.org/10.1007/978-3-319-05633-3_6

short time interval, the vertical jump is considered to be one of the most explosive tasks (Samozino et al. 2008). So, the vertical jump has been widely used for assessing the mechanical power of the lower extremities and various methods have been proposed in the literature to measure it in a field environment (e.g. Bosco et al. 1983; Gray et al. 1962; Samozino et al. 2008, see Chaps. 4 and 5).

Another jump test frequently used for evaluating the mechanical properties of the lower limbs is vertical hopping. Vertical hopping, hereinafter referred to as "hopping", consists of bouncing repeatedly in place on both legs[1] by minimizing the angular displacements at the leg joints, i.e. the hip, knee and ankle joints (Lamontagne and Kennedy 2013). Given the high jump frequency, approximately double compared with the repeated counter-movement jump tests (Bosco 1999), the lower limbs remain more rigid in hopping that during other jumps. This makes it a relevant task for assessing the stiffness of the lower limbs (Kuitunen et al. 2011). Stiffness, which is classically defined as the ability of a body to resist deformation (Serpell et al. 2012), characterizes the elastic behaviour of a structure. It is usually associated with the body's ability to store and return the elastic strain energy, which is a factor that potentially influences the efficiency and performance of human movement (Cavanagh and Kram 1985; Komi and Bosco 1978). There is indeed a growing body of evidence suggesting that lower limb stiffness is implicated in sporting performance (for reviews, see Brazier et al. 2014; Butler et al. 2003; McMahon et al. 2012). For example, it has been found that lower limb stiffness is related to performance in endurance running (Dalleau et al. 1998; Dutto and Smith 2002; Heise and Martin 1998; Kerdok et al. 2002; McMahon et al. 1987), in sprint running (Bret et al. 2002; Chelly and Denis 2001), as well as the long jump (Seyfarth et al. 1999, 2000). Although there is no clear consensus on the topic, it seems that there may be an optimal level of leg stiffness to maximize athletic performance (Arampatzis et al. 2001; McMahon et al. 2012; Pearson and McMahon 2012; Seyfarth et al. 1999). Too much or too little stiffness may be detrimental and possibly induce musculoskeletal injuries (Butler et al. 2003; Flanagan et al. 2008; Williams et al. 2001). Given its implication in sporting performance and injury risk, lower limb stiffness is an important topic in scientific and sporting communities.

In this chapter, we will first present the biomechanical aspects related to the concept and assessment of lower limb stiffness in hopping. Within this context, we will describe the biomechanical model classically used for studying lower limb stiffness, the reference measurement methods, and their limitations. Secondly, we will introduce a simple method for measuring lower limb stiffness in hopping and discuss the application, and limitations, of this method.

---

[1]Although vertical hopping may also be performed on one leg, we will only address aspects related to two-legged hopping in this chapter.

6.2 Lower Limb Stiffness in Hopping

6.2.1 The Mechanical Definition of Stiffness

From a mechanical point of view, stiffness refers to the ability of a body to resist changes in length. For an elastic body,[2] such as a spring, stiffness may be determined from the relationship that exists between the force applied to the body and the change in length produced. This is known as Hooke's law. According to this law, the force needed to deform an elastic body (e.g. for stretching or compressing it) is proportional to the resulting deformation—i.e. the change in length—and the body stiffness. That is:

$$F = K \times \Delta L \qquad (6.1)$$

where F is the force applied (in newtons), K is the stiffness (in newton-metres^{-1}) and ΔL is the change in length (in metres).

Thus, stiffness represents the ratio of the force to the change in length:

$$K = \frac{F}{\Delta L} \qquad (6.2)$$

According to this principle, the higher the stiffness, the higher the required force to deform the body.

6.2.2 Spring-Like Leg Behaviour in Human Hopping

Human hopping is frequently described using a spring-mass model (Blickhan 1989; McMahon and Cheng 1990). This model assimilates the whole-body to an oscillating system that consists of a mass, equivalent to the body mass, supported by a single mass-less spring characterising the behaviour of the lower limbs. At each hop, the spring compresses during the first half of the ground contact phase (as leg joints flex) and lengthens during the second half of the ground contact phase (as leg joints extend), before the subsequent flight phase (Fig. 6.1). The stiffness of the spring, typically referred to as "leg spring", thus reflects the ability of the lower limbs to resist a change in length.

Although the spring-mass model is in appearance a quite simple model, it describes and predicts the mechanics of human bouncing gaits, such as hopping and running, remarkably well (e.g. Blickhan 1989; Farley and Gonzalez 1996; He et al. 1991; McMahon and Cheng 1990). The experimental validation of this model for

---

[2]An elastic body refers to a deformable material body that returns to its original shape and size when the forces causing the deformation are removed.

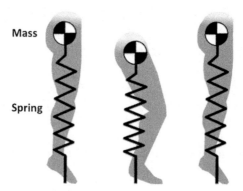

Fig. 6.1 Illustration of the spring–mass model classically used to describe the overall biomechanics of human hopping. In this model, the whole-body is represented by a mass, equivalent to the body mass, and a single massless linear spring representing the behaviour of the lower limbs. At the initial point of ground contact, the leg spring is uncompressed. During the first half of the ground contact phase, the leg spring compresses as the leg joints flex, and it reaches a maximal compression at approximately the middle of the contact phase. During the second half of the ground contact phase, the leg spring lengthens as leg joints extend, before the subsequent aerial phase

hopping has been accomplished mainly through force platform measurements during hopping (Blickhan 1989; Farley et al. 1991; McMahon and Cheng 1990). A force platform is a device that measures the reaction force supplied by the ground, i.e. the ground reaction force. From force platform measurements, it is also possible to compute the displacement of the centre of mass of the body (Cavagna 1975).

The first studies analysing hopping on a force platform show that the vertical force and the vertical centre of mass displacement, which reflects the change in length of the leg spring, exhibit a typical pattern during the ground contact phase (e.g. Blickhan 1989; McMahon and Cheng 1990). Specifically, these studies demonstrate that during the first half of the contact phase, the vertical force increases while the centre of mass displaces downwards, thus reflecting a compression of the leg spring. The maximal vertical centre of mass displacement, i.e. the maximal leg spring compression, is reached approximately at the middle of the contact phase. During the second half of the contact phase, the vertical force decreases while the centre of mass displaces upwards, indicating a lengthening of the spring. In short, during the contact phase both the vertical force signal and the vertical centre of mass displacement behave as a sine wave (Fig. 6.2). This sinusoidal pattern of the force-time signal is characteristic of the behaviour of a linear[3] spring-mass system and thus confirms the spring-like leg behaviour during the ground contact phase of hopping (Blickhan 1989; McMahon and Cheng 1990).

---

[3]The term "linear" means that the deformation of the spring is linearly proportional to the force applied.

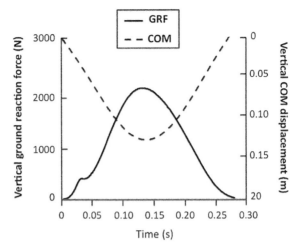

Fig. 6.2 Typical time course of the vertical ground reaction force (GRF) and the vertical displacement of the centre of mass (COM) during the ground contact phase of hopping. The traces were obtained for one subject (mass = 58 kg) hopping at a 2.2 Hz frequency

6.2.3 Modulation of Leg Stiffness in Hopping

The leg spring stiffness represents an overall stiffness of the musculoskeletal system of the lower limbs during the ground contact phase (Farley and Morgenroth 1999). Various anatomical structures contribute to this stiffness such as the muscles, tendons, ligaments, cartilage and bones of the limbs in contact with the ground. Specifically, leg stiffness corresponds to the combination of all the elementary stiffness values of these structures (Butler et al. 2003). Nevertheless, it is worth noting that leg stiffness may be modulated. Indeed, results from the literature reveal that humans are able to adapt leg stiffness to different situations.

Hopping height. Results from the literature show that leg stiffness varies as a function of hopping height (e.g. Farley and Morgenroth 1999; Farley et al. 1991). Precisely, leg stiffness increases when individuals augment the hopping height at a given hopping frequency. For example, Farley and Morgenroth (1999) found that leg stiffness is twice as great when subjects hop at a maximal height ($\approx$9 cm in their study) than at their preferred hopping height ($\approx$3 cm) for a given frequency (2.2 Hz). Preferred hopping height is the height at which an individual naturally hops at a given frequency. In contrast, it is worthwhile to note that when the hopping frequency is not imposed (i.e. freely chosen), leg stiffness does not increase with hopping height (Kuitunen et al. 2011).

Hopping frequency. In addition to hopping height, numerous studies show that hopping frequency also influences leg stiffness (Blickhan 1989; Dalleau et al. 2004; Farley et al. 1991; Granata et al. 2002; Hobara et al. 2010; Rapoport et al. 2003). Indeed, it has been observed that leg stiffness increases linearly with an increase in hopping frequency. Interestingly, it has been observed that when subjects are free to choose their hopping frequency, they spontaneously adopt a preferred frequency of about 2 Hz (Ferris and Farley 1997; Granata et al. 2002; Melvill Jones and Watt 1971).

Contact time. Changing hopping frequency usually leads a variation in ground contact time. Precisely, an increase in hopping frequency is often associated with a decrease in the time spent on the ground (Austin et al. 2003; Chang et al. 2008; Farley et al. 1991; Ferris and Farley 1997; Hobara et al. 2008, 2010; Rapoport et al. 2003). Nevertheless, it is possible to modulate contact time for the same hopping frequency. This modulation of contact time may be made for instance through instructions given to the subjects (Arampatzis et al. 2001; Hobara et al. 2007; Voigt et al. 1998a, b). Hobara et al. (2007) examined the effect of changing contact time on leg stiffness during hopping at the preferred frequency. For this, they compared leg stiffness in two hopping conditions: with preferred contact time ($\approx$238 ms in their study) and with the shortest possible contact time ($\approx$203 ms). In agreement with the findings about the effect of hopping frequency, these authors found that leg stiffness increased when contact time was reduced.

Ground surface properties. Leg stiffness is also influenced by the mechanical properties of the surface underfoot (Farley et al. 1998; Ferris and Farley 1997; Moritz and Farley 2003, 2005; Moritz et al. 2004). Precisely, it has been found that leg stiffness increased when the stiffness of the ground surface was decreased, i.e. when hopping was performed on a softer surface than normal. This adaptation in leg stiffness kept the total stiffness constant, i.e. the stiffness of the leg-surface combination, and thus maintained similar centre of mass motions, in particular in terms of height reached by the centre of mass during the flight phase. However, when the surface is highly damped, humans abandon the spring-mass behaviour (Moritz and Farley 2003, 2005). Indeed, in these conditions, lower limb mechanics need to be altered in order to prevent that the surface dictates the centre of mass motions. When hopping is performed on this type of very damp surface, the lower limbs extend during the first part of the contact phase and then retract in the second part of the stance. This adjustment maintains the centre of mass motion unchanged (Moritz and Farley 2003, 2005).

Loading conditions. Findings from the literature suggest that leg stiffness is sensitive to the loading conditions. Recently, Carretero-Navarro and Marquez (2015) observed that, in a hopping condition at a 3 Hz frequency, subjects increased leg stiffness when carrying an overload (weighted vest) superior to 10% of their body weight (BW). In contrast, these authors did not observe any effect of the load on leg stiffness when the hopping frequency was below 3 Hz or when the load was inferior to 10% of BW. In line with these results, Donoghue and Steele (2009) also found no significant difference in leg stiffness when subjects hopped at a 2.2 Hz frequency with or without an overload of 10% of BW. In short, it seems that the addition of a load may increase leg stiffness during hopping, in particular when the load is superior to 10% of the subject's BW and when the hopping frequency is elevated (≥ 3 Hz).

6.2.4 Mechanisms for Regulating Leg Stiffness in Hopping

To better understand how humans regulate leg stiffness in hopping, numerous studies are investigating stiffness at the joint level (e.g. Farley and Morgenroth 1999; Hobara et al. 2009; Kuitunen et al 2011). Indeed, leg stiffness as a general condition depends on the stiffness at the various joints of the leg, including the hip, knee and ankle joints (Farley and Morgenroth 1999). Stiffer joints will undergo smaller angular displacements during the contact phase, consequently reducing leg spring compression and thus increasing leg stiffness. In most of these studies, leg joints were thought of as springs, referred to as torsional springs, possessing a constant stiffness. Joint stiffness precisely represents the resistance of the joint to change in the angular displacement (Brazier et al. 2014). It is calculated as the ratio of the joint moment to the angular joint displacement. Joint kinematics is usually obtained with a high-speed video camera (Farley and Morgenroth 1999) or a motion capture system (Mrdakovic et al 2014). Joint moment is calculated from inverse dynamics using both kinematic and kinetic (i.e. force platform) data.

Results from studies examining joint stiffness in hopping are quite variable. Some studies show that leg stiffness is mainly regulated by modulating the ankle joint stiffness during hopping at a given frequency (Farley and Morgenroth 1999; Farley et al. 1998). In particular, these authors observed that subjects increased primarily ankle joint stiffness thus augmenting leg stiffness during hopping at 2.2 Hz. In contrast, other studies show that the major determinant for increasing leg stiffness during hopping at the preferred frequency was knee joint stiffness rather than ankle joint stiffness or hip joint stiffness (Hobara et al. 2009; Kuitunen et al. 2011). These conflicting results between studies could be due to differences in task constraints, in particular in the hopping frequency. Indeed, recent studies suggest that the relative contribution of ankle or knee joint stiffness to leg stiffness is frequency dependent (Hobara et al. 2011; Mrdakovic et al. 2014). According to these last studies, the main determinant of leg stiffness switches from knee stiffness to ankle stiffness when the hopping frequency is increased.

The aforementioned studies underline that leg stiffness may be regulated by adjusting stiffness at the joint level. Joint stiffness adjustments may be accomplished by modulating the muscle activity including the level of muscle activation before ground contact (Arampatzis et al. 2001; Hobara et al. 2007) and in the early phase of ground contact (Kuitunen et al 2011), the short-latency stretch reflex response at the instant of touchdown (Komi and Gollhofer 1997; Voigt et al. 1998a, b) and the muscle co-contraction level (Hortobagyi and DeVita 2000). Nevertheless, it should be noted that, in addition to joint stiffness, leg stiffness may also be modulated by altering the joint angle at the instant of touchdown. Indeed, by further extending the leg joints at touchdown, the ground reaction force vector will be more closely aligned with the joints, simultaneously decreasing the joint moments but increasing leg stiffness (Farley and Morgenroth 1999; Farley et al. 1998).

6.2.5 Measurement of Leg Stiffness in Hopping: The Reference Methods

In the literature, lower limb stiffness in hopping has been extensively assessed using a force platform. Various methods based on the spring-mass model have been proposed for calculating stiffness from force platform measurements (for reviews, see Brughelli and Cronin 2008; Butler et al. 2003; Serpell et al. 2012).

One of the simplest and most commonly used methods for calculating leg stiffness in hopping was introduced by McMahon and Cheng (1990). This method consists of calculating leg stiffness by dividing the peak vertical ground reaction force by the maximal vertical displacement of the centre of mass, both occurring approximately in the middle of the ground contact phase:

$$K_R = \frac{F_{max}}{\Delta COM} \quad (6.3)$$

where K_R is the leg stiffness, F_{max} is the peak vertical ground reaction force and ΔCOM is the maximum vertical displacement of the centre of mass.

Vertical displacement of the centre of mass may be computed by a double integration of the vertical centre of mass acceleration (Cavagna 1975), which may be obtained from the vertical ground reaction force using Newton's second law (F = ma). The goal being to determine the displacement, the integration constant is set arbitrarily at zero. The maximal vertical displacement of the centre of mass is then determined from the difference between the maximum and minimum values of the displacement curve. This procedure proposed by Cavagna (1975) for determining the vertical centre of mass displacement from force platform data has been shown to be accurate for a large range of hopping frequencies (Ranavolo et al. 2008).

A second method that has also been widely used for calculating leg stiffness in hopping is one detailed by Farley and Gonzalez (1996). This method requires plotting the vertical ground reaction force against the vertical centre of mass displacement for the ground contact phase, as illustrated in Fig. 6.3. Leg stiffness is then calculated as the slope of the linear regression of this force versus the displacement relationship (Farley and Gonzalez 1996).

A third method for calculating leg stiffness was proposed by McMahon et al. (1987). In this method, leg stiffness is calculated from the total body mass and the natural frequency[4] of the mass–spring system. Natural frequency of oscillation is determined from the vertical velocity of the COM and the duration of the contact and aerial phases, i.e. the contact and flight times, respectively. All these parameters

---

[4]Natural frequency is the frequency at which the spring-mass system oscillates freely, i.e. in the absence of any external force, once set into motion. This natural frequency depends on the mass and the stiffness of the system.

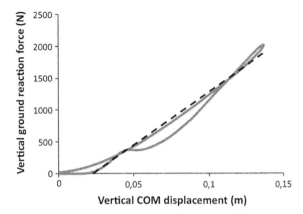

Fig. 6.3 Typical plot of the vertical ground reaction force versus the vertical centre of mass (COM) displacement for the ground contact phase of hopping. The curve was obtained for one subject (mass = 58 kg) hopping at a 2.2 Hz frequency. The slope (dotted line) of this curve represents leg stiffness

are derived from the vertical ground reaction force. Leg stiffness is then calculated by the following formula:

$$K_R = m\omega_0^2 \tag{6.4}$$

where K_R is the leg stiffness, m is the total body mass of the subject and ω_0^2 is the natural frequency of oscillation.

A fourth method for calculating leg stiffness has been described by Cavagna et al. (1988). In this method, stiffness is also calculated from the body mass and the natural frequency of oscillation using the same formula (Eq. 6.4) as McMahon et al. (1987). The sole difference between these two methods resides in the calculation of the natural frequency of oscillation. In their method, Cavagna et al. (1988) used the vertical ground reaction force history for determining the natural frequency of oscillation. More precisely, natural frequency was obtained from the effective contact time, which corresponds to the amount of time that the vertical force is greater than the body weight during the ground contact phase. By assuming that the vertical force-time curve is sinusoidal, one may consider that the effective contact time is equivalent to one-half of a period of oscillation (P/2), where P is equal to the period of oscillation of the spring-mass system. The natural frequency may then be calculated as:

$$\omega_0 = \frac{2\pi}{P} \tag{6.5}$$

where ω_0 is the natural frequency of oscillation and P is the period of oscillation.

Substituting the equation in 6.4, leg stiffness may be directly expressed by the following formula:

$$K_R = m\left(\frac{2\pi}{P}\right)^2 \tag{6.6}$$

where K_R is the leg stiffness, m is the body mass of the subject and P is the period of oscillation.

Studies have compared the leg stiffness values obtained from some of the above-described reference methods (Hébert-Losier and Eriksson 2014; Hobara et al. 2014). Hobara et al. (2014) found no difference in leg stiffness measurements between the method describing leg stiffness as the ratio of the vertical force by the vertical centre of mass displacement (McMahon and Cheng 1990) and the frequency-based method of Cavagna et al. (1988). In contrast, results of the study by Hébert-Losier and Eriksson (2014) suggest that the method proposed by McMahon and Cheng (1990) would be more reliable than the one described by Cavagna et al. (1988). Interestingly, other studies found that the method proposed by McMahon and Cheng (1990) presents a good interday reliability during hopping at 2.2 and 3.2 Hz (Joseph et al. 2013; McLachlan et al. 2006) and a moderate interday reliability during hopping at the preferred frequency (Joseph et al. 2013).

6.2.6 Limitations of the Reference Methods

The various reference methods described in the literature for measuring leg stiffness in hopping present some limitations. One of these limitations is inherent to the spring-mass model, on which these various methods rely. In reality, the human leg does not constitute a perfectly linear spring (Blickhan 1989; Farley and Gonzalez 1996; McMahon and Cheng 1990). Indeed, the lower limb is composed of several joints (mainly hip, knee and ankle) actuated by a set of muscles, which could be considered as a set of variable resistance springs (Bobbert and Casius 2011; Rapoport et al. 2003). However, although the mechanical behaviour of the leg is not entirely characterised by a linear spring, the spring-mass model represents an adequate and widely accepted model for studying leg stiffness in hopping.

In addition to this limitation associated with the model, the reference methods suffer from other limitations that restrict their use in sports, as well as in research. First, these various methods require a force platform, which is an expensive device and thus not readily accessible for most coaches and athletes. Next, the use of these methods is also limited by the data processing required and the computational complexity. To overcome these limitations, we developed a simple method for measuring lower limb stiffness from easily accessible parameters.

6.3 A Simple Method for Measuring Leg Stiffness in Hopping

6.3.1 Theoretical Foundations of the Method

The simple method proposed here relies on the general assumption that the vertical force-time signal during the ground contact phase of hopping may be modelled by a simple sine function (Cavagna et al. 1988; McMahon and Cheng 1990), which is described by the following equation:

$$F(t) = F_{max} \times \sin\left(\frac{\pi}{T_C} \times t\right) \quad (6.7)$$

where F_{max} is the maximal vertical force (in N), T_C is the contact time (in s) and represents the half period of the sine wave.

By supposing that the area under the sine curve is equivalent to the impulse of the vertical force signal during the contact time, it is possible to calculate the maximal vertical force as follows (see Dalleau et al. 2004 for details):

$$F_{max} = m \times g \times \frac{\pi}{2} \times \left(\frac{T_F}{T_C} + 1\right) \quad (6.8)$$

where m is the mass of the individual (in kg), g is the gravitational acceleration ($g = 9.81$ m s^{-1}) and T_F is the flight time.

The maximal vertical centre of mass displacement, ΔCOM (in m), during the contact phase may be obtained from the following equation:

$$\Delta COM = -\frac{F_{max}}{m}\frac{T_C^2}{\pi^2} + g\frac{T_C^2}{8} \quad (6.9)$$

Stiffness being the ratio of the maximal vertical force to the maximal vertical centre of mass displacement, it may be calculated as:

$$K_N = \frac{m \times \pi(T_F + T_C)}{T_C^2\left(\frac{T_F+T_C}{\pi} - \frac{T_C}{4}\right)} \quad (6.10)$$

Thus, leg stiffness may be calculated from just the following parameters: the subject's body mass (m), the contact (T_C) and flight (T_F) times.

6.3.2 Experimental Validation of the Method

This simple method has been validated against a reference method relying on force platform data. For this, an experiment was conducted over two sessions and enrolled eight subjects. In the first session, subjects performed submaximal two-legged hops with a large range of frequencies. They were instructed to hop on a force platform covered by a single contact mat for 10 s at frequencies ranging from 1.8 to 4.0 Hz (with increments of 0.2 Hz). In the second session, subjects performed ten maximal two-legged hops with their hands on their hips and their legs as straight as possible. In both sessions, the data recorded by the force platform (Kistler, Switzerland) and the contact mat (Mayser, Germany) was sampled at 500 Hz. The contact mat detected foot contact with the ground and thus provided the contact and flight times during hopping. The leg stiffness estimated by the simple method (K_N) was calculated from these contact and flight times using Eq. 6.10. This estimated stiffness was compared with the stiffness measured by the reference method based on force platform data. The reference stiffness (K_R) was calculated as the ratio of the peak vertical ground reaction force to the maximal vertical displacement of the centre of mass (McMahon and Cheng 1990).

Results show that the absolute differences between both methods ranged from 0.1 kN m^{-1} (i.e. 0.2% of K_R, obtained at 2.8 Hz) to 4.6 kN m^{-1} (i.e. 7.2% of K_R, obtained at 3.6 Hz; Table 6.1) during the submaximal hopping test. The absolute mean difference (all frequencies included) between both methods was 1.8 kN m^{-1}, i.e. 3.8% of K_R.

Table 6.1 Mean values and standard deviation (SD) of leg stiffness obtained from the reference method (K_R) and the simple method (K_N) for each hopping frequency

Hopping frequency (Hz)	K_R (kN m^{-1}) Mean ± SD	K_N (kN m^{-1}) Mean ± SD	Mean difference (kN m^{-1})	Limits of agreement Mean ± 2SD
1.8	21.8 ± 11.6	23.3 ± 10.2	1.5	−7.719 to 3.993
2	29.2 ± 12.9	28.1 ± 9.2	−1.1	−5.83 to 1.614
2.2	33.9 ± 10.6	33.4 ± 8.5	−0.5	−6.086 to 1.098
2.4	40.0 ± 10.1	37.7 ± 6.3	−2.3	−6.284 to 2.44
2.6	42.1 ± 7.8	41.8 ± 5.5	−0.3	−7.047 to 3.621
2.8	45.0 ± 4.5	44.9 ± 5.7	−0.1	−8.434 to 4.47
3	52.8 ± 6.7	50.2 ± 4.8	−2.5	−8.56 to 5.636
3.2	54.9 ± 8.2	52.3 ± 5.8	−2.5	−8.794 to 5.878
3.4	58.6 ± 4.2	55.2 ± 5.8	−3.4	−15.222 to 9.15
3.6	63.8 ± 8.5	59.3 ± 5.7	−4.6	−8.565 to 6.779
3.8	62.4 ± 9.9	61.8 ± 6.0	−0.6	−7.121 to 5.859
4	68.4 ± 6.7	66.8 ± 5.8	−1.6	−5.927 to 6.693

The mean difference and the limits of agreement between both methods are also presented

Furthermore, a high significant correlation was found between K_N and K_R in the submaximal hopping test (r = 0.94; $p < 0.001$). A strong correlation was also found between K_N and K_R in the maximal hopping test. In the maximal hopping test, the mean stiffness (± standard deviation) obtained by the simple method was 30.6 ± 10.0 and 34.0 ± 13.1 kN m^{-1} for the reference method. Taken together, these results suggest that the simple method is valid for assessing leg stiffness in submaximal and maximal hopping conditions.

6.3.3 Advantages and Limitations of the Method

The simple method enables the measurement of leg stiffness from simple parameters: the body mass of the individual, and contact and flight times. As compared with force platform methods, minimal data is stocked, data processing and calculation are greatly simplified, and no calibration is required. This method may thus be particularly advantageous for assessing leg stiffness in a field environment, as well as in laboratory conditions. In addition, the fact that this method is simple and inexpensive makes it accessible to most coaches and athletes.

Nevertheless, it must be recognized that the simple method presents some limitations. First, given that this method relies on the spring-mass model, the limitation associated with this model (see 6.2.6) also applies to the simple method.

A second limitation is associated with the modelling of the vertical force-time curve by a sine function. For a pure spring-mass system, the vertical force should theoretically have a bell shaped curve during the ground contact phase (McMahon and Cheng 1990; McMahon et al. 1987), which is not completely the case in reality. Compared to this bell shaped curve, the sine shaped model presents, for an equal impulse amount, a lower maximal force (Dalleau et al. 2004). This approximation of the vertical force signal curve by a sine wave may thus limit the accuracy of the simple method. In particular, this may explain the slight underestimation of stiffness obtained by this method as compared to the reference stiffness when hopping frequency is above 1.8 Hz (see Table 6.1). This tendency of the simple method to slightly underestimate leg stiffness has also been observed by another study (Hobara et al. 2014), following the initial validation of the method.

A third limitation of this simple method is related to the fact that the input parameters, i.e. body mass, contact and flight times, have to be accurately measured in order to ensure the accuracy of leg stiffness measurements. Morin et al. (2005) adapted this simple method for measuring leg stiffness in running (see Chap. 8 for details). By conducting a sensitivity analysis, these authors noted that the parameter with the greatest influence on stiffness was the contact time. Specifically, variation of the contact time influenced stiffness in a proportion of about 1:2, which means that a change of 10% in this parameter leads to a change of 20% in leg stiffness. Regarding the other parameters, it was found that body mass influenced stiffness in a proportion of about 1:1 and even less for flight time. These findings draw our attention to the necessity of accurately measuring these input parameters when using the simple method. Several systems currently allow measuring contact and flight times with accuracy, like for example contact mats, optical measurement systems (e.g. Optojump^{TM}), accelerometers and high-speed cameras. Readers may refer to Chap. 8 for detailed information about available devices and technologies for measuring these temporal parameters.

6.3.4 Application of the Method

The aim of developing this simple method was to facilitate leg stiffness assessment so that it can be carried out in a field environment by coaches or athletes themselves. Furthermore, this method has also been developed for use in research, particularly to study the regulation of stiffness in the optimization of sports performance. Numerous studies suggest that leg stiffness is related to performance, in particular in sports activities involving the stretch-shortening cycle, such as running and jumping (Brazier et al. 2014; Butler et al. 2003). However, there is no clear consensus on the optimal amount of stiffness for performance. Some studies show that higher levels of stiffness are beneficial to performance (Chelly and Denis 2001; Dalleau et al. 1998). Conversely, other studies suggest that lower levels of stiffness may improve the storage and utilization of elastic strain energy and thus enhance performance (Laffaye et al. 2005). Furthermore, some authors state that there is an optimal level of stiffness and that increasing this level would not necessarily enhance performance (Seyfarth et al. 2000). Hence, the relationship between stiffness and performance is not fully understood.

Within this context, we used the simple method to investigate the relationship between leg stiffness and another parameter of performance, i.e. the reactive power, in highly trained athletes including sprinters and ski racers (Dalleau et al. 2007). Reactive power, also called mechanical power, was measured by the method developed by Bosco et al. (1983). For this, subjects were instructed to perform maximal hops during a certain period (7 s in this study). Reactive power was calculated simply from contact and flight times using the following formula:

$$P_R = \frac{g^2 T_F T_T}{4 T_C} \qquad (6.11)$$

where P_R is the reactive power (in W kg^{-1}), g is the gravitational acceleration ($g = 9.81$ m s^{-1}), T_F is the flight time, T_C is the contact time and T_T is the total time ($T_F + T_T$).

Results show that Senegalese sprinters exhibited less leg stiffness (21.6 ± 3.7 kN m^{-1}) than both Italian sprinters (34.9 ± 5.3 kN m^{-1}) and ski racers (33.0 ± 5.4 kN m^{-1}), while no significant difference was found between Italian sprinters and ski racers ($p > 0.05$). Regarding the reactive power, we observed that Italian sprinters produced a higher reactive power (64.1 ± 4.6 W kg^{-1}) than Senegalese sprinters (49.5 ± 7.5 W kg^{-1}) and skiers

Table 6.2 Mean values and standard deviation of leg stiffness obtained from the simple method (in maximal hopping) for athletes from different sport disciplines

Athletes	Leg stiffness (kN m^{-1})	Body mass (kg)
National level male sprinters	31.7 ± 4.0	72.8 ± 7.9
National level male discus throwers	40.7 ± 2.0	96.2 ± 8.5
National level male hammer throwers	39.4 ± 2.1	97.2 ± 12.2
National level male shot putters	42.0 ± 2.4	107.0 ± 23.7

(50.0 ± 7.1 W kg^{-1}). In contrast, there was no significant difference in the reactive power between the Senegalese sprinters and skiers ($p > 0.05$). A significant correlation between leg stiffness and reactive power was found in sprinters ($r^2 = 0.68$; $p < 0.001$, Italian and Senegalese sprinters combined), but not in skiers ($p > 0.05$).

These results suggest that leg stiffness is related to the reactive power in athletes trained to produce power in similar conditions. Maximal hopping and sprinting may indeed be considered as activities particularly involving the stretch-shortening cycle, contrary to ski racing. Thus, probably due to the specificity of their training, sprinters seem to be more apt to take advantage of the stretch-shortening cycle. With similar stiffness, Italian sprinters are able to develop higher power than ski racers. With less stiffness, Senegalese sprinters produce power similar to skiers. These findings stress that the sport speciality or the type of training influences leg stiffness control. Other studies have confirmed that the type of training influences the regulation of leg stiffness. For example, Hobara et al. (2008) examined the differences between power-trained athletes and distance runners during hopping at 1.5 and 3 Hz. They found that the power athletes exhibited greater leg stiffness ($\approx$38 kN m^{-1} at 1.5 Hz and $\approx$62 kN m^{-1} at 3 Hz) than the endurance-trained athletes ($\approx$28 kN m^{-1} at 1.5 Hz and $\approx$48 kN m^{-1} at 3 Hz) at both hopping frequencies. By way of indication, other leg stiffness values obtained from the simple method (maximal hopping) for athletes from different sports disciplines are displayed in the following Table 6.2.

6.4 Conclusion

The method presented in this chapter constitutes a valid method for assessing leg stiffness in submaximal and maximal hopping conditions. This method calculates leg stiffness from only the individual's body mass and the contact and flight times during hopping, which may be obtained with basic technical equipment. This simple method may be particularly advantageous for assessing leg stiffness in a field environment, as well as in laboratory conditions.

References

Arampatzis A, Schade F, Walsh M, Bruggemann GP (2001) Influence of leg stiffness and its effect on myodynamic jumping performance. J Electromyogr Kinesiol 11(5):355–364

Austin GP, Tiberio D, Garrett GE (2003) Effect of added mass on human unipedal hopping at three frequencies. Percept Mot Skills 97(2):605–612

Blickhan R (1989) The spring-mass model for running and hopping. J Biomech 22:1217–1227

Bobbert MF, Richard Casius LJ (2011) Spring-like leg behaviour, musculoskeletal mechanics and control in maximum and submaximum height human hopping. Philos Trans R Soc Lond B Biol Sci 366(1570):1516–1529

Bosco C (1999) Strength assessment with the Bosco's test. Italian Society of Sports Science, Rome, Italy

Bosco C, Luhtanen P, Komi PV (1983) A simple method for measurement of mechanical power in jumping. Eur J Appl Physiol Occup Physiol 50(2):273–282

Brazier J, Bishop C, Simons C, Antrobus M, Read PJ, Turner AN (2014) Lower extremity stiffness: effects on performance and injury and implications for training. Strength Conditioning J 36(5):103–112

Bret C, Rahmani A, Dufour AB, Messonnier L, Lacour JR (2002) Leg strength and stiffness as ability factors in 100 m sprint running. J Sports Med Phys Fitness 42(3):274–281

Brughelli M, Cronin J (2008) A review of research on the mechanical stiffness in running and jumping: methodology and implications. Scand J Med Sci Sports 18(4):417–426

Butler RJ, Crowell HP, Davis IM (2003) Lower extremity stiffness: implications for performance and injury. Clin Biomech 18(6):511–517

Carretero-Navarro G, Márquez G (2015) Effect of different loading conditions on leg stiffness during hopping at different frequencies. Sci Sports 31(2):e27–e31

Cavagna GA (1975) Force platforms as ergometers. J Appl Physiol 39(1):174–179

Cavagna GA, Franzetti P, Heglund NC, Willems P (1988) The determinants of the step frequency in running, trotting and hopping in man and other vertebrates. J Physiol 399:81–92

Cavanagh PR, Kram R (1985) Mechanical and muscular factors affecting the efficiency of human movement. Med Sci Sports Exerc 17(3):326–331

Chang YH, Roiz RA, Auyang AG (2008) Intralimb compensation strategy depends on the nature of joint perturbation in human hopping. J Biomech 41(9):1832–1839

Chelly SM, Denis C (2001) Leg power and hopping stiffness: relationship with sprint running performance. Med Sci Sports Exerc 33(2):326–333

Dalleau G, Belli A, Bourdin M, Lacour JR (1998) The spring-mass model and the energy cost of treadmill running. Eur J Appl Physiol Occup Physiol 77(3):257–263

Dalleau G, Belli A, Viale F, Lacour JR, Bourdin M (2004) A simple method for field measurements of leg stiffness in hopping. Int J Sports Med 25(3):170–176

Dalleau G, Rahmani A, Verkindt C (2007) Relationship between power and musculotendinous stiffness in high level athletes. Sci Sports 22(2):110–116

Donoghue O, Steele L (2009) Acute effects of hopping with weighted vest on vertical stiffness. In: ISBS-Conference Proceedings Archive 1(1)

Dutto DJ, Smith GA (2002) Changes in spring-mass characteristics during treadmill running to exhaustion. Med Sci Sports Exerc 34(8):1324–1331

Farley CT, Blickhan R, Saito J, Taylor CR (1991) Hopping frequency in humans: a test of how springs set stride frequency in bouncing gaits. J Appl Physiol 71(6):2127–2132

Farley CT, Gonzalez O (1996) Leg stiffness and stride frequency in human running. J Biomech 29(2):181–186

Farley CT, Houdijk HH, Van Strien C (1985) Louie M (1998) Mechanism of leg stiffness adjustment for hopping on surfaces of different stiffnesses. J Appl Physiol 85(3):1044–1055

Farley CT, Morgenroth DC (1999) Leg stiffness primarily depends on ankle stiffness during human hopping. J Biomech 32(3):267–273

Ferris DP, Farley CT (1997) Interaction of leg stiffness and surfaces stiffness during human hopping. J Appl Physiol 82(1):15–22

Flanagan EP, Galvin L, Harrison AJ (2008) Force production and reactive strength capabilities after anterior cruciate ligament reconstruction. J Athl Train 43(3):249–257

Granata KP, Padua DA, Wilson SE (2002) Gender differences in active musculoskeletal stiffness. Part II. quantification of leg stiffness during functional hopping tasks. J Electromyogr Kinesiol 12(2):127–135

Gray R, Start K, Glenross D (1962) A test of leg power. Res Q 33:44–50

He JP, Kram R, McMahon TA (1991) Mechanics of running under simulated low gravity. J Appl Physiol 71(3):863–870

Hébert-Losier K, Eriksson A (2014) Leg stiffness measures depend on computational method. J Biomech 47(1):115–121

Heise GD, Martin PE (1998) "Leg spring" characteristics and the aerobic demand of running. Med Sci Sports Exerc 30(5):750–754

Hobara H, Inoue K, Kobayashi Y, Ogata T (2014) A comparison of computation methods for leg stiffness during hopping. J Appl Biomech 30(1):154–159

Hobara H, Inoue K, Muraoka T, Omuro K, Sakamoto M, Kanosue K (2010) Leg stiffness adjustment for a range of hopping frequencies in humans. J Biomech 43(3):506–511

Hobara H, Inoue K, Omuro K, Muraoka T, Kanosue K (2011) Determinant of leg stiffness during hopping is frequency-dependent. Eur J Appl Physiol 111(9):2195–2201

Hobara H, Kanosue K, Suzuki S (2007) Changes in muscle activity with increase in leg stiffness during hopping. Neurosci Lett 418(1):55–59

Hobara H, Kimura K, Omuro K, Gomi K, Muraoka T, Iso S, Kanosue K (2008) Determinants of difference in leg stiffness between endurance- and power-trained athletes. J Biomech 41(3):506–514

Hobara H, Muraoka T, Omuro K, Gomi K, Sakamoto M, Inoue K, Kanosue K (2009) Knee stiffness is a major determinant of leg stiffness during maximal hopping. J Biomech 42(11):1768–1771

Hortobagyi T, DeVita P (2000) Muscle pre- and coactivity during downward stepping are associated with leg stiffness in aging. J Electromyogr Kinesiol 10(2):117–126

Joseph CW, Bradshaw EJ, Kemp J, Clark RA (2013) The interday reliability of ankle, knee, leg, and vertical musculoskeletal stiffness during hopping and overground running. J Appl Biomech 29(4):386–394

Joyce D, Lewindon D (2016) Sports injury prevention and rehabilitation: integrating medicine and science for performance solutions. Routledge, New York

Kerdok AE, Biewener AA, Mcmahon TA, Weyand PG, Herr HM (2002) Energetics and mechanics of human running on surfaces of different stiffnesses. J Appl Physiol (1985) 92(2):469–478

Komi PV, Bosco C (1978) Utilization of stored elastic energy in leg extensor muscles by men and women. Med Sci Sports 10(4):261–265

Komi PV, Gollhofer A (1997) Stretch reflex can have an important role in force enhancement during SSC-exercise. J Appl Biomech 13:451–460

Kuitunen S, Ogiso K, Komi PV (2011) Leg and joint stiffness in human hopping. Scand J Med Sci Sports 21(6):e159–e167

Laffaye G, Bardy BG, Durey A (2005) Leg stiffness and expertise in men jumping. Med Sci Sports Exerc 37(4):536–543

Lamontagne M, Kennedy MJ (2013) The biomechanics of vertical hopping: a review. Res Sports Med 21(4):380–394

McLachlan KA, Murphy AJ, Watsford ML, Rees S (2006) The interday reliability of leg and ankle musculotendinous stiffness measures. J Appl Biomech 22(4):296–304

McMahon JJ, Comfort P, Pearson SJ (2012) Lower limb stiffness: effect on performance and training considerations. Strength Conditioning J 34(6):94–101

McMahon TA, Cheng GC (1990) The mechanics of running: how does stiffness couple with speed? J Biomech 23(Suppl 1):65–78

McMahon TA, Valiant G, Frederick EC (1987) Groucho running. J Appl Physiol 62(6): 2326–2337

Melvill Jones G, Watt DG (1971) Observations on the control of stepping and hopping movements in man. J Physiol 219(3):709–727

Morin JB, Dalleau G, Kyrolainen H, Jeannin T, Belli A (2005) A simple method for measuring stiffness during running. J Appl Biomech 21(2):167–180

Moritz CT, Farley CT (2003) Human hopping on damped surfaces: strategies for adjusting leg mechanics. Proc Biol Sci 270(1525):1741–1746

Moritz CT, Farley CT (2005) Human hopping on very soft elastic surfaces: implications for muscle pre-stretch and elastic energy storage in locomotion. J Exp Biol 208(Pt 5):939–949

Moritz CT, Greene SM, Farley CT (2004) Neuromuscular changes for hopping on a range of damped surfaces. J Appl Physiol (1985) 96(5):1996–2004

Mrdakovic V, Ilic D, Vulovic R, Matic M, Jankovic N, Filipovic N (2014) Leg stiffness adjustment during hopping at different intensities and frequencies. Acta Bioeng Biomech 16(3):69–76

Pearson SJ, McMahon J (2012) Lower limb mechanical properties: determining factors and implications for performance. Sports Med 42(11):929–940

Ranavolo A, Don R, Cacchio A, Serrao M, Paoloni M, Mangone M, Santilli V (2008) Comparison between kinematic and kinetic methods for computing the vertical displacement of the center of mass during human hopping at different frequencies. J Appl Biomech 24(3):271–279

Rapoport S, Mizrahi J, Kimmel E, Verbitsky O, Isakov E (2003) Constant and variable stiffness and damping of the leg joints in human hopping. J Biomech Eng 125(4):507–514

Samozino P, Morin JB, Hintzy F, Belli A (2008) A simple method for measuring force, velocity and power output during squat jump. J Biomech 41(14):2940–2945

Serpell BG, Ball NB, Scarvell JM, Smith PN (2012) A review of models of vertical, leg, and knee stiffness in adults for running, jumping or hopping tasks. J Sports Sci 30(13):1347–1363

Seyfarth A, Blickhan R, Van Leeuwen JL (2000) Optimum take-off techniques and muscle design for long jump. J Exp Biol 203(Pt 4):741–750

Seyfarth A, Friedrichs A, Wank V, Blickhan R (1999) Dynamics of the long jump. J Biomech 32(12):1259–1267

Voigt M, Chelli F, Frigo C (1998a) Changes in the excitability of soleus muscle short latency stretch reflexes during human hopping after 4 weeks of hopping training. Eur J Appl Physiol Occup Physiol 78(6):522–532

Voigt M, Dyhre-Poulsen P, Simonsen EB (1998b) Modulation of short latency stretch reflexes during human hopping. Acta Physiol Scand 163(2):181–194

Williams DS, 3rd, McClay IS, Hamill J (2001) Arch structure and injury patterns in runners. Clin Biomech 16(4):341–347

Chapter 7
A Simple Method for Measuring Force, Velocity, Power and Force-Velocity Profile of Upper Limbs

Abderrahmane Rahmani, Baptiste Morel and Pierre Samozino

Abstract Upper limb abilities can be assessed from different kinds of exercises (e.g., cranking, push-up, and medicine ball put test). Since the bench press is a very common exercise used in training routines by most athletes in many sports, its interest in the scientific literature raised in the past two decades. As presented in previous chapters during jumping or cycling, from several bench presses performed against different loads, coaches and athletes can simply and accurately define their upper limb force-velocity (F-v) profile. They can estimate their theoretical maximal force (F_0), velocity (V_0) and power (P_{max}). The aim of this chapter is to present the important points that must be taken into account to optimize the use of the bench press as a routine testing. In a first part, this chapter will focus on the importance of taking into consideration all the mechanical inertia involved in the bench press exercise. Not considering the upper limb mass in the calculation of the force produced during the exercise implies an underestimation in the F-v profile that can reach 30% for P_{max}. This could conduct to incorrect choices in the optimal load during training and thus limit the performance improvement. In the second part of this chapter, we present a simple mechanical model of the bench press to study the importance of the upper limb acceleration in the estimation of the force produced. The moving system (i.e., lifted mass and upper limbs) is modeled from rigid segments and the force can be determined thanks to four simple measurements: the vertical displacement of the lifted load, the elbow angle measured using a goniometer, the arm and forearm lengths and the constant horizontal position of the

---

A. Rahmani (✉) · B. Morel
Laboratory "Movement, Interactions, Performance", Department of Sport Sciences,
Faculty of Sciences and Technologies, Le Mans University, EA 8 4334, Le Mans, France
e-mail: abdel.rahmani@univ-lemans.fr

P. Samozino
Laboratoire Inter-universitaire de Biologie de la Motricité, Université de Savoie Mont Blanc,
Campus Scientifique, 73000 Le Bourget du Lac, Chambéry, France
e-mail: pierre.samozino@univ-smb.fr

© Springer International Publishing AG 2018
J.-B. Morin and P. Samozino (eds.), *Biomechanics of Training and Testing*,
https://doi.org/10.1007/978-3-319-05633-3_7

hand on the barbell. The validity of this model has been confirmed through experimental data obtained from a force platform. An important point is that the kinematics and kinetics of this model allow demonstrating that the acceleration of the moving system is similar to the one of the barbell. Finally, based on the previous statement, the last part of this chapter presents a simple method for assessing force, velocity and power during a ballistic bench press performed on a traditional guided barbell, based on the Newtonian laws and only three simple parameters: (i) upper limb mass estimated as 10% of the body mass; (ii) barbell flight height recorded with a nylon cable tie and (iii) push-off distance measured with a measuring tape. Consequently, coaches and athletes could accurately determine their F-v profile and extrapolate reliable mechanical parameters (F_0, v_0, Sfv and P_{max}) in order to maximize upper limbs performance and manage training programs in field conditions.

7.1 Introduction

The previous chapters present various simple methods to evaluate muscle power capabilities and F-v profile of lower limbs during ballistic push-off. The majority of studies focusing on lower limbs have been conducted using jumping exercises. While these exercises represent key movements for many sporting activities, upper limbs also have their importance in many activities (e.g., throwing, hitting, and rowing). Different methods have been proposed to assess the upper limb anaerobic abilities including all-out cranking exercises (Vanderthommen et al. 1997; Driss et al. 1998), medicine ball put test (Stockbrugger and Haennel 2001, 2003) and arm jump (Laffaye et al. 2014). While the validity and reliability of these tests have been attested, the bench press has the advantage to be one of the most common exercises used in training routines by most athletes in many sports. Indeed, bench press is an optimal training movement to increase the anterior trunk (*pectoralis* major and minor), arms (*triceps brachii*) and shoulders (anterior and medial deltoid) (Wilson et al. 1989; Barnett et al. 1995). An increasing interest to the use of the bench press exercise as a simple test for assessing upper limb strength is also observed in the scientific literature (Pearson et al. 2007; Padulo et al. 2012; Buitrago et al. 2013; Sreckovic et al. 2015; García-Ramos et al. 2016).

Two sub-types of bench press exist: the traditional bench press during which the barbell must remain in the hands of the athlete at the end of the propulsive phase of the movement (i.e., the barbell is voluntarily decelerated in order to not throw it), and the ballistic bench press (also called ballistic bench throw) during which the barbell is accelerated during the whole push-off phase inducing a flight phase, as during jumping. Whatever the considered bench press, the muscular parameters of the upper limbs are usually determined using force platforms (Rahmani et al. 2009; Young et al. 2015) or kinematic systems, such as optical encoders (Rambaud et al. 2008; Jandačka and Vaverka 2009) or linear transducers (Garnacho-Castaño et al. 2014; Sreckovic et al. 2015; García-Ramos et al. 2016).

Considering similar loads, the ballistic bench press allows the development of higher values of force, velocity, power and muscle activation in comparison to the traditional one (Newton et al. 1996). The deceleration phase occurring during the traditional bench press exercise seems to be responsible for these results (Cormie et al. 2011). Indeed, Sánchez-Medina et al. (2010) reported that when light and medium loads are lifted during a traditional bench press, the deceleration at the end of the movement is greater than what would be expected with the unique effect of gravity. The net force applied to the barbell (which is the mechanical output measured) may so underestimate the force produced by the agonist muscles due to the activation of antagonist muscles, which apply force in the opposite direction to the load motion in order to stop the movement (Jarić et al. 1995). However, the ballistic bench press becomes inadequate when performed with high loads. In this case, the movement is not ballistic anymore and the barbell cannot be thrown. Despite these differences, the force-velocity profiles of both traditional and ballistic bench press are linear, allowing the determination of the parameters evoked in the previous chapters (the theoretical maximal force, velocity and power, and a fortiori the individual Sfv). Then, the choice of the type of bench press used for training or testing depends on the goal to achieve. Ballistic bench press is preferred during power training as athletes are able to generate higher values of velocity, force and then power with light to moderate loads. Ballistic bench press can also be considered as more representative of ecological ballistic movements. The traditional bench press is by definition the only one that can be used for high loads and so to evaluate the one repetition maximum (1-RM).

This chapter will focus on the upper limb evaluation from the bench press exercise. First, we will discuss from previous experimental data the importance of taking the upper limb mass into account. Then, a mechanical model of the bench press exercise will be presented to highlight the importance of considering the involved limb segment (i.e., arm and forearm). Finally, a simple method will be detailed to assess the upper limb abilities thanks to only three simple parameters (upper limb mass, barbell flight height and push-off distance) that are easy to measure outside a laboratory and without specific devices.

7.2 The Force, Velocity, Power Mechanical Profile

7.2.1 Importance of the Upper Limb Inertia During the Bench Press

As previously mentioned, whatever the considered bench press exercise (i.e., traditional vs. ballistic), the force-velocity (F-v) and power-velocity (P-v) profiles fit linear and second polynomial models, respectively (Fig. 7.1; Rambaud et al. 2008; Sreckovic et al. 2015; García-Ramos et al. 2016). The F-v relationship and explosive maximal power are widely used parameters when studying the

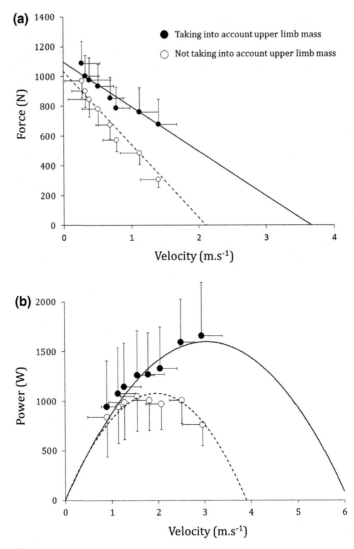

Fig. 7.1 Mean force–velocity (**a**) and power–velocity (**b**) relationships taking (open circles) or not taking (filled circles) upper-limb mass into account

mechanical characteristics of muscles or muscle groups. In explosive events like throwing, lower and upper-limb force and maximal power have been evidenced to contribute to the final performance (Bourdin et al. 2010). Determining precisely the maximal power is then important to organize the athlete's training.

It should be kept in mind that mostly all studies used kinematic systems to investigate the muscular power. These systems enable muscular power assessment in terms of lifted load displacement during an exercise. From an external load, and once known its displacement and the time to reach it, mean power is estimated using the Newtonian laws. In order to assess kinematic parameters, the whole mechanical system inertia (i.e., mass of the lifted load plus the inertia of the levers or involved body segments) must be carefully determined to precisely calculate the load at which the power training is optimized (Rambaud et al. 2008). Several authors have shown that the force produced during single-joint extension of the lower limbs is underestimated if lever arm and leg inertia are not taken into account (Winter et al. 1981; Nelson and Duncan 1983; Rahmani et al. 1999). This can lead to an underestimation of the maximal power, maximal force, and maximal velocity extrapolated from the F-v and P-v relationships (Rahmani et al. 1999).

In several bench press studies (Cronin et al. 2000; Shim et al. 2001; Izquierdo et al. 2002; Cronin and Henderson 2004; Sánchez-Medina et al. 2014; García-Ramos et al. 2016), the force was calculated on the basis of load only (i.e., without taking into account the total inertia of the system load plus upper limb mass). This was done "to evaluate the simplest possible approach that can be used for routine testing" (García-Ramos et al. 2016). However, this implies that the upper limb mass and the effort required to accelerate it were neglected, which can, as discussed above, lead to an underestimation of maximal power production. This methodological bias would explain why mean maximal force values obtained in team sport players using kinematic devices (Izquierdo et al. 2002; Cronin et al. 2003) are lower than those measured with a force platform (Wilson et al. 1991a, b, 1994; Murphy et al. 1994).

In order to illustrate this fact, a previous study (Rambaud et al. 2008) aimed at comparing the force calculated from a kinematic encoder to that simultaneously measured with a force platform fixed under the bench (Fig. 7.2). The traditional bench press exercise was done under a guided horizontal barbell. Then, we hypothesized that forces produced on the anteroposterior and mediolateral axis could be neglected. The instantaneous velocity and acceleration of the barbell were calculated from successive displacement time-derivatives for each lift. Instantaneous force (F, in N) was calculated as follows:

$$F = M(a+g) + F_f \quad (7.1)$$

where M is the considered moving mass, g is the gravitational acceleration (9.81 m s^{-2}), a is the calculated acceleration (m s^{-2}), and F_f is the friction force determined by a freefall test added to the concentric phase. F was determined by taking only the lifted load into account ($F_{peak}b$) or the total moving mass (lifted load plus upper limb mass estimated from Winter's anthropometric tables (Winter 2009) ($F_{peak}t$). The instantaneous power (in W) was calculated as the product of force and velocity at any given time.

The mean mass of the upper limbs represented about 10% of the total body mass of the subject. Since the lifted masses ranged from 7 to 74 kg, neglecting the upper limb mass has obviously an impact on the force calculation. When the upper limb mass was ignored, the force calculated with the kinematic device, regardless the lifted load, was significantly lower than the one measured with the force platform

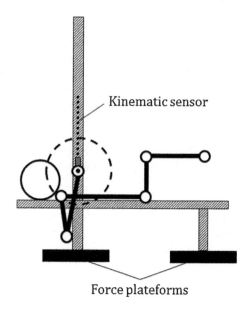

Fig. 7.2 A picture of the guided horizontal barbell used during the bench press exercise

(Fig. 7.3). This underestimation was greater for lighter loads because the relative contribution of the upper limb mass to the total inertia decreased as the lifted load increased (from 54 to 10% at 7 and 74 kg, respectively). Considering the upper limb mass in the force calculation, there was no difference between the forces directly measured with the force platform and those calculated from the Newtonian laws. Forces were significantly correlated (r = 0.91; $p < 0.001$), close to the identity line (Fig. 7.4).

7.2.2 Consequence of the Upper Limb Inertia on the Force-Velocity Profile

Considering or not the upper limb mass, the F-v is significantly linear (Fig. 7.1a, r = 0.75–0.98, $p < 0.05$), and the P-v is significantly described by a second order polynomial regression (Fig. 7.1b, r = 0.88–0.99, $p < 0.05$). Neglecting the upper limb mass has an impact on the extrapolated force-velocity parameters. The underestimation of F_0, V_0, P_{max} and V_{opt} when the upper limb is neglected is equal to 6, 41, 32 and 35%, respectively. Even if the difference is significant, the underestimation concerning F_0 is weak. This is due to the relative lower contribution of the upper limb inertia to the total inertia for the heavy lifted loads. The underestimation of P_{max} is more problematic since this parameter is obtained at light lifted loads. The theoretical optimal loads corresponded to a lifted load of 36 kg when the upper limb is considered, whereas this optimal load was about twice lower (15 kg) when the upper limb mass is neglected. This may have

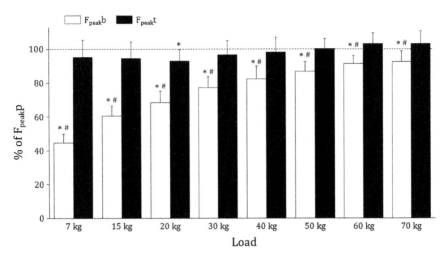

Fig. 7.3 Peak force values calculated from the kinematic sensor taking ($F_{peak}t$; in black) and not taking ($F_{peak}b$ in grey) upper limb mass into account and expressed in percentage of the peak force values measured by the force platform, ($F_{peak}P$, reference method). Asterisk: Significantly different from 100% $F_{peak}p$ ($P < 0.001$); Hash: $F_{peak}b$ significantly different from $F_{peak}t$ ($P < 0.001$)

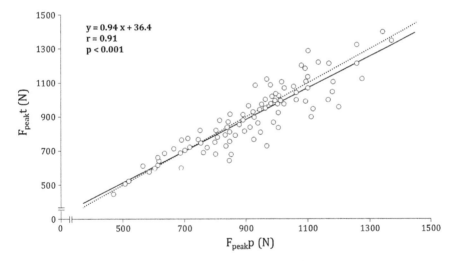

Fig. 7.4 Relationship between force peak values measured with the force platform ($F_{peak}P$) and peak force values calculated from data obtained with the kinematic device taking upper limb inertia into account ($F_{peak}t$). The dotted line represents the identity line

important consequences in explosive training since this kind of training are based on the optimal load at which the maximal power should be addressed (Caiozzo et al. 1981; Kanehisa and Miyashita 1983; Kaneko et al. 1983).

7.3 A Simple Model of the Bench Press Exercise

7.3.1 Importance of the Shoulder During the Bench Press

In the context of the evaluation, the previous section insists on the importance of the upper limb inertia. However, the model used in the previous study considered the upper limb as a punctual mass, moving only vertically. Rambaud et al. (2008) did not give any information on the upper limb acceleration, and its importance in the force estimation. The bench press exercise involved two joints (the elbow and the shoulder) and several muscular groups (*pectoralis* major and minor, *triceps brachii*, anterior and medial deltoid) which are progressively involved. To dissociate each segment participating in the movement (i.e., arm and forearm) is thus capital to appreciate their respective impact in the whole movement kinematics and dynamics. A first approach was to consider a model of the bench press including only the elbow joint (i.e., the shoulder and the wrist were supposed still) (Fig. 7.5). The force could be determined thanks to four simple measurements: the vertical displacement of the lifted load recorded by using a kinematic device, the elbow angle measured using a goniometer, the arm and forearm lengths estimated with the Winter's table (Winter 2009) and the constant horizontal position of the hand on the barbell measured with a tape. However, the computer simulation of the movement evidenced that the sum of the arm and forearm lengths does not reach the maximal height at which le load is lifted (personal data). This implies that the shoulder must be included in the upper limb model (Fig. 7.5c).

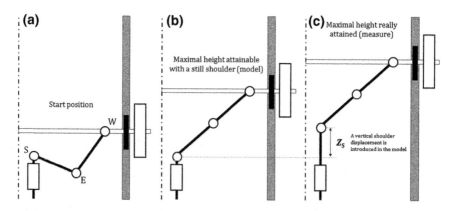

Fig. 7.5 Mechanical model of the upper limb: **a** start position. With the shoulder S supposed still, **b** the bar cannot reach the maximal height experimentally measured. A shoulder vertical displacement must be introduced to the model (**c**). E: elbow, W: wrist

7.3.2 A Simple Model Based on Three Segments: Shoulder, Arm and Forearm

Description of the model. Since the bench press exercise was performed with a guided horizontal barbell, actions of the two upper limbs are assumed to be symmetrical. The model had three degrees of freedom. Two revolute joints were used to model the shoulder and elbow rotations, and the vertical shoulder displacement (Z_S) was represented by introducing a prismatic joint (Fig. 7.6a). The position of the subject's hands was noted (x_0, Z). The coordinate x_0 represented the horizontal position of the hand, which was constant because the movement was performed under a vertically guided barbell. Z was the vertical displacement of the barbell and Z_0 was the vertical position of the hand at rest relative to the horizontal axis.

The absolute angles of the upper arm (θ_a) and forearm ($\theta_a + \theta_f$) were expressed relatively to the horizontal axis. θ_f was calculated from the angle measured between the upper arm and the forearm as $\theta_f = 180 - \theta$, where the anatomic angle of the elbow θ was measured by goniometry.

Inverse kinematic model. An inverse kinematic model q was used to calculate the joint coordinates θ_a (in rad) and Z_S (in m) derived from the vertical displacement Z and the elbow angle θ_f:

$$q = \begin{bmatrix} \theta_a \\ \theta_f \\ Z_S \end{bmatrix} \quad (7.2)$$

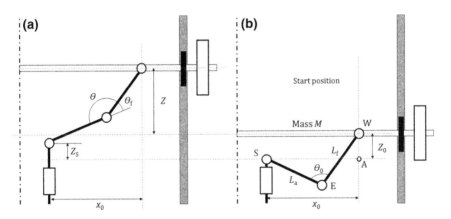

Fig. 7.6 a Mechanical model of the upper limb during the bench press exercise with 3 limb segments linked by 2 revolute joints and a prismatic joint: x_0, horizontal position of the wrist; Z_0, initial vertical position of the wrist relative to the horizontal axis; Z, vertical displacement of the lifted mass; L_a, upper arm length; L_f, forearm length; θ: elbow angle; θ_a, absolute upper arm angle; θ_f, forearm angle relative to the arm position. **b**. Initial position of the subject: S, shoulder; E, elbow; W, wrist; A, orthogonal projection of the wrist on the horizontal axis; θ_0, initial elbow angle

θ_a and Z_S were calculated from the hand coordinates which was written as followed:

$$x_0 = L_a \cos\theta_a + L_f \cos(\theta_a + \theta_f) \qquad (7.3)$$

$$Z + Z_0 - Z_S = L_a \sin\theta_a + L_f \sin(\theta_a + \theta_f) \qquad (7.4)$$

where L_a is the length of the upper arm (in m) and L_f is the length of the forearm (in m), both estimated from Winter's table (Winter 2009), Z_0 is the initial vertical position of the hand.

The absolute angle of the arm θ_a (in rad) is derived from Eq. 7.3:

$$\theta_a = \tan^{-1}\left(\frac{Bx_0 + A\sqrt{C - x_0^2}}{Ax_0 + B\sqrt{C - x_0^2}}\right) \qquad (7.5)$$

where $A = L_a + L_f \cos\theta_f$, $B = -L_f \sin\theta_f$, $C = A^2 + B^2$.

The vertical displacement of the shoulder is derived from Eq. 7.4. Firstly, we need to calculate Z_0. This can be done geometrically from the rest position (Fig. 7.6b). Applying the Pythagoras' theorem in the triangle SAW (A is a virtual point allowing to express the distance SW from the known distance), we can write:

$$x_0^2 + Z_0^2 = SW^2 \qquad (7.6)$$

The SW side of the triangle SEW can then be expressed as:

$$SW^2 = L_a^2 + L_f^2 - 2L_a L_f \cos\theta_0$$

From Eqs. 8.4 and 8.6, Z_0 can be written as:

$$Z_0 = \sqrt{L_a^2 + L_f^2 - 2L_a L_f \cos\theta_0 - x_0^2} \qquad (7.7)$$

Z_S is then equal to:

$$Z_S = Z + Z_0 - \sqrt{C - x_0^2} \qquad (7.8)$$

For details see Appendix A in (Rahmani et al. 2009).

Acceleration of the combined center of mass. In this model, the human body is considered as two distinct mechanical rigid systems: (i) the moving system composed of the lifted mass M, the upper limbs (arms and forearms, the hand is not considered) and the shoulders (the mass of the shoulders is neglected); (ii) the resting system composed of the trunk, the head and the lower limbs, which are considered to remain fixed during the bench press exercise. This latter system is not considered in the force calculation.

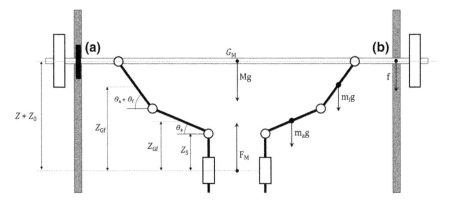

Fig. 7.7 a Diagram of the vertical position of centre of mass of the arm (Z_{Ga}), forearm (Z_{Gf}) and the barbell (Z_{GM}); **b** Diagram of the external forces applied to the limbs and barbell system in one hand during the bench press exercise. F_M, force produced by the subject; f, friction forces; $m_a g$, $m_f g$ and Mg, weights of the upper arm, forearm and lifted mass, respectively; θ_a, absolute angle relative to the horizontal axis; $\theta_a + \theta_f$, absolute angle between the upper arm and the forearm

To determine the vertical position Z_G of the combined center of mass of the lifted mass, arms and forearms, it is necessary to calculate the vertical position of each center of mass of the element of the moving system (i.e., Z_{GM}, Z_{Ga} and Z_{Gf} the vertical position of the lifted mass, the arm and the forearm, respectively) (Fig. 7.7a). Z_{GM}, Z_{Ga} and Z_{Gf} can be written as:

$$Z_{G_M} = Z_S + Z_0 \tag{7.9}$$

$$Z_{G_a} = Z_S + a_a \sin \theta_a \tag{7.10}$$

$$Z_{G_f} = Z_S + L_a \sin \theta_a + a_f \sin(\theta_a + \theta_f) \tag{7.11}$$

where a_a and a_f are the position of the center of mass of the arm and the forearm relatively to the proximal joint, respectively. a_a and a_f were estimated from Winter's table (Winter 2009).

Then, Z_G is expressed as:

$$Z_G = \frac{M(Z+Z_0) + 2m_a Z_{G_a} + 2m_f Z_{G_f}}{m + 2m_a + 2m_f} \tag{7.12}$$

where m_a and m_f are the arm and forearm masses, respectively, determined from the Winter's anthropometric tables (Winter 2009). Z_G is twice derivated to calculate the acceleration of the combined center of mass acceleration $\ddot{Z}_G$.

Force calculations. The force F_M produced at the shoulder during the bench press is then determined from the mechanical model (Fig. 7.7b) and expressed as:

$$F_M = M\ddot{Z} + 2m_a\ddot{Z}_{G_a} + 2m_f\ddot{Z}_{G_f} + (M + 2m_a + 2m_f)g + F_f \tag{7.13}$$

where $\ddot{Z}$, $\ddot{Z}_{G_a}$ and $\ddot{Z}_{G_f}$ are the accelerations of the lifted mass M, the arm and forearm segments, respectively and F_f the friction forces. m_a and m_f were multiplied by 2 to take the two upper limbs into account, assuming that the movement was symmetric. Z_{GM}, Z_{Ga} and Z_{Gf} were twice derivated to determine the accelerations $\ddot{Z}$, $\ddot{Z}_{G_a}$ and $\ddot{Z}_{G_f}$, respectively.

7.3.3 Kinematic Parameters

The displacement-time courses of the lifted mass Z, the moving system center of mass Z_G, the center of mass of the arm Z_{Ga} and forearm Z_{Gf} and the shoulder Z_S were identical but not equal to Z (Fig. 7.8). For a given lifted mass, the vertical difference between Z and Z_G was constant throughout the bench press exercise. Consequently, the vertical velocity and acceleration of the combined center of mass and the lifted mass were identical during the bench press exercise (Fig. 7.9). The acceleration determined from the model followed the one measured with the force platform, as it is the case during squat exercise performed under a guided barbell (Rahmani et al. 2000). The difference at the end of the curve is mainly due to the software treatment (see Rahmani et al. 2009 for details). However, this part of the curve corresponds to the end of the vertical displacement, when upper limbs are

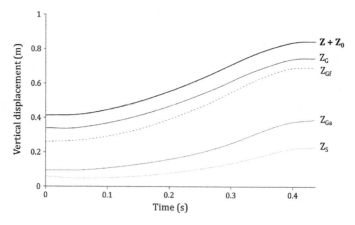

Fig. 7.8 Typical vertical displacement-time curves for the lifted mass (Z + Z_0), the global centre of mass Z_G, the forearm centre of mass Z_{Gf}, the upper arm centre of mass Z_{Ga}, and the shoulder (Z_S) during a bench press exercise with a weight of 44 kg

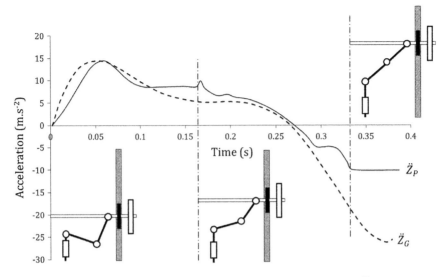

Fig. 7.9 Typical acceleration-time curves measured by the kinematic device ($\ddot{Z}_G$) and the force platform ($\ddot{Z}_P$)

stretched and decelerate the barbell. This part is out of the pushing phase and is not considered in the force calculation. In addition, the difference between Z and Z_G decreases with the increase of the lifted mass. The heavier the lifted mass, the shorter the distance between the centers of mass of the system and of the lifted mass. This is due to the position of the center of mass of the moving system, which is always located close to the greatest mass (i.e., the lifted mass). We can assume that for lifted loads heavier than 74 kg, the vertical displacement of Z and Z_G would be superimposed.

Regarding the vertical displacement Z and Z_{Gf}, the difference between them is constant (0.016 ± 0.01 m) for the whole displacement-time curve, whatever the subject or the lifted mass. This result indicates that the elbow extension, mainly performed by the *triceps brachii* at the end of the movement, is too short to influence the centre of mass displacement. The movement of the forearm can then be considered as essentially a translation movement.

The major part of the bench press exercise is due to the arm rotation, performed by the pectoralis major and the anterior deltoid. This is illustrated by the vertical displacement of the arm (Z_{Ga}) and the shoulder (Z_S). Differences between Z and both Z_S and Z_{Ga} followed the same profile whatever the subject or the lifted mass. These differences increase progressively during 65% of the total displacement, describing the removal of the lifted mass with both arm and shoulder. After that, these differences remain constant until the end of the exercise. This instant of the movement corresponds to the alignment of the arms with the forearms. This result was observed for all subjects, whatever the lifted mass.

7.3.4 Kinetic Parameters—Validation of the Model

As mentioned above, the acceleration of the moving system and the one computed from the force platform are identical (Fig. 7.9). Consequently, there is no significant difference between the force calculated from the model (F_M) and the one directly measured with the force platform (F_P). The absolute difference between the two values was less than 2.5% considering all the measurements for each lifted mass. The validity of the model is also supported by a coefficient of variation lower than 1% between the two methods. In addition, whatever the lifted mass, F_M is significantly correlated to F_P ($r = 0.99$, $p < 0.001$), with a regression slope not different from unity, and the y-intercept of the linear regression not statistically different from 0 (Fig. 7.10).

In order to be complete, F_M is not statistically different from the force estimated by Rambaud et al. (2008) using a kinematic device (see Sect. 7.2; Table 7.1). An inverse dynamical model is easily constructed using the present model with the experimental results, allowing the determination of joint forces and torques. For this, determination of the acceleration of the arm and forearm is necessary. This kind of model could easily be used by sport scientists to identify the relative importance of each muscle group during bench press exercise, improving the understanding of upper limb injury occurrence and allowing assessing actual rehabilitation program efficiency.

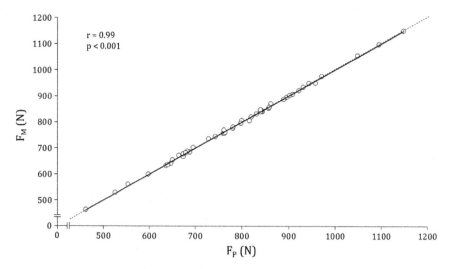

Fig. 7.10 Relationship between the force calculated from the model (F_M) and the force directly measured with the force platform (F_P). The dotted line represents the identity line

Table 7.1 Mean values (± standard deviation) of the force measured with the force platform (F_{PF}), and those estimated with the bench press model (F_M) and using a kinematic device (F_K) (as presented Sect. 7.2)

Mass (kg)	F_{PF} (N)	F_M (N)	F_K (N)
24	621 ± 99	619 ± 97	620 ± 95
34	694 ± 95	697 ± 95	698 ± 96
44	805 ± 85	804 ± 85	804 ± 86
54	829 ± 108	829 ± 109	827 ± 105
64	875 ± 102	875 ± 103	875 ± 101
74	942 ± 91	943 ± 91	943 ± 93

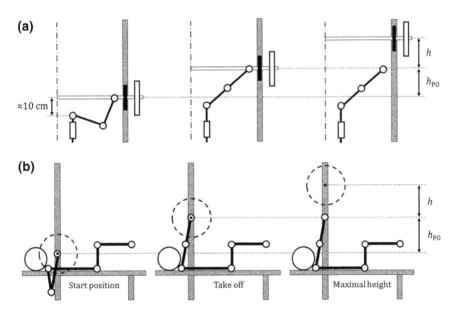

Fig. 7.11 The three key positions during a bench press throw performed with a guided barbell and the two distances (h, h_{PO}) used in the proposed computation. **a** Frontal view (only one upper limb is represented). **b** Sagittal view

7.4 A Simple Method for Measuring Force, Velocity and Power During the Bench Press Exercise

7.4.1 Theoretical Bases and Equations

This method (Rahmani et al. 2017) is based on the simple method developed by Samozino et al. 2008 during squat jump (see Chap. 4). So, only three simple parameters are required for these computations: the mass of the studied system (i.e., upper limbs plus lifted mass), the vertical displacement during the freefall phase (h) and the vertical push-off distance (h_{po}) extracted from a ballistic bench press (Fig. 7.11). The previous assumptions are applied to this method: the acceleration of the barbell is representative of the studied system's acceleration and the mass of the upper limbs are taken into account in the calculation of the force produced.

Since the movement is performed on a guided barbell machine with a friction force, the acceleration during the freefall (a_{ff}) is not the gravitational acceleration. Assuming the friction force is constant during the freefall, a_{ff} can be estimated with the second law of Newton as:

$$a_{ff} = \frac{g.m_b + F_f}{m_b} \qquad (7.14)$$

Substituting g by a_{ff} in Eqs. 4 and 8 of Samozino et al. (2008) gives:

$$\overline{F} = m_{ul+b} \times \frac{g.m_b + F_f}{m_b} \times \left(\frac{h}{h_{po}} + 1\right) \qquad (7.15)$$

$$\overline{v} = \sqrt{\frac{\frac{g.m_b + F_f}{m_b} \times h}{2}} \qquad (7.16)$$

In this method, h is measured with a nylon cable tie fixed around the rail of the guided barbell machine (Fig. 7.12) which allows a reading of the highest height attained by the barbell. This tie is moved upwards along the rail as the barbell is thrown by the athletes, but stays still in its maximal height position as the barbell moves downwards. Finally, h_{po} is measured as the difference between the initial position of the barbell (i.e., in contact with the security catches) and the maximal height attained at the end of the push-off (Fig. 7.11). All the dimensions can be measured using a non-flexible tape, with 0.1 cm accuracy.

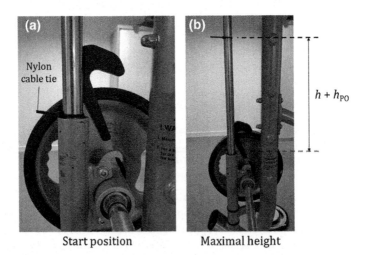

Fig. 7.12 Picture of the nylon cable tie fixed around the rail of the guided barbell machine at the rest position (**a**) and after the flight phase (**b**)

Calculating force and velocity following Eqs. 7.15 and 7.16 for bench press performed at different additional loads gives different points of the force-velocity relationship: the higher the moving mass (upper limbs mass + barbell mass), the higher the force and the lower the velocity, as during squat jump. F-v curves were then extrapolated to obtain F_0 and v_0, which corresponds to the intercepts of the F–v curve with the force and the velocity axis, respectively. The slope of the F-v linear relationship (Sfv) was also considered for further analysis. Values of maximal power of the power-velocity relationship (P_{max}) were calculated as previously validated (Vandewalle et al. 1987; Samozino et al. 2012):

$$P_{max} = \frac{F_0 \times v_0}{4} \quad (7.17)$$

7.4.2 Validation of the Method

The validity of the computation method was established by comparing (i) $\bar{F}$ and $\bar{v}$ obtained from the computation method to those simultaneously measured with an accelerometer (Myotest® Pro; Myotest SA, Sion, Switzerland) fixed on the barbell; and (ii) the mechanical parameters extrapolated from the F-v relationships (i.e., F_0, V_0, Sfv, and P_{max}) obtained from these two methods. In this study, twelve healthy and physically active males performed two ballistic bench presses at different loads (30, 40, 50, 60 and 70% of body mass). Regarding the results, the validity of the method is supported by the almost perfect relationships observed for the force (r = 0.95, $p < 0.001$) and the velocity (r = 0.89, $p < 0.001$) estimated by the two methods. The equations of the regression lines were not different from that of the identity line. The magnitude of the correlation observed for the force was in line with those observed during squat and countermovement jumps (r from 0.95 to 1) (Samozino et al. 2008; Giroux et al. 2014; Jiménez-Reyes et al. 2017). For the velocity, the coefficient of correlation is slightly lower than those obtained for the force, again as previously observed during squat jumps (0.87–0.94) (Giroux et al. 2014). Nevertheless, there is no difference between the force and the velocity measured by the two methods (systematic bias is around 30 N for the force, and 0.07 m s^{-1} for the velocity; CV% < 10%). In addition, very high between-trials reliability was found for a given load by the intraclass coefficient of correlation (ICC) higher than 0.8 for force and velocity, which is in line with those reported in previous studies focusing on ballistic bench press (Alemany et al. 2005) and classical bench press exercise (Comstock et al. 2011; Garnacho-Castaño et al. 2014). The CV% here obtained suggested adequate absolute reliability (i.e., <10%) for $\bar{F}$ and $\bar{v}$ (ranging from 0.8 to 1.7% and 1.4 to 6.3%, respectively), in agreement with previous studies mentioned above. Thus, these results evidenced high between-trials reliability for the computation method.

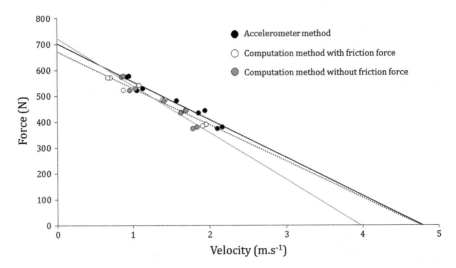

Fig. 7.13 Typical traces of F-v relationships obtained from the accelerometer method (black), the computation method taken all the mechanical inertia into account (white) and the computation method neglecting the friction forces (grey)

Finally, in agreement with the accelerometer method, the relation between force and velocity estimated by the simple method was well described by a negative linear relationship (Fig. 7.13), as it was previously shown for classical bench press (Rambaud et al. 2008; García-Ramos et al. 2015) and bench throw (Sreckovic et al. 2015; García-Ramos et al. 2016). The validity of the simple method was supported by the strong correlations of the force-velocity slope ($r^2 = 0.99$, $p < 0.001$), F_0 ($r^2 = 0.93$, $p < 0.001$), v_0 ($r^2 = 0.59$, $p < 0.05$), and P_{max} ($r^2 = 0.87$, $p < 0.001$; all not different from unity).

7.4.3 Limits of the Method

To validate the method, the force and velocity estimated from Eqs. 7.15 and 7.16 was compared to the measure simultaneously obtained with an accelerometer, instead of a force platform (referred as the 'gold standard'). This choice was done because the accelerometer directly measures the motion of the lifted barbell, including the flight phase. The use of a force platform presents two main disadvantages. Firstly, during ballistic bench throws, the moving system (upper limbs and lifted mass) is split in two separate systems at the release. This makes difficult to track the true moment of release in opposition to what happened during squat jump (the force platform signal is null). Secondly, a force platform monitors all the reaction forces occurring during the movement, also those produced from "parasite contractions" (e.g., lower limbs movement, abdominal muscles contraction), making difficult to estimate the net force applied to the moving system. Using an accelerometer has the advantage to determine precisely at which time the barbell is thrown and to estimate only the force developed by the upper limbs to accelerate the system (Comstock et al. 2011).

Another limitation concerns the necessity to estimate the friction force due to the guided barbell system into account. Since the bench presses were performed under a vertical guided barbell, it seems important to take all the mechanical parameters into account, and also the friction force (F_f) due to the guided barbell system. We acknowledge that some coaches may not be familiar with the procedure to determine F_f but, unfortunately, by neglecting F_f, errors on the F-v parameters could occur, even if more modern machines present lower friction forces. The force-velocity profiles including or not F_f in the computations for a typical individual are depicted in Fig. 7.13. Considering the whole experimental population, V_0 is overestimated by 16 ± 6%, F_0 is underestimated by 5 ± 5% and the Sfv is underestimated by 25 ± 10%. Such a bias is important to consider during bench press since the lifted mass (including the upper limb mass) is low in comparison to the one lifted during a loaded squat. Another way to say it is that the proportion of the friction force is high when considering the sum of the forces applied to the system during a bench press. In the present study, the ratio between F_f and the total lifted mass ranges between 24% (for the lightest lifted load) and 10% (for the heaviest one). The variation of the ratio with the lifted loads explains why the estimation of the slope is the most affected when F_f is neglected. In comparison, if the studied movement was a squat, this ratio would be ranged between 5% (for the heaviest one) and 6% (for the lightest lifted load) which can explain why F_f is generally neglected in this case. Fortunately, the acceleration of the friction losses (a_{Ff}) can easily be determined by measuring the time during which the barbell is falling on a given displacement d. It can be assumed that the friction forces are constant during the falling test. So, a_{Ff} could also be considered as constant and is equal to:

$$a_{Ff} = \frac{2 \cdot d}{t^2} \quad (7.18)$$

The time t can be easily measured using a smartphone, including the build-in camera with a 240-Hz sampling rate, which allows to measure the time with a sufficient accuracy and d with a non-flexible measuring tape.

7.4.4 Practical Applications

The model used here is identical to the one previously proposed on jump (see Chap. 4) and present the same practical application as those discussed in these studies (i.e., maximizing power production). In the same way, (García-Ramos et al. 2016) observed that the F_0 is strongly correlated with the 1-RM measured during a bench press exercise (r = 0.92–0.94). The force-velocity relationship is then useful to assess the upper-body maximal capabilities to generate force, velocity, and power.

Warm up. As indicated in Chap. 5, after 5–10 min of a typical general warm-up (e.g., running or cycling), specific warm-up has to include ballistic bench throws with a progressive increase in the intensity (for example: 10 submaximal reps at 20 kg, 8 at 30 kg, 6 at 40 kg, 4 at 50 kg, 3 at 60 kg, 2 at 70 kg). Throwing the barbell with the upper limb is not a "natural" movement, and apprehension of this type of movement should be eliminated to ensure the best conditions for assessing objectively the force and power abilities of an individual. Obviously a familiarization session should be scheduled if athletes are not accustomed to ballistic bench press.

Push-off distance. A major concern should be addressed to correctly determine the push-off distance (h_{po}). Participants laid supine on the bench. The barbell was positioned across their chest at nipple level above the *pectoralis major*, supported by the lower mechanical stops of the measurement device ($\approx$5 cm above the chest). Participants held the barbell choosing the most comfortable position. This handgrip was determined during the warm-up and must be marked on the barbell with tape to ensure reproducibility. h_{po} was measured as the difference between the initial position of the barbell (i.e., in contact with the security catches) and the maximal height attained after the push-off (Fig. 7.11). To reach this latter distance, participants were asked to tense maximally their upper limb, including the shoulder antepulsion. The movement should be done as smoothly as possible and the back should stay in contact with the bench.

Starting position. During the validation of the method, participants were asked to cross their legs to standardize the position and to avoid the influence of the ground reaction force resulting from lower limbs pushing on the floor. This is recommended during testing but is not mandatory during real training which can be done as desired. That being said, since the method is based on the h_{po} measured under a guided barbell, the force, velocity and power determination remains identical whatever the considered starting position. A practical point that should be taken into account may concern the starting position of the barbell. This one is generally positioned across the individual's chest at nipple level above the *pectoralis major* ($\approx$5 cm above the chest). It could be preferable to start the ballistic bench press from a higher height (a comfortable height such as 10 or 15 cm). In this way, the athlete is in a more comfortable and optimal position to overcome the greatest inertia mainly encountered at the beginning of the movement (especially, for the heaviest lifted loads or for participants who are not accustomed to ballistic bench press).

Individualization of training. Considering strength training programs, the simple method proposed here for the bench press can be used to compare athletes, to monitor, and to individualize training from their F-v profile and the requirements of the task. Figure 7.14 shows very high differences in F-v relationships among young

7 A Simple Method for Measuring Force, Velocity, Power ...

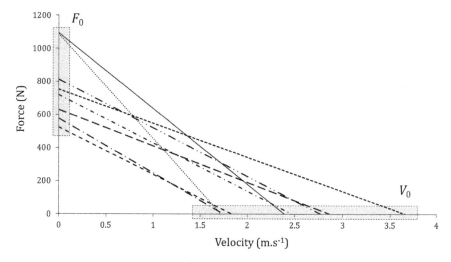

Fig. 7.14 Force-velocity relationships obtained using the simple method on young high level French basket-ball players who performed bench throws at different loads. The grey areas highlight the high inter-individual variability in maximal force (F_0) and velocity (V_0) capabilities

high-level basket-ball players of the same age with a higher variability in V_0 than in F_0 values. We can observe that players who presented the highest maximal force (F_0) were not those who produced the highest force at high movement velocities (V_0). When strength training aims at improving ballistic upper limb extensions (e.g., to improve explosive passes or long distance shots), it is more efficient to focus on increasing V_0 than F_0, notably given that a basket ball weighs ∼600 g. So, in this context, players who present high F_0 values need velocity based strength training. Conversely, players with high V_0 values should follow training focusing on the entire F-v spectrum. This highlights the large interest for sport practitioners to determine F-v profiles for individualizing strength training of upper limbs.

7.5 Conclusion

This chapter presents simple methods to evaluate the upper limb force during bench press exercise. A particular attention should be brought to the different inertia involved during the bench press exercise. To neglect the upper limb mass in the force calculation leads to an underestimation of the force, and a fortiori of the maximal power an individual can produce. This has the consequence of underestimating the optimal load at which the training program should be performed.

A simple model of the bench press exercise demonstrates that the acceleration of the moving system (lifted mass plus upper limb mass) is similar to the acceleration of the barbell. Consequently, the inverse kinetic model demonstrated a great validity in the force estimation thanks to four simple measurements: the vertical displacement of the lifted load, the elbow angle, the lengths of the arm and forearm and the horizontal position of the hand. The comparison of the force calculated from this model and the one measured with a force platform shows no significant difference between the two scores.

This leads to evaluate a simple method to assess the force, velocity and power produced during a ballistic bench press performed outside a laboratory, thanks to the Newtonian laws and only three simple parameters: (i) the upper limb mass estimated as 10% of the body mass, (ii) the barbell flight height recorded thanks to a nylon cable tie and (iii) the push-off distance measured with a measuring tape. The method developed for jumping exercise is also valid to estimate the mechanical properties of the upper limb with only slight adaptations (i.e., taking into account the upper limb mass and friction forces). Consequently, coaches and athletes could accurately determine their F-v profiles, extrapolate reliable mechanical parameters (F_0, V_0, Sfv and P_{max}) in order to maximize upper limbs performance and manage and individualize training programs in field conditions.

References

Alemany JA, Pandorf CE, Montain SJ, Castellani JW, Tuckow AP, Nindl BC (2005) Reliability assessment of ballistic jump squats and bench throws. J Strength Cond Res 19:33–38. https://doi.org/10.1519/14783.1

Barnett C, Kippers V, Turner P (1995) Effects of variations of the bench press exercise on the EMG activity of five shoulder muscles. J Strength Cond Res 9:222–227. https://doi.org/10.1519/00124278-199511000-00003

Bourdin M, Rambaud O, Dorel S, Lacour JR, Moyen B, Rahmani A (2010) Throwing performance is associated with muscular power. Int J Sports Med 31:505–510. https://doi.org/10.1055/s-0030-1249622

Buitrago S, Wirtz N, Yue Z, Klienider H, Mester J (2013) Mechanical load and physiological responses of four different resistance training methods in bench press exercise. J Strength Cond Res 27:1091–1100

Caiozzo VJ, Perrine JJ, Edgerton VR (1981) Training-induced alterations of the in vivo force-velocity relationship of human muscle. J Appl Physiol 51:750–754

Comstock BA, Solomon-Hill G, Flanagan SD, Earp JE, Luk HY, Dobbins KA, Dunn-Lewis C, Fragala MS, Ho J-Y, Hatfield DL, Vingren JK, Denegar CR, Volek JS, Kupchak BR, Maresh CM, Kraemer WJ (2011) Validity of the Myotest in measuring force and power production in the squat and bench press. J Strength Cond Res 25:2293–2297

Cormie P, McGuigan MR, Newton RU (2011) Developing maximal neuromuscular power: Part 2 training considerations for improving maximal power production. Sport Med 41:125–146. https://doi.org/10.2165/11538500-000000000-00000

Cronin JB, Henderson ME (2004) Maximal strength and power assessment in novice weight trainers. J Strength Cond Res 18:48–52

Cronin JB, McNair PJ, Marshall RN (2000) The role of maximal strength and load on initial power production. Med Sci Sports Exerc 32:1763–1769. https://doi.org/10.1097/00005768-200010000-00016

Cronin JB, McNair PJ, Marshall RN (2003) Force-velocity analysis of strength-training techniques and load: implications for training strategy and research. J Strength Cond Res 17:148. https://doi.org/10.1519/1533-4287(2003)017<0148:FVAOST>2.0.CO;2

Driss T, Vandewalle H, Monod H (1998) Maximal power and force-velocity relationships during cycling and cranking exercises in volleyball players: correlation with the vertical jump test. J Sports Med Phys Fitness 38:286–293

García-Ramos A, Jaric S, Padial P, Feriche B (2015) Force-velocity relationship of upper-body muscles: traditional vs. ballistic bench press. J Appl Biomech 1–44. doi:https://doi.org/10.1123/jab.2015-0162

García-Ramos A, Jaric S, Padial P, Feriche B (2016) Force-velocity relationship of upper body muscles: traditional versus ballistic bench press. J Appl Biomech 32:178–185. doi:https://doi.org/10.1123/jab.2015-0162

Garnacho-Castaño MV, López-Lastra S, Maté-Muñoz JL (2014) Reliability and validity assessment of a linear position transducer. J Sport Sci Med 14:128–136

Giroux C, Rabita G, Chollet D, Guilhem G (2014) What is the best method for assessing lower limb force-velocity relationship? Int J Sports Med 36:143–149. https://doi.org/10.1055/s-0034-1385886

Izquierdo M, Häkkinen K, Gonzalez-Badillo JJ, Ibanez J, Gorostiaga EM (2002) Effects of long-term training specificity on maximal strength and power of the upper and lower extremities in athletes from different sports. Eur J Appl Physiol 87:264–271. https://doi.org/10.1007/s00421-002-0628-y

Jandaĉka D, Vaverka F (2009) Validity of mechanical power output measurement at bench press exercise. J Hum Kinet 21:33–40. https://doi.org/10.2478/v10078

Jarić S, Ropret R, Kukolj M, Ilić DB (1995) Role of agonist and antagonist muscle strength in performance of rapid movements. Eur J Appl Physiol Occup Physiol 71:464–468. https://doi.org/10.1007/BF00635882

Jiménez-Reyes P, Samozino P, Pareja-Blanco F, Conceiçao F, Cuadrado-Penafiel V, Gonzales-Badillo JJ, Morin JB (2017) Validity of a simple method for measuring force-velocity-power profile in countermovement jump. Int J Sports Physiol Perform 12:36–43. https://doi.org/10.1123/IJSPP.2015-0484

Kanehisa H, Miyashita M (1983) Specificity of velocity in strength training. Eur J Appl Physiol Occup Physiol 52:104–106. https://doi.org/10.1007/BF00429034

Kaneko M, Fuchimoto T, Toji H, Suei K (1983) Training effect of different loads on the force-velocity relationship and mechanical power output in human muscle. Scand J Sport Sci 5:50–55

Laffaye G, Collin JM, Levernier G, Padulo J (2014) Upper-limb power test in rock-climbing. Int J Sports Med 35:670–675. https://doi.org/10.1055/s-0033-1358473

Murphy AJ, Wilson GJ, Pryor JF (1994) Use of the iso-inertial force mass relationship in the prediction of dynamic human performance. Eur J Appl Physiol Occup Physiol 69:250–257. https://doi.org/10.1007/BF01094797

Nelson SG, Duncan PW (1983) Correction of isokinetic and isometric torque recordings for the effects of gravity. A clinical report. Phys Ther 63:674–676

Newton RU, Kraemer WJ, Hakkinen K, Humphries BJ, Murphy AJ (1996) Kinematics, kinetics, and muscle activation during explosive upper body movements. J Appl Biomech 12:37–43

Padulo J, Mignogna P, Mignardi S, Tonni F, D'Ottavio S (2012) Effect of different pushing speeds on bench press. Int J Sports Med 33:376–380. https://doi.org/10.1055/s-0031-1299702

Pearson S, Cronin J, Hume P, Slyfield D (2007) Kinematic and kinetics of the bech press and bench pull exerises in a strength trained sporting population. Symp A Q J Mod Foreign Lit, 27–30

Rahmani A, Belli A, Kostka T, Dalleau G, Bonnefoy M, Lacour J-R (1999) Evaluation of knee extensor muscles under non-isokinetic conditions in elderly subjects. J Appl Biomech 15:337–344. https://doi.org/10.1123/jab.15.3.337

Rahmani A, Dalleau G, Viale F, Hautier CA, Lacour J-R (2000) Validity and reliability of a kinematic device for measuring the force developed during squatting. J Appl Biomech 16:26–35

Rahmani A, Rambaud O, Bourdin M, Mariot JP (2009) A virtual model of the bench press exercise. J Biomech 42:1610–1615. https://doi.org/10.1016/j.jbiomech.2009.04.036

Rahmani A, Samozino P, Morin J-B, Morel B (2017) A simple method for assessing upper limb force-velocity profile in bench press. Int J Sports Physiol Perform 1–23. doi:https://doi.org/10.1123/ijspp.2016-0814

Rambaud O, Rahmani A, Moyen B, Bourdin M (2008) Importance of upper-limb inertia in calculating concentric bench press force. J Strength Cond Res 22:383–389. https://doi.org/10.1519/JSC.0b013e31816193e7

Samozino P, Morin JB, Hintzy F, Belli A (2008) A simple method for measuring force, velocity and power output during squat jump. J Biomech 41:2940–2945. https://doi.org/10.1016/j.jbiomech.2008.07.028

Samozino P, Rejc E, Di Prampero PE, Belli A, Morin JB (2012) Optimal force-velocity profile in ballistic movements-Altius: Citius or Fortius? Med Sci Sports Exerc 44:313–322. https://doi.org/10.1249/MSS.0b013e31822d757a

Sánchez-Medina L, Perez CE, Gonzalez-Badillo JJ (2010) Importance of the propulsive phase in strength assessment. Int J Sports Med 31:123–129. https://doi.org/10.1055/s-0029-1242815

Sánchez-Medina L, González-Badillo JJ, Pérez CE, Pallarés JG (2014) Velocity- and power-load relationships of the bench pull vs bench press exercises. Int J Sports Med 35:209–216. https://doi.org/10.1055/s-0033-1351252

Shim a L, Bailey ML, Westings SH (2001) Development of a field test for upper-body power. J Strength Cond Res 15:192–197. doi:https://doi.org/10.1519/1533-4287(2001)015<0192:DOAFTF>2.0.CO;2

Sreckovic S, Cuk I, Djuric S, Nedeljkovic A, Mirkov D, Jaric S (2015) Evaluation of force–velocity and power–velocity relationship of arm muscles. Eur J Appl Physiol 115:1779–1787. https://doi.org/10.1007/s00421-015-3165-1

Stockbrugger B, Haennel RG (2001) Validity and reliability of a medicine ball explosive power test. J Strength Cond Res 15:431–438. https://doi.org/10.1519/1533-4287(2001)015<0431:VAROAM>2.0.CO;2

Stockbrugger B, Haennel R (2003) Contributing factors to performance of a medicine ball explosive power test: a comparison between jump and nonjump athletes. J Strength Cond Res 17:768–774. https://doi.org/10.1519/1533-4287(2003)017<0768:CFTPOA>2.0.CO;2

Vanderthommen M, Francaux M, Johnson D, Dewan M, Lewyckyj Y, Sturbois X (1997) Measurement of the power output during the acceleration phase of all-out arm cranking exercise. Int J Sports Med 18:600–606

Vandewalle H, Peres G, Heller J, Panel J, Monod H (1987) Force-velocity relationship and maximal power on a cycle ergometer - Correlation with the height of a vertical jump. Eur J Appl Physiol Occup Physiol 56:650–656. doi:https://doi.org/10.1007/BF00424805

Wilson G, Elliott BC, Kerr GK (1989) Bar path and force profile characteristics for maximal and submaximal loads in the bench press. Int J Sport Biomech 21:450–462. https://doi.org/10.1123/ijsb.5.4.390

Wilson GJ, Elliott BC, Wood GA (1991a) The effect on performance of imposing a delay during a stretch-shorten cycle movement. Med Sci Sport Exerc 23:364–370. https://doi.org/10.1249/00005768-199103000-00016

Wilson GJ, Wood G a, Elliott BC (1991b) Optimal stiffness of series elastic component in a stretch-shorten cycle activity. J Appl Physiol 70:825–833

Wilson GJ, Murphy AJ, Pryor JF (1994) Musculotendinous stiffness: its relationship to eccentric, isometric, and concentric performance. J Appl Physiol 76:2714–2719

Winter DA (2009) Biomechanics and motor control of human movement. Wiley, New York

Winter DA, Wells RP, Orr GW (1981) Errors in the use of isokinetic dynamometers. Eur J Appl Physiol Occup Physiol 46:397–408. https://doi.org/10.1007/BF00422127

Young KP, Haff GG, Newton RU, Gabbett TJ, Sheppard JM (2015) Assessment and monitoring of ballistic and maximal upper-body strength qualities in athletes. Int J Sports Physiol Perform 10:232–237. https://doi.org/10.1123/ijspp.2014-0073

Part III
Running

Chapter 8
A Simple Method for Measuring Lower Limb Stiffness During Running

Jean-Benoit Morin

Abstract During running, the lower limbs musculo-tendinous system behaves like a spring-mass system. From an external mechanical point of view, the complexity of the running motion (that involves force production, motion and coordination at the hip, knee and ankle levels, to ensure propulsion and balance) can be well described by the behavior of a single, linear spring-mass system. The vertical ground reaction force over time during contact on the ground shows a sine-wave behavior, and the vertical downward displacement of the center of mass a concomitant u-shape behavior. In this model the relationship between compressive force in the spring and change of length is linear and the slope of this relationship is equivalent to the stiffness of the spring. Since this model was first developed in the late 1980s, it has been used as a practical approach to study human and animal running mechanics, mainly with ground reaction forces measurements (force plates or instrumented treadmill) and high-speed motion analysis. This chapter presents the theoretical bases of this model, and a simple field method to compute the main spring-mass variables during running. This method is based on a sine-wave modeling of the ground reaction force trace and requires only five simple input variables: running velocity, the body mass of the runner, their lower limb length and contact and aerial times. In this chapter, we also detail how to measure these input variables, and examples of use of the sine-wave method in research and training practice.

> Science replaces the complicated visible with the simple invisible
>
> Jean Perrin, Physicist (1870–1942)

Jean-Benoit Morin (✉)
Laboratory of Human Motricity, Education Sport and Health,
Université Côte d'Azur, 261 Route de Grenoble, 06205 Nice, France
e-mail: jean-benoit.morin@unice.fr

© Springer International Publishing AG 2018
J.-B. Morin and P. Samozino (eds.), *Biomechanics of Training and Testing*,
https://doi.org/10.1007/978-3-319-05633-3_8

8.1 Introduction

Running is something millions of humans do on a weekly, if not daily basis. Thousands of years ago, our ancestors used running to hunt their own food, and probably to escape predators. Nowadays, people run for exercise and health purposes, to train for competitions, or just for the sole pleasure of running. Sit on a bench at a playground for a few minutes and you'll observe that young kids almost never walk around. They run, even though, contrary to cycling or other motor skills, they never really learned how to run. As pointed out in a famous book (McDougall 2010) and paper in Nature (Bramble and Lieberman 2004), we are "born to run". This activity is natural and most runners are likely not aware of the immensely complex mechanisms that allow them to "just run". Running can be defined as propelling one's body mass on the ground from one leg stance to another. It is a "single-leg forward bouncing" motion, and it implies very complex adjustments in balance, muscle activity and limbs coordination, force production by the muscles, and force transmission to the skeletal system by the tendons, plus harmonious flexions and extensions of the three main joints of the lower limb (hip, knee, ankle) in the sagittal plane of motion.[1] When running, this machinery organizes itself so quickly that it allows typical running step rate of about 3 steps per second, with ground contact phases of usually less than 300 ms. And this step cycle happens roughly 10.000 times during a one-hour jog at 10 km h^{-1}.

The neuromuscular, tendinous and osteo-articular features of the running motion are in fact so complex, that the best engineers in the world could design and build running biped[2] or quadruped[3] robots only a few years ago. Put this into perspective with other technical prowess such as landing humans on the moon (1969), and you'll have a better idea of how complex this apparently simple movement is. For these reasons, even though runners have long been subjects to observation and study (e.g. ancient Greeks paintings representing running athletes), the scientific study of human running mechanics is a very young discipline. In order to run a macroscopic approach and "see the forest before the tree", biomechanists and physiologists used simple integrative models to describe and analyze biped locomotion (Dickinson et al. 2000). The model used to describe and study running mechanics is the spring-mass model (SMM).

In this chapter, we will define this model, the typical measurements involved, and a simple method we developed to compute the mechanical features of this model. We will then show examples of use and applications, and discuss the main

---

[1] Although running involves motions of the body in all three planes of space, we will limit our approach in this chapter to the sagittal plane of motion.

[2] See for instance the running robot Phides from the Delft University of Technology: http://www.3me.tudelft.nl/en/about-the-faculty/departments/biomechanical-engineering/research/dbl-delft-biorobotics-lab/bipedal-robots/.

[3] See for instance the M.I.T Cheetah Robot: http://biomimetics.mit.edu.

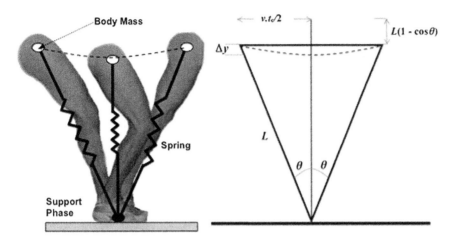

Fig. 8.1 Left: schematic representation of the spring mass model of running. The centre of mass trajectory can be roughly described by that of the white circle at the great trochanter. Right: simplified geometry of the spring-mass system during the support phase. L: lower limb length at initial contact; θ: angle swept by the system during the forward motion for each half of the support phase. Assuming a constant running velocity (v in m s^{-1}), the horizontal displacement of the centre of mass during the support phase is equal to $v \cdot tc/2$, with tc the contact time in s

technologies that can be used to measure the mechanical inputs of this simple method.

8.2 The Spring-Mass Model of Running

The SMM (Fig. 8.1) models the runner as a mass (equal to the total body mass) loading a linear spring that represents the lower limb (thigh, leg and foot segments, hip, knee and ankle joints). At each step, as the runner lands on the ground, the mass compressed the spring during its downward motion, and then the spring extends in the second part of the ground contact phase, before the subsequent aerial phase. Contrary to rebound vertical jump and hopping (see Chap. 6), the SMM for running implies an oscillation forward during this compression-extension cycle.

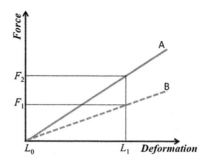

Fig. 8.2 Schematic force-deformation relationship for two perfect linear springs. A is stiffer than B (and in turn B is more compliant than A). When a given compression force is applied (e.g. F_1), the resulting deformation of spring A is lower than that of spring B. For the same change in length L_1, a higher compression force is necessary for spring A than for spring B. Stiffness for spring A is computed as $k_A = F_2/L_1$ and for spring B as $k_B = F_1/L_1$. In this example, $k_A > k_B$

8.2.1 Spring Stiffness

Mechanically, the stiffness[4] of a deformable system is the ratio of the force applied to deform this system and the resulting deformation (change of length). In the case of a pure linear spring, the relationship between the compression force applied (F in N) and the resulting deformation (ΔL in m) is linear, and the slope of this linear relationship is the stiffness (k in N m^{-1}) of the system (Fig. 8.2). If a linear spring has a stiffness of k = 1000 N m^{-1}, applying a force of 10 N (i.e. equivalent to the weight on earth of a mass of $\approx$1 kg) will be necessary to compress it by 1 cm.

8.2.2 Spring-Mass Behavior During Bouncing and Running

A major finding of the first studies of human jumping and running movements with force plates on which subjects jumped and ran (Blickhan 1989; McMahon and Cheng 1990), is that the recordings of vertical GRF over time showed sine-wave like behavior. As complex as the human running motion may be, the vertical GRF behavior during the support phase (sine-wave) and the aerial motion of the center of mass during the aerial phase (parabolic) follow quite simple mathematical behaviors. The vertical GRF over time during a running support phase can be well fitted by the following equation (Fig. 8.3):

---

[4]Although it has an indirect relationship, this stiffness is distinct from what is sometimes referred to as "muscle stiffness" or "joint stiffness". The latter usually indicate a loss of range of motion and the possibly associated pain.

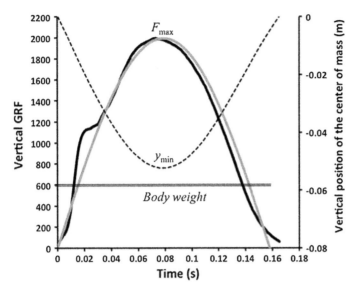

Fig. 8.3 Vertical GRF measured with a force plate (2000 Hz sampling rate) for a male runner of 62 kg running at a constant speed of 5 m s^{-1}. The grey trace shows the sine-wave modelling of the vertical GRF for this stance phase. The dashed line shows the vertical trajectory of the centre of mass

$$F(t) = F_{max} \sin\left(\frac{\pi}{tc} t\right) \quad (8.1)$$

Knowing the vertical GRF data and the runner's body mass m allows one to compute the vertical displacement of the center of mass during the support phase using the laws of motion (Cavagna 1975). Then, using the simplified stance phase geometry shown in Fig. 8.1, the change in leg length during contact can be computed (McMahon and Cheng 1990; Farley and González 1996). Then, as shown in Fig. 8.3, a plot of the vertical GRF against the leg length changes shows a pretty linear trace. On this basis, the SMM for running has been used to analyze running locomotion mechanics in humans, but also in biped and quadruped animals (Alexander 1991, 2003, 2004; Farley et al. 1993; Kram and Dawson 1998).

8.2.3 Spring-Mass Stiffness in Running: Definitions and Assumptions

Since the lower limb compression and length change follow an overall out-of-phase behavior during the stance phase, a simple way to compute the lower limb stiffness (k_{leg}) is to divide the maximal compression force by the corresponding maximal leg length change (Farley and González 1996). These two values correspond to the maximal values of the linear relationship between force and change in length shown in Fig. 8.2:

$$k_{leg} = F_{max} \cdot \Delta L^{-1} \quad (8.2)$$

In addition, in order to describe the vertical motion of the center of mass and to focus on the vertical bouncing mechanism only, the vertical stiffness (k_{vert}) has been proposed (Farley et al. 1993; Farley and González 1996; Farley and Ferris 1998). It is computed as the ratio of the maximal compression (ground reaction) force to the maximal downward displacement of the center of mass (which corresponds to the position y_{min} in Fig. 8.3):

$$k_{vert} = F_{max} \cdot \Delta y^{-1} \quad (8.3)$$

The SMM for running is a simple, integrative model with two main mechanical features (and their sub-components): k_{leg} and k_{vert}. As mentioned in introduction, these mechanical variables are integrative and encompass numerous and complex neuromuscular and mechanical phenomena simultaneously characterizing the running system (Farley and Ferris 1998). As for all models aiming at simplifying biological phenomena to better describe and interpret them, it implies several assumptions:

- the lower limb is not a perfect, linear spring. It is made of three main segments (thigh, leg, foot) that basically allow the running motion based on joint angular motions (hip, knee, ankle). The joint angular stiffness has been discussed among other methods to quantify lower limb stiffness (Brughelli and Cronin 2008a), and some authors proposed to use the term "quasi-stiffness" (Latash and Zatsiorsky 1993). However, the overall behavior and the mechanical outputs of this complex, multi-segmental system are very close to those of a linear spring-mass system in both running and vertical bouncing (see Chap. 6).
- the basis postulate of the SMM and of the simple method presented below is that the vertical GRF-time signal is close to a sine-wave equation during running. As discussed in the limitation section, this is true for a large spectrum of running speeds (i.e. from about 10 to about 25 km h^{-1}), for which the overall quality of the mathematical fitting is high (Morin et al. 2005). This assumption is not valid at very low running speeds (<8 km h^{-1}, personal unpublished observations), and has recently been challenged for top running speeds (>36 km h^{-1}) in world-class sprinters (Clark and Weyand 2014).
- the SMM and the computation of stiffness that will be detailed in the next paragraph are based o the fact that the maximal compression force (F_{max}) and the minimal length of the lower limb (that corresponds to y_{min} in Fig. 8.3, and that is used to compute the leg length change ΔL) are reached simultaneously. As shown in the typical example of Fig. 8.3, the delay is very short between these two instants (Silder et al. 2015).

- the two main assumptions that are made when using the SMM for running are related to the geometrical representation of the lower limb motion during the stance (Fig. 8.1). First, the point of force application onto the ground is assumed to be fixed during the stance. This is not the case in reality, since the center of pressure moves by about 10–20 cm under the foot during contact. This has been shown to influence the stiffness computations (Bullimore and Burn 2006) and a correction could be applied, with an estimation of the length of this "forward translation of the point of force application" based on running velocity (Lee and Farley 1998). The important point here is that, in case of intra-subject comparisons or inter-subjects comparisons using the same computation method (training process or other similar investigations), this assumption does not challenge the results obtained. Second, the angular sector swept by the lower limb during running is assumed to be symmetrical around the vertical mid-stance position (Fig. 8.1), and the length of the lower limb at foot strike L is assumed to be equal to the reference value measured from the great trochanter to the ground in anatomical position. This simplification has been discussed and video analyses confirmed that the lower limb length at foot strike is overestimated in the classical SMM (Arampatzis et al. 1999). The main point in our opinion is that this assumption of a symmetrical angle swept during the support phase makes the SMM invalid in accelerated/decelerated running, and during uphill/downhill running.

Using the SMM to describe and analyze running mechanics supposes acknowledgement of these assumptions, and our point is that although they are due to the simplification of a very complex reality, they basically do not challenge the observations made in training or research contexts (see below in the applications section) provided all measurements and comparisons are made consistently with the same procedures and computations.

8.2.4 Reference Methods and Typical Values

In addition to simple anthropometrics (body mass and leg length) and running speed, computing stiffness using the SMM requires measurement of the vertical GRF and position of the center of mass over time. Although synchronizing a high-frequency video analysis with GRF data can lead to accurate measurement of center of mass position over time, both F_{max} and Δy can be derived from the sole GRF signal using a method detailed by Cavagna (1975). Briefly, this method uses the laws of mechanics to compute vertical acceleration of the center of mass over time, and then velocity and position are obtained by integrating this acceleration signal (Fig. 8.4).

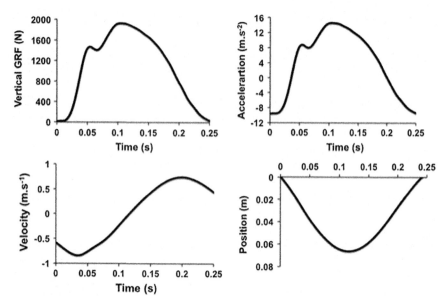

Fig. 8.4 The vertical GRF data (upper-left panel) are measured on an instrumented treadmill (1000 Hz sampling rate) for a male runner of 70 kg running at a constant speed of 3.33 m s^{-1}. Vertical acceleration of the centre of mass (upper-right panel) is computed as $a(t) = F(t)/m - g$ with m the subject's body mass and g the acceleration due to gravity (9.81 m s^{-2} on Earth). Then, centre of mass velocity (lower-left) and position (lower right) over time are obtained by integration of the acceleration signal, as proposed by Cavagna (1975)

Then, ΔL is computed from the lower limb geometry presented in Fig. 8.1 as follows (Farley et al. 1993; Farley and González 1996; Farley and Ferris 1998):

$$\Delta L = L - \sqrt{L^2 - \left(\frac{vtc}{2}\right)^2} + \Delta y \tag{8.4}$$

Note that L can be estimated from subject height H using the anthropometrical model of Winter (1979):

$$L = 0.53H \tag{8.5}$$

Ground reaction force data are typically measured with force-plate systems embedded into the running ground (e.g. Slawinski et al. 2008b; Rabita et al. 2011) or with instrumented treadmills (e.g. Dutto and Smith 2002).

The numerous studies that reported SMM stiffness values in human running consistently show that k_{leg} is basically constant over a wide range of running velocities, whereas k_{vert} increases with running speed (Fig. 8.5). Interestingly, this has also been observed in a wide range of animals such as dogs, goats, or kangaroos

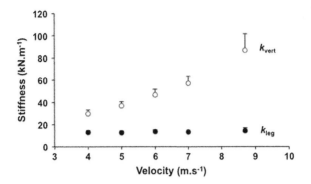

Fig. 8.5 Vertical and leg stiffness typical values for a group of male middle-distance runners at 4, 5, 6, 7 m s^{-1}, and at their maximal velocity. Data were measured with a ground-embedded force plate

[Farley et al. (1993) studied several animals ranging from ≈0.1 kg rats to ≈130 kg horses].

Although it is consistent over a wide range of running speeds (which shows how the runner's lower limb system adapts to the velocity constraint) (Brughelli and Cronin 2008b), k_{leg} has been shown to change under a few specific conditions.

Step frequency and contact time. First, when runners change their step frequency, their k_{leg} changes (Farley and González 1996; Morin et al. 2007): when step frequency increases, k_{leg} increases and vice versa. This is in fact mainly due to the change in contact time that occurs when step frequency is increased or decreased (Morin et al. 2007). We observed, by manipulating separately step frequency and contact time at a constant running velocity, that contact time was the main determinant of k_{leg} changes in such conditions. A factor of about 2 was found between the percent changes in k_{leg} and in contact time. For instance, if contact time was decreased by 10%, k_{leg} was increased by about 20–25%, and vice versa (Morin et al. 2007).

Running surface. Another external factor that induces changes in k_{leg} is the stiffness of the running surface. As for hopping, running on a softer or harder-than-normal surface is associated with higher and lower k_{leg}, respectively (Ferris and Farley 1997; Ferris et al. 1998, 1999). The neuromuscular system is able to adjust stiffness very fast to adapt to changes in running surface (Ferris et al. 1999). This clear effect of running surface stiffness on the SMM mechanics might be of importance when considering the surface of running tracks (McMahon and Greene 1978, 1979), and the consequences on energetics (Kerdok et al. 2002).

Loading-unloading. Although it is not a crucial point in typical running practice, loading (i.e. unloading or reduced gravity and additional loading) has an effect on the SMM mechanics. Under reduced gravity (unloading) conditions, He et al. (1991) observed a tendency towards decreased k_{leg} only under extreme unloading (0.2 g), and Silder et al. (2015) recently showed that loading was associated with an increase in leg stiffness. This change in k_{leg} was associated with an increased knee flexion, and an overall crouched running pattern that was termed "Grouch Running" by McMahon et al. (1987).

Fatigue. Finally, the consequences of fatigue in running on the SMM mechanics have been investigated, with contrasting results depending on the intensity-duration of the exercise considered. Indeed, fatigue induced by short (sprint, typically up to 10–12 s), intermediate (middle-distance, about 5–15 min) or very long (ultra-endurance, longer than 2–4 h) is associated with different changes in the spring-mass behavior. Note that as fatigue in running might be associated in real practice with a decrease in running velocity, the studies discussed below include testing at similar fixed running velocities, or changes that are beyond the expected relationship between k_{vert} and running velocity as illustrated in Fig. 8.5.

Repeated sprints. The main effect of fatigue after repeated sprints of about 5 s, with short recovery time (about 25 s), which is typical from the physical demand of team sports such as soccer or rugby, is that both k_{vert} and k_{leg} decrease (Girard et al. 2011a, b, c, 2015, 2016). This is mainly associated with reduced step frequency, longer contact time and a greater downward displacement of the center of mass during contact. In other words, repeated maximal-intensity sprints with short-duration recovery leads to an impairment of the spring-mass behavior and the overall ability of athletes to bounce on the ground effectively, which might in turn alter both their sprint performance and their ability to do complex movements that are crucial in team sports: changing direction, accelerating, jumping and cutting. Concerning longer sprints, we found similar results (except a non-significant decrease in k_{leg}) when asking athletes to perform four all-out 100-m sprints with 120 s passive recovery (Morin et al. 2006). Finally, Hobara et al. 2010 (Morin et al. 2006) observed such decreases in both vertical and leg stiffness over an all-out 400-m sprint.

Middle-distance. Contrary to very short (sprint) and very long (ultra-endurance, see below) exercises, the changes in SMM mechanics induced by intermediate-duration maximal efforts, which are typical from middle-distance running or Olympic-distance triathlon, are not consistent among studies. Indeed, some studies report no change in spring-mass behavior on average for the group of subjects tested (Hunter and Smith 2007; Slawinski et al. 2008b; Le Meur et al. 2013) whereas others observe alterations, with, as for sprinting, a decrease in both types of stiffness (Dutto and Smith 2002). Some other studies report changes for one type of stiffness and not for the other (Rabita et al. 2011; Girard et al. 2013). Such contrasting outcomes may result from differences in the testing protocols (constant velocity until exhaustion vs. time trial over a given distance), moment and velocity of testing (pre-post vs. during the running effort), or more likely to variable individual responses to the effects of fatigue, as discussed by Hunter and Smith (2007).

Ultra-endurance. With the increasing popularity of "ultra-marathon" and "ultratrail" races, i.e. races longer than 42 km (up to more than 300 km) with a considerable amount of uphill and downhill running,[5] research has recently focused on the neurophysiological and biomechanical consequences of such demanding running efforts. Studies consistently show that the consequence of fatigue on the spring-mass behavior in this context is an overall change toward a higher or unchanged leg stiffness and a higher vertical stiffness, mainly due to a lower downward displacement of the center of mass during contact and a higher step frequency (Morin et al. 2011a, b; Degache et al. 2013, 2016). Our hypothesis was that this change could be a way for subjects to run with a smoother and thus safer running pattern, i.e. reducing the aerial time and overall impact faced at each step, to potentially reduce the muscle, joint and tendinous pain in the legs at each step (Morin et al. 2011b; Millet et al. 2012). Interestingly, similar adaptations have been observed when muscle structure and/or functions are altered in other contexts: after an intense downhill run inducing delayed-onset muscle soreness (Morin et al. 2011b; Millet et al. 2012), after a quadriceps muscle biopsy (Morin et al. 2009) or with ageing (Cavagna et al. 2008). Note that the changes in spring-mass mechanics were not related to the substantial decrease in muscle force capability observed in these ultra-endurance studies (Martin et al. 2010; Millet et al. 2011; Morin et al. 2012b).

8.2.5 Limitations of the Reference Methods

As detailed in the beginning of this chapter, the spring-mass model has been used to describe and predict the mechanical behavior of runners in various contexts. Although it has been discussed and criticized, it is an overall accepted integrative model of human running locomotion. However, in addition to the basic limitations of the model listed above, the classical reference methods present some limitations to a wider use in both research sports practice:

- first and foremost, these reference methods require costly specific devices to measure GRF: force plates or instrumented treadmills. These are typical research laboratory devices and although some running clinics or physiotherapy centers are equipped with such devices, it is not the case of most running coaches or athletes.
- a consequence is that, except for rare research protocols (e.g. Slawinski et al. 2008b; Rabita et al. 2011), running mechanics and spring-mass behavior are studied in laboratory conditions with subjects running for one or two steps on a ground-embedded force plate, or for several steps on an instrumented treadmill. In both cases, the measurement protocol only simulates real-life running practice.

---

[5]See for instance one of the most famous, with 160 km and more than 9000 m of positive elevation change: http://www.ultratrailmb.com.

- finally, as earlier discussed, data processing is required to compute k_{vert}, k_{leg} and their components F_{max}, Δy and ΔL from GRF data, which prevents many athletes and practitioners to use the SMM.

In order to lift these limitations, we proposed a simple field method to compute spring-mass mechanics from a few variables of the running pattern that are easier to obtain in typical training practice.

8.3 A Simple Method for Measuring Stiffness During Running

8.3.1 Theoretical Bases and Equations

The simple method is based on the assumption that the vertical GRF over time has a sine-wave behavior during running, as shown in Fig. 8.3. Then, on the basis of the sine-wave Eq. (8.1), and according to the laws of motion [for detailed computations, see Morin et al. (2005)], the final equations of the method are the following:

$$k_{leg} = F_{max} \cdot \Delta L^{-1} \tag{8.2}$$

and

$$k_{vert} = F_{max} \cdot \Delta y^{-1} \tag{8.3}$$

with

$$F_{max} = mg\frac{\pi}{2}\left(\frac{ta}{tc}+1\right) \tag{8.6}$$

$$\Delta y = -\frac{F_{max}}{m}\frac{tc}{\pi^2} + g\frac{tc^2}{8} \tag{8.7}$$

and

$$\Delta L = L - \sqrt{L^2 - \left(\frac{vtc}{2}\right)^2} + \Delta y \tag{8.4}$$

The five input variables necessary to compute these SMM characteristics are the runner's body mass m (in kg) and leg length L (in m), the running velocity v (in m s^{-1}) and the contact and aerial times (tc and ta in s).

8.3.2 Validation of the Method

Two protocols were performed in order to validate the simple method. For the first protocol, 8 male runners performed 30-s running bouts at 7 different velocities ranging from 3.33 to 6.67 m s^{-1} on an instrumented treadmill (ADAL, HEF Tecmachine, Andrézieux-Bouthéon, France, Fig. 13.4). Mechanical data of vertical GRF were sampled at 1000 Hz for 10 consecutive steps and statistical comparisons were made between the reference and the proposed method for all steps at each running velocity. For the second protocol, 10 male trained middle-distance runners volunteered to run at 5 different velocities (4, 5, 6, 7 m s^{-1} and their individual maximal sprinting velocity) on a force plate (Kistler, Switzerland) installed into the running track. Data from one step at each velocity were sampled at 1800 Hz and compared between the reference and the proposed methods.

Overall, the method showed an acceptable to very good concurrent validity [for full details and statistics, see Morin et al. (2005)]. For k_{vert} and k_{leg}, we obtained a mean absolute bias between methods of 0.12% (ranging from 1.53% at 6.67 m s^{-1} to 0.07% at 6.11 m s^{-1}) and of 6.05% (from 9.82% at 3.33 m s^{-1} to 3.88% at 6.67 m s^{-1}) respectively during treadmill running. For the overground running condition, the bias was 2.30% (from 3.64% at 5 m s^{-1} to 0.25% at 6 m s^{-1}) for k_{vert} and 2.54% (from 3.71% at 5 m s^{-1} to 1.11% at the maximal velocity) for k_{leg}. In addition, the linear regressions between reference and proposed methods were significant and high ($R^2 = 0.89$–0.98) for both k_{vert} and k_{leg} in the treadmill and overground protocols. The other mechanical variables of the SMM (F_{max}, Δy and ΔL) also showed low absolute bias (0.67–6.93%, Table 8.1).

In addition to this initial validation, some studies brought additional support to the reliability and interest of this sine-wave method (Coleman et al. 2012; Pappas et al. 2014).

8.3.3 Input Variables and Importance of Contact Time

As detailed in the equations above, the simple method presented is based on five inputs that must be accurately measured (body mass, running velocity, lower limb length, contact and flight times). In order to estimate the relative importance (weight) of each of these variables in the final calculation of k_{leg} and k_{vert}, we did a sensitivity analysis that simulated changes in each of these five variables while keeping all other variables unchanged. The results of this analysis revealed what variable(s) had the greatest influence on the final stiffness values, and thus the variable(s) that would require the most careful and accurate measurement. From reference values, all variables were changed by up to +10 or −10%, and the resulting stiffnesses were computed.

This theoretical analysis clearly shows that, ceteris paribus, the contact time has the greatest influence on k_{leg} computation, and that contact and aerial times have a

Table 8.1 Mean absolute bias between reference values of the SMM mechanical variables and the values computed with the proposed method

	Variable	Δy (cm)	ΔL (cm)	F_{max} (kN)	k_{vert} (kN m^{-1})	k_{leg} (kN m^{-1})
TREADMILL PROTOCOL	Reference	5.37 ± 1.02	20.2 ± 3.0	2.05 ± 0.34	37.7 ± 8.8	10.4 ± 2.34
	Simple Method	5.20 ± 0.91	20.0 ± 3.0	1.91 ± 0.32	37.7 ± 8.9	9.75 ± 2.19
	Bias (%)	3.28 ± 1.10	0.93 ± 0.43	6.93 ± 2.52	0.12 ± 0.53	6.05 ± 3.02
FORCE PLATE PROTOCOL	Reference	4.71 ± 1.48	16.2 ± 1.7	2.13 ± 0.21	51.4 ± 21.5	13.3 ± 1.9
	Simple Method	4.60 ± 1.33	16.1 ± 1.7	2.06 ± 0.24	50.2 ± 20.4	13.0 ± 2.54
	Bias (%)	2.34 ± 2.42	0.67 ± 1.09	3.24 ± 2.08	2.30 ± 1.63	2.54 ± 1.16

Mean absolute bias is computed as follows: Bias = |(model-reference)/reference| · 100

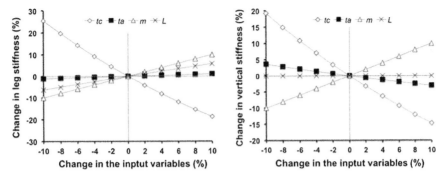

Fig. 8.6 Sensitivity analysis of the simple method proposed. The mechanical inputs of the model plotted against the corresponding leg (left panel) and vertical (right panel) variations

substantial influence on vertical stiffness. For instance, a 10% change in contact time will result, in theory, in a 20–25% change in k_{leg} or k_{vert}. In practice, this means that if a runner has a 10% shorter contact time (e.g. 180 vs. 200 ms), he will run with a 25% higher k_{leg}, and vice versa. As seen in the applications below, such changes can result from training, fatigue, injury, or a voluntary change in the running pattern.

In order to experimentally test the results of this theoretical simulation, we performed a study in which we were able to control and isolate the effects of step frequency and contact time on stiffness during running (Morin et al. 2007). We asked subjects to run on an instrumented treadmill at 3.33 m s^{-1} in several conditions including higher and lower step frequencies set by an audio tone. Given the direct relationship between contact time and step frequency, we also asked the subjects to run with shorter and longer contact times while following their preferred step frequency that was imposed by the audio tone. In the latter conditions, we could isolate the effect of contact time on spring-mass stiffness, independently from the associated effect of step frequency (Farley and González 1996), as we did in the theoretical simulation presented in Fig. 8.6. The results of this experimental study basically confirmed the significant and substantial effect of contact time on k_{leg}, and the overall slope of the linear relationship between tc and k_{leg} was about 2, which is consistent with the 1:2–2.5 relative weight of tc on k_{leg} illustrated in Fig. 8.6. This shows the fundamental importance of an accurate measurement of contact time when using this method.

8.3.4 Limits of the Method

Assumptions of the spring-mass model. The simple method presented here is based on the use of the spring-mass model of running. Therefore, all the limitations and assumptions associated with the use of this model (see Sect. 8.2.4) also apply to the use of this method.

Running conditions: acceleration, speed and slope. The linear spring-mass model is based on a geometrical consideration of the lower limb during the stance phase that is mainly characterized by a symmetry of the angle swept before and after mid-stance (Fig. 8.1) (Blickhan 1989; McMahon and Cheng 1990; Farley and González 1996; Bullimore and Burn 2006). Consequently, all experimental conditions that clearly induce a running pattern in which this basic assumption of the model is infringed lead to questionable data. Although all inputs of the simple method used may be measured in any running condition, the spring-mass variables computed (e.g. vertical and leg stiffness) do not make much sense if the basic postulates of the model are not respected. These conditions include in particular non-constant velocity (i.e. accelerated or decelerated) and incline (uphill or downhill) running. In these cases, the angle swept by the center of mass is not symmetrical around the mid-stance point (center of mass above the center of pressure) and thus Eq. (8.4) and others do not give correct values. Finally, the simple method presented here has been originally validated (Morin et al. 2005) for running speeds ranging from 12 to about 28 km h^{-1}. Some measurements performed after this publication (see Sect. 8.5.2) showed that this validity was still acceptable at lower speeds of 10 km h^{-1}, but less at even lower speeds (8 km h^{-1}, close to the walk-run transition speed). Finally, a recent study (Clark and Weyand 2014) discussed the validity of the sine-wave modeling approach used here at top running speeds in world-class sprinters (i.e. higher than 36 km h^{-1}) and showed that the ground reaction force measured on an instrumented treadmill deviated from the sine-wave trace in such extreme conditions.

Accuracy of input variables measurement. As shown in the theoretical simulation of the model (Fig. 8.6) and in our experimental study (Morin et al. 2007), the measurement of contact and aerial time during running is paramount for the overall accuracy of the stiffness data obtained with this simple method. Although of lower importance, running velocity should also be measured as accurately as possible. Finally, anthropometric variables of body mass and lower limb length should also be measured accurately. For the latter, an accurate pair of scales and a measuring tape is sufficient. The most important factor to ensure accuracy and reliability of the data obtained when using this method will therefore be the technologies used for measuring contact and aerial time, and running velocity.

8.4 Technologies and Input Measurements

As detailed in the previous sections, the devices and technologies used to obtain the main mechanical inputs of the model (contact, aerial time and running velocity) must be accurate enough to ensure the validity of stiffness computations. For instance, since a typical contact time lasts about 0.2 s (ranging from about 0.1 s in high-speed sprinting to about 0.3 s in low speed jogging), a device operating at a

Fig. 8.7 Typical measurement setting with a contact/pressure mat (left, Gaitrite™) installed on the running floor, or an optical device (middle and right, Optojump™) installed on the track during a 100-m sprint or a pole-vault run-up

sampling frequency of 100 Hz will have a time resolution of 0.01 s, which represents a potential 5% error for measuring a 0.2 s event. This error drops to 2.5% for a 200-Hz device and reaches 10% for a 50-Hz device. Note that standard television footage and video cameras operate most of the time at lower amounts of frames per second.

Force plates and instrumented treadmills. Although one can expect researchers and practitioners to use the standard (reference) method (see Sect. 8.2.3) for spring-mass model analysis of running when force plates or instrumented treadmills are available, these devices may be used to measure contact and aerial times. These devices usually operate at sampling frequencies of 1000 Hz or more, ensuring accurate measurements. Thus, provided that the running velocity is accurately measured (treadmill sensor or timing gates for force plate measurements), these contact and aerial time data can lead to accurate computations of spring-mass variables using the present method.

Contact mat and optical devices set on the ground. A practical way to measure contact and flight time, as well as running velocity, is to set a contact/pressure mat (e.g. Gaitrite™), or an optical device (e.g. Optojump™) directly on the ground (Fig. 8.7). The main advantage is that the input variables are obtained accurately without interfering with the runner's technique since the runner carries no sensor whatsoever and (free overground running condition). The other interest is that measurements can be performed directly on the training or competition site, and not necessarily at the laboratory, and for middle-distance or sprint running, direct track measurements are possible, as illustrated in Fig. 8.7 and Sect. 8.5.

Footswitches, pressure soles and accelerometers set at the foot. A practical way to measure the time of foot contact on the ground (and thus the aerial time) during

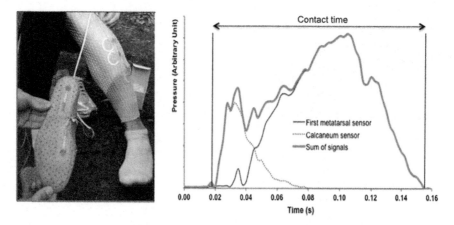

Fig. 8.8 Left: force sensing resistors taped on the insole of a running shoe, at the calcaneum and head of the 1st metatarsal. Right: the pressure signals (sampling rate of 400 Hz) are summed and synchronized with surface electromyography measurements during a running step at 12 km h^{-1}. Light wires and data acquisition devices are carried by the athlete thanks to special clothing and medical stockings

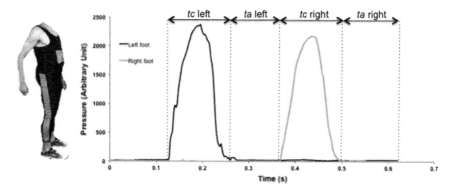

Fig. 8.9 Left: light wires and data acquisition devices are carried by the athlete thanks to special clothing and medical stockings. Right: contact and aerial times measured using footswitches (sampling rate of 400 Hz) during two successive sprint steps (top speed of about 10 m s^{-1})

running is to have pressure sensors directly taped on the foot of the runner, or for more comfort on the shoe insole. This setting has been used in clinical evaluation of the walking gait with footswitches pasted on the subjects' toes (Aminian et al. 2002), and we used it in our field sprint studies (see Sect. 8.5, Morin et al. 2006), as shown in Fig. 8.8. Despite their lack of accuracy as to the amount of pressure applied, these footswitches (Force Sensing Resistor type) are light, very thin, and not expensive. Their measurement of temporal features of the running step (contact and aerial times, Fig. 8.9) is very accurate and reliable (as shown by our personal data of comparison with instrumented treadmill data). The other advantage is that

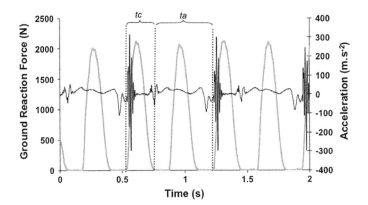

Fig. 8.10 Typical signals of raw acceleration and ground reaction force obtained during a 16 km h^{-1} run (forefoot strike pattern). The accelerometer was tightly taped on the external side of the shoe, at the 5th metatarsal head (as in Fig. 12.1)

they may be connected and synchronized with other devices such as muscle activity sensors (surface electromyography) as we used in a sprint study (Slawinski et al. 2008a, Fig. 8.8).

Using a similar approach, some manufacturers and authors proposed to install small and light accelerometers onto the shoe, as shown in Fig. 12.1 (Hoyt et al. 1994, 2000; Weyand et al. 2001; Giandolini et al. 2014). Although valid and accurate signals may be obtained with this methodology (see for instance Fig. 8.10), the raw data acquisition, storage and processing are usually more complex than with footswitches, should only contact and aerial times be measured.

Sport watch and accelerometers set at the trunk. Recently, another type of systems have been produced that can measure contact, aerial times and the associated running step spatio-temporal variables. These are based on accelerometers placed at the chest (e.g. Garmin HRM-Run™ monitor, Fig. 8.11) or at the hip (e.g. Myotest™). Basically, these light sensors are carried onto the runner's body, not far from the center of mass. In theory, as soon as the runner leaves the ground from one contact to the next one, the center of mass has a vertical acceleration equal to the gravitational acceleration. This is no longer the case as soon as the runner touches the ground a few hundreds of a second later. So should the device be correctly calibrated and accurate enough (and body mass, lower limb length and running speed be known), contact and aerial times can be measured. In turn, applying the laws of falling bodies and/or the equations presented in this chapter, all other spring-mass variables can be computed from this simple basis[6] (Gindre et al. 2016) (Fig. 8.12). To our knowledge, only one study has been published about the

---

[6]Patent: Flaction P, Quievre J, Morin JB (2013) PATENT. Integrated portable device and method implementing an accelerometer for analyzing biomechanical parameters of a stride. US20130190657.

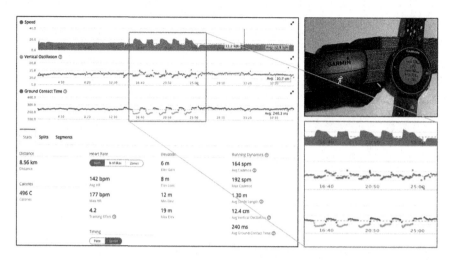

Fig. 8.11 Some sport watches such as the Garmin Forerunner 620™ include an accelerometer (Garmin HRM-Run™) carried at the chest (right). Contact time and other mechanical variables are then monitored for all steps of a run (left, example for a standard jog), allowing detailed follow-up of running mechanics, and calculation of the spring-mass variables using the simple method presented here. In this typical running session, five bouts of 1' run at maximal aerobic speed (20 km h^{-1}) and the associated changes in contact time and vertical oscillation of the center of mass appear in the red squares

validity of the Garmin HRM-Run™ monitor (Watari et al. 2016). The results of this study show a valid measurement of ground contact time, which is in line with our own testing (unpublished data) that shows an acceptable validity and reliability compared to a instrumented treadmill and Optojump™ data. The Myotest™ Run accelerometer device has been tested in two similar studies, that overall conclude it is reproducible, and valid for step rate measurements, although significant differences with reference devices have been found for contact and aerial times (Gouttebarge et al. 2015; Gindre et al. 2016). These devices are therefore very interesting for training and long-term monitoring purposes in a sports context provided comparisons between data are made with the same devices. Their ease of use, comfort, and low cost compared to other technologies are a clear advantage.

Video analysis. As for devices installed on the running ground, a simple and direct way to measure the duration of contact and aerial phases in running is to film and record the steps, provided the number of frames per second is high enough to ensure accuracy. This has been done in some studies, especially in real performance context. For instance, (Le Meur et al. 2013) used a camera (Casio™ Exilim EX-F1) with a 300 fps slow motion mode to measure contact and aerial times, and study how spring-mass variables changed with fatigue and race tactics during an international triathlon competition. Standard television footage (not more than 100 fps, typically around 30 or 60) have been used to estimate spring-mass variables in world-class athletes from top competition recordings (Taylor and Beneke 2012). However, we think, as explained above, that these conditions do not allow accurate

Fig. 8.12 Left: the Myotest™ accelerometer device is carried at the lower end of the trunk during running. Right: running mechanics and spring-mass variables (e.g. vertical stiffness) are displayed after a "run check" test

and reliable measurements of contact time, and in turn lead to questionable values of spring-mass variables using the simple method presented here.

Smartphone app. Some smartphones now feature cameras that allow for slow motion modes with up to 240 fps (Apple iPhone 6™ and subsequent models). So if one can film and record the steps of a runner in a way that allows clear visual detection of foot ground contact beginning and end, this high filming rate could lead to accurate measurements. This is the basis of a new app named "Runmatic"[7] that calculates all spring-mass variables, as well as the left-right asymmetry and monitors how their changes over time (long-term monitoring), from contact, flight time, running speed and body mass and height inputs. As shown in Fig. 8.13, the runner's feet should be filmed over several steps, and then by just clicking on the screen as foot ground contact and take-off are identified visually, all contact and aerial times can be computed. In turn, all spring-mass variables can be derived and left-right asymmetries calculated.

The validation of this app consisted in comparing the contact and aerial times as measured by two observers (inter-observer reliability) to reference data measured with an opto-electronic device (Optojump™) set on a treadmill for 8 consecutive steps at speeds ranging from 10 to 18 km h^{-1} in two subjects (about 100 points

---

[7]https://itunes.apple.com/us/app/runmatic/id1075902287?l=es&ls=1&mt=8.

Fig. 8.13 The iPhone app "Runmatic" uses the high-speed video slow motion mode to identify the contact and aerial times. Then, running pattern mechanics are computed and displayed on the screen

compared in total). Overall, the results show a very high validity and reliability of the app measurements compared to the opto-electronic device (Balsalobre-Fernandez et al. 2017).

8.5 Applications

The main objective of developing, validating and publishing the simple method presented in this chapter was to use it in both research and sports training practice. Over the past decade, we did so, and we were honored to see other academics or coaches adding this method to their running analysis toolbox. Here we will detail two examples of application of this method to investigate running mechanics and spring-mass behavior in very different contexts: the maximal-speed (sprint) and maximal-duration (ultra-endurance) conditions.

8.5.1 Sprint Running

Due to the fact that a human body running at a top speed of 10 m s^{-1} or more (Usain Bolt reached a maximal velocity beyond 12 m s^{-1} during his 9.58 s 100-m World Record) covers about 10 m in one single second, studying sprint running mechanics has long been a challenge. Therefore, as detailed in Chap. 10, sprint instrumented treadmill have been used, and overground experiments have long been limited to studying one or two steps on force plates. In an original experiment we performed in 2004, we were able to calculate the main spring-mass variables of physical education students during an entire 100-m sprint using the simple method presented here (Morin et al. 2006). Subjects' mass and lower limb length were measured using classical devices, running speed was measured with a radar device operating at 35 Hz (Stalker ATS, Applied Concepts, Plano, TX) and finally contact and aerial times were measured at a sampling rate of 400 Hz with footswitches taped on the insoles of the running shoes. At the time of the experiment, we could not record and save pressure data with the currently common flash memory cards or similar technologies, so raw data were transmitted via cable to a PC stored in the backpack of an experimenter who was driving a motorbike side-by-side with the runner. This pilot was skilled enough to exactly follow the runner's acceleration during the 100-m.

This experiment allowed us to show the first spring-mass data during actual overground sprint running. Table 8.2 shows the main variables for the 20-m sections of the dash except the 20–40 m one since subjects were acceleration during this section.

In the same experiment, we investigated the effects of fatigue on spring-mass mechanics on order to better know (i) how the human body changes its spring-mass characteristics under the effect of fatigue due to the repetition of maximal-intensity tasks and (ii) potentially orient the training practice in sports that involve long and/or repeated sprints. Table 8.2 shows the main changes observed with fatigue when repeating four 100-m with 2 min of passive rest. Along with the expected decrease in performance and mean and maximal running speed, the two components of the spring-mass model that showed the most substantial changes were the vertical oscillation of the center of mass during stance and the vertical stiffness. The changes observed were of greater magnitude than what could have been anticipated from the velocity-k_{vert} relationship (Fig. 8.5) and suggest that the ability to limit the downward displacement of the center of mass during the stance and the loss of k_{vert} under fatigue might be a performance factor in such conditions (close to what happens during repeated sprints and changes of direction over team sports such as soccer, rugby, basketball or handball).

Another field experiment was performed in 2011 with this time elite sprinters including the 2010 European Champion on 100-m, 200-m and 4 × 100-m relay (Fig. 8.7 right). During this study (Morin et al. 2012a), we measured sprinters spring-mass behavior over their maximal velocity phase (data not published) using Optojump™ rails set on the track along the 40–60 m section of the 100-m.

Table 8.2 Mean ± SD for the main running mechanics and spring-mass variables over a 100-m run in 8 trained physical education students

Variable	20–40 m	40–60 m	60–80 m	80–100 m	Changes with fatigue between the first and fourth 100-m (%)
t_c (ms)	111 ± 14	108 ± 15	110 ± 12	113 ± 11	+14.7 ± 0 7.2*
t_v (ms)	137 ± 13	141 ± 13	142 ± 11	146 ± 16	+3.87 ± 4.19
v (m s^{-1})	7.93 ± 0.34	8.33 ± 0.34	8.24 ± 0.24	7.89 ± 0.40	−11.6 ± 3.1*
k_{leg} (kN m^{-1})	19.7 ± 4.8	19.8 ± 5.2	18.9 ± 3.5	19.5 ± 4.3	−9.53 ± 9.62
k_{vert} (kN m^{-1})	93.9 ± 14.6	98.3 ± 16.6	93.8 ± 10.2	89.6 ± 9.8	−20.6 ± 7.9*
F_{max} (kN)	2.54 ± 0.25	2.63 ± 0.27	2.60 ± 0.20	2.60 ± 0.21	−4.88 ± 3.67
Δy (cm)	2.8 ± 0.4	2.8 ± 0.4	2.8 ± 0.3	3.0 ± 0.3	+21.2 ± 9.4*
ΔL (cm)	13.8 ± 2.7	14.2 ± 2.8	14.5 ± 2.3	14.2 ± 2.2	+6.88 ± 7.78

The changes (in %) between the first and fourth 100-m show how variables change with fatigue
*Indicates a significant change ($P < 0.05$) as shown by an analysis of variance

Table 8.3 Spring-mass variables averaged for six consecutive steps between the 40 and 60-m marks of a 100-m sprint test in a world-class individual (personal best time of 9.92 s)

	t_c (ms)	t_v (ms)	Δy (cm)	k_{vert} (kN m^{-1})	k_{leg} (kN m^{-1})
Left 1	102	131	2.44	116	14.3
Right 1	104	132	2.51	112	13.1
Left 2	96	138	2.38	127	16.7
Right 2	100	133	2.41	119	14.3
Left 3	98	136	2.40	123	15.6
Right 3	101	131	2.41	118	13.3
Average all	100	134	2.43	119	14.5
Coefficient of variation (%)	2.9	2.2	1.8	4.4	9.5
Average left	99	135	2.41	122	15.5
Average right	104	133	2.51	120	14.3

The running velocity was constant at about 11.0 m s^{-1} during these steps. The athlete used athletics spike shoes and starting-blocks

Table 8.3 shows the data for several consecutive steps of the world class athlete tested during this experiment.

Note that in addition to an accurate study of the athlete's mechanics and a long-term follow-up should measurements be repeated over a season, this method allows deeper analyses such as left-right asymmetry computation and pre-post injury testing to improve the return-to-sport and rehabilitation processes. These are some of our current projects, with for instance a recent study on anterior cruciate ligament surgery (Mazet et al. 2016). We also considered other sports in which sprint running is key, and we could for instance study the final phase of the run-up in elite pole-vaulters including the current World Record holder (Cassirame et al. 2015), and the effects of elevated tracks on spring-mass behavior, run-up terminal velocity and performance in elite athletes.

8.5.2 Ultra-endurance

Although ultra-endurance events (e.g. road 24-h or 100-km, mountain 100-mile, etc....) may seem totally opposite to short maximal sprints such as a 100-m, one common feature is the difficulty to perform on-site experiments to study running mechanics. We initiated the study of very long running mechanics on an instrumented treadmill with a 24-h protocol (Morin et al. 2011a). Then, in order to gain more insights into actual ultra-trail practice, an increasingly popular activity, we wanted to investigate the consequence of a typical race (Ultratrail du Mont Blanc®, 160 km, about 9000 m positive elevation change). The study design (Morin et al. 2011b) was a pre-post measure with subjects asked to run at a constant speed on the ground (Fig. 8.7) of a laboratory located close to the start-finish line of the race. In order to remove any effect of running velocity on the data computed, they were asked to follow a constant 12 km h^{-1} pace while running on the pressure mat. Out of the 34 subjects initially recruited, 22 finished the race and 18 could be tested (4 were unable to complete the running bout at 12 km h^{-1} because of pain and/or fatigue). Table 8.4 shows that the main changes in this ultra-endurance fatigue context are opposite to those observed in sprint running (Table 8.2). In particular, subjects after the mountain ultramarathon showed lower aerial time with unchanged contact time whereas fatigue after repeated sprints induced greater contact times with almost unchanged aerial times. In addition, the vertical displacement of the center of mass during the running stance was greater after four straight 100-m sprints, whereas it was reduced after a 160-km mountain ultramarathon.

These contrasting results show that the consequences of fatigue on running biomechanics are very different depending on the duration and intensity of the exercise considered. We interpreted these results as a decrease in the ability to push the ground and resist the impact and high forces requested to "bounce" back as fatigue develops in sprint running. On the contrary, the changes in spring-mass behavior observed after the 160-km ultratrail were consistent with those observed

Table 8.4 Mean ± SD for the main running mechanics and spring-mass variables pre- and post-mountain ultra-marathon

Variable	Pre ultratrail	Post ultratrail	Changes with fatigue between pre- and post-race (%)
t_c (ms)	249 ± 16	252 ± 17	+1.40 ± 7.21
t_v (ms)	118 ± 19	96 ± 22	−18.5 ± 17.4*
k_{leg} (kN m^{-1})	9.87 ± 1.45	9.44 ± 1.10	−3.71 ± 8.78
k_{vert} (kN m^{-1})	25.1 ± 2.3	26.6 ± 3.3	+5.64 ± 11.7
F_{max} (kN)	2.32 ± 0.16	2.17 ± 0.16	−6.30 ± 7.03*
Δy (cm)	6.6 ± 0.05	5.9 ± 0.08	−11.6 ± 10.5*
ΔL (cm)	16.9 ± 1.4	16.4 ± 1.4	−2.90 ± 7.55

Running time ranging from 25 to 43 h in the subjects tested
*Indicates a significant change ($P < 0.05$) as shown by a t-test for paired samples

over a twice longer race, the Tor des Geants® (Italy) during which we performed a similar study (Degache et al. 2016), and interpreted as a way for subjects to run with less hip, knee and ankle flexion (and the associated eccentric muscle actions) and "bounce" less on the ground. This allowed them to reduce the amount of braking impulse necessary at each step to brake the downward motion of the body as it lands onto the ground, and also probably reduce the associated amount of joint, tendinous and muscular pain at each step. This "smoother" and "safer" running pattern observed consistently in ultra-endurance experiments was not correlated with the massive decrease in knee extensors or ankle plantar flexors strength (Millet et al. 2011), and we suggested it could be a strategy used (consciously or not) by runners to "save the legs" and cover the complete distance (Millet et al. 2012). Interestingly, very similar changes as those observed after ultra-endurance races (especially mountain races) were also observed when the muscle structure or function is altered: after intense downhill running (Chen et al. 2007, 2009), a muscular biopsy of the *vastus lateralis* muscle (Morin et al. 2009) or with ageing (Karamanidis and Arampatzis 2005; Cavagna et al. 2008). In comparison, a series of maximal short treadmill sprints inducing a substantial loss of strength did not result in any change in the constant velocity running pattern (Morin et al. 2012b).

In conclusion, although laboratory devices and protocols initially allowed the validation and bases of the spring-mass model analysis for running, experiments are now possible in real-life exercise and sport conditions. This simple model has been refined and challenged for some specific conditions over time (e.g. very high sprinting velocity), but it overall allows a sound, macroscopic analysis of running biomechanics, encompassing the resulting output of the complex neuromuscular and osteo-tendinous machineries.

References

Alexander RM (1991) Energy-saving mechanisms in walking and running. J Exp Biol 160:55–69
Alexander RM (2003) Principles of animal locomotion. Princeton University Press, Princeton, NJ
Alexander RM (2004) Bipedal animals, and their differences from humans. J Anat 204:321–330
Aminian K, Najafi B, Büla C et al (2002) Spatio-temporal parameters of gait measured by an ambulatory system using miniature gyroscopes. J Biomech 35:689–699
Arampatzis A, Brüggemann GP, Metzler V (1999) The effect of speed on leg stiffness and joint kinetics in human running. J Biomech 32:1349–1353
Balsalobre-Fernández C, Agopyan H, Morin J-B (2017) The validity and reliability of an iPhone app for measuring running mechanics. J Appl Biomech 33:222–226
Blickhan R (1989) The spring-mass model for running and hopping. J Biomech 22:1217–1227
Bramble DM, Lieberman DE (2004) Endurance running and the evolution of Homo. Nature 432:345–352
Brughelli M, Cronin J (2008a) A review of research on the mechanical stiffness in running and jumping: methodology and implications. Scand J Med Sci Sports 18:417–426
Brughelli M, Cronin J (2008b) Influence of running velocity on vertical, leg and joint stiffness: modelling and recommendations for future research. Sports Med 38:647–657
Bullimore SR, Burn JF (2006) Consequences of forward translation of the point of force application for the mechanics of running. J Theor Biol 238:211–219

Cassirame J, Sanchez H, Morin J-B (2015) Elevated track in pole vault, what advantage for run-up determinants of performance? In: 33rd International conference on biomechanics in sports, at Poitiers France. Poitiers, pp 1–5

Cavagna GA (1975) Force platforms as ergometers. J Appl Physiol 39:174–179

Cavagna GA, Legramandi MA, Peyré-Tartaruga LA (2008) Old men running: mechanical work and elastic bounce. Proc Biol Sci 275:411–418

Chen TC, Nosaka K, Tu J-H (2007) Changes in running economy following downhill running. J Sports Sci 25:55–63

Chen TC, Nosaka K, Lin M-J et al (2009) Changes in running economy at different intensities following downhill running. J Sports Sci 27:1137–1144

Clark KP, Weyand PG (2014) Are running speeds maximized with simple-spring stance mechanics? J Appl Physiol 117:604–615

Coleman DR, Cannavan D, Horne S, Blazevich AJ (2012) Leg stiffness in human running: Comparison of estimates derived from previously published models to direct kinematic–kinetic measures. J Biomech 45:1987–1991

Degache F, Guex K, Fourchet F et al (2013) Changes in running mechanics and spring-mass behaviour induced by a 5-hour hilly running bout. J Sports Sci 31:299–304

Degache F, Morin J-B, Oehen L et al (2016) Running mechanics during the world's most challenging mountain ultramarathon. Int J Sports Physiol Perform 11:608–614

Dickinson MH, Farley CT, Full RJ et al (2000) How animals move: an integrative view. Science 288:100–106

Dutto DJ, Smith GA (2002) Changes in spring-mass characteristics during treadmill running to exhaustion. Med Sci Sports Exerc 34:1324–1331

Farley CT, González O (1996) Leg stiffness and stride frequency in human running. J Biomech 29:181–186

Ferris DP, Farley CT (1997) Interaction of leg stiffness and surfaces stiffness during human hopping. J Appl Physiol 82:15–22–4

Farley CT, Ferris DP (1998) Biomechanics of walking and running: center of mass movements to muscle action. Exerc Sport Sci Rev 26:253–285

Farley CT, Glasheen J, McMahon TA (1993) Running springs: speed and animal size. J Exp Biol 185:71–86

Ferris DP, Louie M, Farley CT (1998) Running in the real world: adjusting leg stiffness for different surfaces. Proc R Soc B Biol Sci 265:989–994

Ferris DP, Liang K, Farley CT (1999) Runners adjust leg stiffness for their first step on a new running surface. J Biomech 32:787–794

Giandolini M, Poupard T, Gimenez P et al (2014) A simple field method to identify foot strike pattern during running. J Biomech 47:1588–1593

Gindre C, Lussiana T, Hebert-Losier K, Morin J-B (2016) Reliability and validity of the Myotest® for measuring running stride kinematics. J Sports Sci 34:664–670

Girard O, Mendez-Villanueva A, Bishop D (2011a) Repeated-sprint ability—Part I. Sport Med 41:673–694

Girard O, Micallef J-P, Millet GP (2011b) Changes in spring-mass model characteristics during repeated running sprints. Eur J Appl Physiol 111:125–134

Girard O, Racinais S, Kelly L et al (2011c) Repeated sprinting on natural grass impairs vertical stiffness but does not alter plantar loading in soccer players. Eur J Appl Physiol 111:2547–2555

Girard O, Millet G, Slawinski J et al (2013) Changes in running mechanics and spring-mass behaviour during a 5-km time trial. Int J Sports Med 34:832–840

Girard O, Brocherie F, Morin J-B et al (2015) Comparison of four sections for analyzing running mechanics alterations during repeated treadmill sprints. J Appl Biomech 31:389–395

Girard O, Brocherie F, Morin J-B, Millet GP (2016) Running mechanical alterations during repeated treadmill sprints in hot *versus* hypoxic environments. A pilot study. J Sports Sci 34:1190–1198

Gouttebarge V, Wolfard R, Griek N et al (2015) Reproducibility and validity of the myotest for measuring step frequency and ground contact time in recreational runners. J Hum Kinet 45: 19–26

He JP, Kram R, McMahon TA (1991) Mechanics of running under simulated low gravity. J Appl Physiol 71:863–870

Hobara H, Inoue K, Gomi K, et al (2010) Continuous change in spring-mass characteristics during a 400 m sprint. J Sci Med Sport 13:256–261. doi:10.1016/j.jsams.2009.02.002

Hoyt RW, Knapik JJ, Lanza JF et al (1994) Ambulatory foot contact monitor to estimate metabolic cost of human locomotion. J Appl Physiol 76:1818–1822

Hoyt DF, Wickler SJ, Cogger EA (2000) Time of contact and step length: the effect of limb length, running speed, load carrying and incline. J Exp Biol 203:221–227

Hunter I, Smith GA (2007) Preferred and optimal stride frequency, stiffness and economy: changes with fatigue during a 1-h high-intensity run. Eur J Appl Physiol 100:653–661

Karamanidis K, Arampatzis A (2005) Mechanical and morphological properties of different muscle-tendon units in the lower extremity and running mechanics: effect of aging and physical activity. J Exp Biol 208:3907–3923

Kerdok AE, Biewener AA, McMahon TA et al (2002) Energetics and mechanics of human running on surfaces of different stiffnesses. J Appl Physiol 92:469–478

Kram R, Dawson TJ (1998) Energetics and biomechanics of locomotion by red kangaroos (*Macropus rufus*). Comp Biochem Physiol B Biochem Mol Biol 120:41–49

Latash ML, Zatsiorsky VM (1993) Joint stiffness: Myth or reality? Hum Mov Sci 12:653–692

Le Meur Y, Thierry B, Rabita G et al (2013) Spring-mass behaviour during the run of an international triathlon competition. Int J Sports Med 34:748–755

Lee CR, Farley CT (1998) Determinants of the center of mass trajectory in human walking and running. J Exp Biol 201:2935–2944

Martin V, Kerherve H, Messonnier LA et al (2010) Central and peripheral contributions to neuromuscular fatigue induced by a 24-h treadmill run. J Appl Physiol 108:1224–1233

Mazet A, Morin J-B, Semay B et al (2016) Modifications du pattern mécanique de course dans les suites d'une plastie du ligament croisé antérieur. Sci Sports 31:219–222

McDougall C (2010) Born to run: the hidden tribe, the ultra-runners, and the greatest race the world has never seen. Profile Books Ltd., London, UK

McMahon TA, Greene PR (1978) Fast running tracks. Sci Am 239:148–163

McMahon TA, Greene PR (1979) The influence of track compliance on running. J Biomech 12:893–904

McMahon TA, Cheng GC (1990) The mechanics of running: how does stiffness couple with speed? J Biomech 23(Suppl 1):65–78

McMahon TA, Valiant G, Frederick EC (1987) Groucho running. J Appl Physiol 62:2326–2337

Millet GY, Tomazin K, Verges S et al (2011) Neuromuscular consequences of an extreme mountain ultra-marathon. PLoS ONE 6:e17059

Millet GY, Hoffman MD, Morin JB (2012) Sacrificing economy to improve running performance– a reality in the ultramarathon? J Appl Physiol 113:507–509

Morin JB, Dalleau G, Kyröläinen H et al (2005) A simple method for measuring stiffness during running. J Appl Biomech 21:167–180

Morin J-B, Jeannin T, Chevallier B, Belli A (2006) Spring-mass model characteristics during sprint running: correlation with performance and fatigue-induced changes. Int J Sports Med 27:158–165

Morin JB, Samozino P, Zameziati K, Belli A (2007) Effects of altered stride frequency and contact time on leg-spring behavior in human running. J Biomech 40:3341–3348

Morin J-B, Samozino P, Féasson L et al (2009) Effects of muscular biopsy on the mechanics of running. Eur J Appl Physiol 105:185–190

Morin J-B, Samozino P, Millet GY (2011a) Changes in running kinematics, kinetics, and spring-mass behavior over a 24-h run. Med Sci Sport Exerc 43:829–836

Morin JB, Tomazin K, Edouard P, Millet GY (2011b) Changes in running mechanics and spring– mass behavior induced by a mountain ultra-marathon race. J Biomech 44:1104–1107

Morin J-B, Bourdin M, Edouard P et al (2012a) Mechanical determinants of 100-m sprint running performance. Eur J Appl Physiol 112:3921–3930

Morin J-B, Tomazin K, Samozino P et al (2012b) High-intensity sprint fatigue does not alter constant-submaximal velocity running mechanics and spring-mass behavior. Eur J Appl Physiol 112:1419–1428

Pappas P, Paradisis G, Tsolakis C et al (2014) Reliabilities of leg and vertical stiffness during treadmill running. Sport Biomech 13:391–399

Rabita G, Slawinski J, Girard O et al (2011) Spring-mass behavior during exhaustive run at constant velocity in elite triathletes. Med Sci Sport Exerc 43:685–692

Silder A, Besier T, Delp SL (2015) Running with a load increases leg stiffness. J Biomech 48:1003–1008

Slawinski J, Dorel S, Hug F et al (2008a) Elite long sprint running: a comparison between incline and level training sessions. Med Sci Sport Exerc 40:1155–1162

Slawinski J, Heubert R, Quievre J et al (2008b) Changes in spring-mass model parameters and energy cost during track running to exhaustion. J Strength Conditioning Res 22:930–936

Taylor MJ, Beneke R (2012) Spring mass characteristics of the fastest men on earth. Int J Sports Med 33:667–670

Watari R, Hettinga B, Osis S, Ferber R (2016) Validation of a torso-mounted accelerometer for measures of vertical oscillation and ground contact time during treadmill running. J Appl Biomech 32:306–310

Weyand PG, Kelly M, Blackadar T et al (2001) Ambulatory estimates of maximal aerobic power from foot -ground contact times and heart rates in running humans. J Appl Physiol 91:451–458

Winter D (1979) Biomechanics of human movement. Wiley, NewYork

Chapter 9
A Simple Method for Determining Foot Strike Pattern During Running

Marlene Giandolini

Abstract Runners and especially long-distance runners face various overuse injuries, such as Achilles tendinopathy, shin splints or stress fractures. These overuse injuries are caused in parts by the repetitive load application on musculoskeletal structures. Otherwise, the load repartition in running is narrowly related to the foot strike pattern adopted. Three foot strike patterns are typically identified: heel strike, midfoot strike and forefoot strike. Depending the foot strike pattern adopted, the intensity and location of loads applied on the runner's body differ and so indirectly the risk to sustain certain overuse injuries. As a consequence, identifying runner foot strike pattern profile is of interest in order to prevent injuries by giving training and/or footwear recommendations according to the runner-specific injury history. This chapter aims at presenting a simple method for determining foot strike pattern in situ and continuously. This method requires two accelerometers to record heel and forefoot accelerations. From these signals, the delay between the heel and metatarsals strikes, represented by their respective peak acceleration, is measured. From this temporal variable, the percentage of each foot strike pattern over a race or a training session is evaluated. By synchronizing accelerometers with GPS unit, this method also permit to study the interaction between fatigue, slope, surface, etc., and the foot strike pattern profile for a given runner. This simple method is of interest for fundamental research but also for clinicians, coaches or running shoe manufacturers.

9.1 Introduction: Why Evaluate Foot Strike Pattern?

Running kinematics of an individual can be characterized by a large panel of variables relative to either the flight phase or the contact phase, such as flight and contact times, step frequency, stride length, joints' range of motion, leg compression,

---

M. Giandolini (✉)
Amer Sports Innovation and Sport Sciences Lab, Salomon SAS,
14 Chemin Des Croiselets, 74370 Metz-Tessy, France
e-mail: marlene.giandolini@salomon.com

etc. Among them, the type of foot strike pattern has recently become of great interest in the research area. Three foot strike techniques have been identified: a rearfoot strike (RFS), in which the point of the first contact with the ground is the heel or the rear third part of the sole and in which the midfoot and forefoot parts do not contact the ground at foot strike; a forefoot strike (FFS), in which the point of the first contact was the forefoot or the front half of the sole and in which the heel does not contact the ground at foot strike but slowly goes down to touch the ground during midstance; a midfoot strike (MFS), in which the heel and the ball of the foot land quasi-simultaneously, the point of the first contact can be thus either the rearfoot or forefoot parts (Altman and Davis 2012; Hasegawa et al. 2007). Basically, the RFS is the predominant foot strike pattern but the parts of RFS, MFS and FFS runners depend on the running activity considered, and so are dependent of the running speed and the distance to cover. For instance, in 800–1500 m races, 27% of RFS runners, 42% of MFS runners and 31% of FFS runners were identified (Hayes and Caplan 2012). In half-marathon, two studies observed about 75–89% of RFS runners, 4–24% of MFS runners and 1–2% of FFS runners (Hasegawa et al. 2007; Larson et al. 2011). Regarding marathon races, there are about 88–94% of rearfoot strikers, 3–5% of midfoot strikers and 1% of forefoot strikers (Kasmer et al. 2013a; Larson et al. 2011). The reason for which the foot strike technique has aroused such keen interest is multifactorial. Indeed, the foot strike pattern used could be a significant factor in the etiology of certain types of running-related injuries and may influence performance, making its identification necessary.

9.1.1 Foot Strike Pattern and Running-Related Injuries

Epidemiological investigations have reported that patella femoral pain syndrome, Achilles tendinopathy, medial tibial stress syndrome (shin splints), and tibial stress fractures are among the most common injuries sustained by runners (Lopes et al. 2012; Taunton et al. 2002). Regarding the etiology of these injuries, it has been suggested that patella femoral pain syndrome could be partly attributed to excessive hip adduction (Morgenroth et al. 2014), that Achilles tendinopathy and shin splints could be induced in part by high traction force of plantar flexors (Cibulka et al. 1994; Kader et al. 2002; McCrory et al. 1999; Moen et al. 2009), and that tibial stress fracture could be caused by repetitive shocks and high rate of loading (Ewers et al. 2002; Warden et al. 2014; Zadpoor and Nikooyan 2011). Numerous studies observed that adopting a FFS results in a large decrease in vertical loading rate measured from vertical ground reaction force (Divert et al. 2005; Giandolini et al. 2013; Kulmala et al. 2013; Lieberman et al. 2010; Shih et al. 2013). More recently, decreased vertical peak tibial acceleration and lower content within the impact-related frequency range, i.e. from 12 to 20 Hz (Shorten and Winslow 1992), has been observed when running with a FFS compared to a RFS (Gruber et al. 2014). Further, lower pressure at the patella-femoral joint with a FFS compared to a RFS was recently shown (Kulmala et al. 2013; Vannatta and Kernozek 2014).

This lowering in knee joint stress could be related to a decrease in hip adduction during the contact phase (Davis 2005; Kulmala et al. 2013; McCarthy et al. 2014; Vannatta and Kernozek 2014). Therefore, one can suggest that a FFS could be more advantageous to reduce impact severity and the strain applied at the knee joint by lowering the rate of loading, minimizing the power of impact frequency content and limiting the hip frontal plane motion. This may thus have implications in the risk of joint and bone injuries. Otherwise, forefoot striking necessitates a more plantar flexed ankle at initial contact, as a matter of fact inducing a higher plantar flexors activity during the stride cycle (Giandolini et al. 2013; Shih et al. 2013). This increases the traction force applied by plantar flexor muscles on the Achilles tendon and the tibial periosteum (Cibulka et al. 1994; Kader et al. 2002; McCrory et al. 1999; Moen et al. 2009). Interestingly, it was observed that Achilles tendon load increases with a FFS compared to a RFS (Almonroeder et al. 2013; Kulmala et al. 2013). It can be therefore reasonably assumed that adopting a FFS would increase the risk of Achilles tendinopathy and shin splints, if the runner is not well adapted. The foot strike pattern influences the intensity and location of the stress applied on the runner's body, and as a consequence may affect the prevalence of certain types of running-related injuries. Thus, the foot strike technique might have implications in the clinic field as well as in the footwear industry area.

9.1.2 Foot Strike Pattern and Performance

The influence of the foot strike pattern on factors of performance is largely discussed. Findings differ and at this time no one can establish a consistent conclusion whether or not the foot strike technique affects performance. Regarding both in situ or lab experiments, results vary radically. For instance, Kasmer et al. (2013a) reported that the part of RFS runners during a marathon decreased as the race performance increased, and Santos-Concejero et al. (2014) recently reported a better running economy when adopting a MFS or FFS. Contrastingly, Ogueta-Alday et al. (2013) found the inverse result: rearfoot strikers would be more economical than midfoot or forefoot strikers. Otherwise, Larson et al. (2011) observed no significant relationship between the foot strike of runners and their marathon performance, and no difference in running economy, oxygen consumption or carbohydrate oxidation when running with a RFS or with a FFS were observed as well (Perl et al. 2012; Gruber et al. 2013). Finally, Hasegawa et al. (2007) highlighted that the better ranked runners during an half-marathon race exhibited shorter contact time, kinematic specificity in forefoot strikers. There is therefore many discrepancies in conclusions about the effect of the foot strike pattern or its related kinematics on performance, and one cannot refute the possibility of a relationship between these both parameters.

9.2 Existing Methods to Evaluate Foot Strike Pattern: Definitions and Limits

Basically, experimenters aiming at identifying foot strike pattern during running use either a kinematic or kinetic method. Indeed, measuring foot-to-ground angle by 2D motion analysis or assessing the position of the center of pressure at initial contact by force or pressure measures are hitherto the two most common methods used.

9.2.1 Measurement of Foot Strike Angle by 2D Motion Analysis

The kinematic approach consists in assessing foot strike angle, i.e. the angle between running surface and plantar surface at initial contact in the sagittal plane. This method can be applied either in lab on treadmill, commonly by setting retroflective markers at heel and 5th metatarsal onto the shoe or foot (Daoud et al. 2012; Lieberman et al. 2010), or in real practice, by recording runners passing at one or several points of a race (Hasegawa et al. 2007; Kasmer et al. 2013a; Larson et al. 2011). While the lab application permits to assess an almost unlimited number of steps on treadmill, the field application permits to evaluate the foot strike pattern in real practice taking hence into account external factors (surface, slope, speed variation, fatigue, competition, etc.) but assessing only one to three steps per runner. Another issue which should be mentioned is that the reliability of the foot strike's identification is likely affected by the sample frequency used. When using high-speed video cameras and setting retroflective markers onto the subject's foot, the initial contact and the foot strike angle are easily and precisely detected and measured, respectively. From this measurement, the foot strike pattern can be classified as RFS, MFS or FFS using for instance the Altman and Davis' criteria (FFS $< -1.6° <$ MFS $< 8° <$ RFS) (Altman and Davis 2012). However, using $\sim$120- or 240-Hz cameras, basically in in situ experiments, makes the touchdown's detection unclear. The identification of the foot strike pattern becomes hence quite subjective and experimenter-dependent. Note that to compensate this drawback, the step-by-step motion analysis could be repeated by two or three experimenters (Kasmer et al. 2013a; Lieberman et al. 2010).

9.2.2 Measurement of Foot Strike Index from Kinetics

The kinetic approach consists in assessing by force or pressure measurements the position of the center of pressure at landing relative to the foot length (Cavanagh and Lafortune 1980). From this *foot strike index*, steps are identified as RFS, MFS and FFS using the criteria proposed by Cavanagh and Lafortune (1980): a RFS is described by a foot strike index lower than 33%, a MFS by an index ranging from 34 to 66%, and a FFS by an index higher than 67%. As for the kinematic method, the use of the foot strike index in identifying the running pattern is limited by either the number of steps analyzed for field measurements or the conditions in which the running kinematic is observed for lab measurements.

Analyzing as many steps as possible in situ is necessary when one aims at investigating the foot strike pattern since running kinematics can be affected by many external factors, such as slope (Buczek and Cavanagh 1990), speed (Hayes and Caplan 2012), surface (Muller et al. 2012), and fatigue (Hasegawa et al. 2007; Kasmer et al. 2013a, b; Larson et al. 2011; Morin et al. 2011a, b). This can make the running pattern more or less variable. Considering the aforementioned limits of kinematic and kinetic methods, an acceleration-based approach has been recently proposed permitting the foot strike pattern's assessment from the sampling of an almost unlimited number of steps in field conditions.

9.3 A Novel Field Method Based on Acceleration Measurements

The acceleration-based method consists in recording heel and metatarsals vertical acceleration and measuring the delay between the heel peak acceleration and the metatarsals peak acceleration describing the heel-to-ground contact and the forefoot-to-ground contact, respectively.

9.3.1 Material and Data Analysis

Two accelerometers are firmly fixed on runner's shoe at the heel and metatarsals. Two examples of setting are shown in Fig. 9.1. The heel accelerometer is set at the back of the shoe above the midsole. The metatarsals accelerometer is placed on the external side of the shoe at the head of the 5th metatarsal above the midsole (Fig. 9.1a). The forefoot accelerometer can also be placed on the dorsal surface of the foot above metatarsals (Fig. 9.1b). Comparing both metatarsals accelerometer's placements (i.e. on the external side or on the instep), there is a 3-ms difference in the measured delay between heel and metatarsals strikes. So when using the instep placement, three milliseconds must be subtracted from the delay measured. Small and lightweight mono-axial accelerometers are preferable first to do not disturb the runner, and second to reduce measurement errors. Indeed, overshot peak accelerations were observed using 6-grams skin-mounted accelerometer compared to those of a bone-mounted accelerometer (Hennig and Lafortune 1988). High tension in the system of attachment and setting accelerometers on rigid and non-deformable parts of the shoe will also improve signals quality.

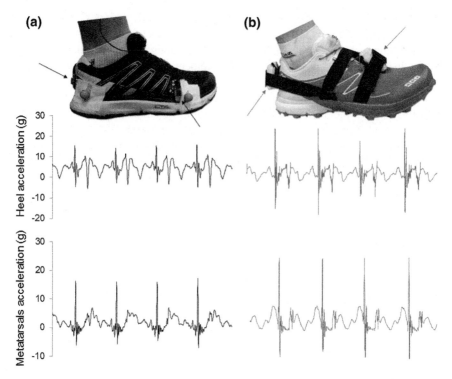

Fig. 9.1 Heel and metatarsals accelerometers' placements. Panel **a** accelerometers (ADXL150, Analog Device, USA) and placement used for the validation of the simple method (Giandolini et al. 2014). Panel **b** accelerometers (Agile Fox, Hikob, France) and placement used for field applications (Giandolini et al. 2015). For each placement, typical raw acceleration signals were presented

The two accelerometers must be time-synchronized. Accelerations should be sampled at a minimum of 1000 Hz in order to permit an accurate detection of the heel and metatarsals strikes. A 30 Hz high-pass filter is applied to acceleration signals to easily detect acceleration peaks by removing the active (<2 Hz) and impact (12–20 Hz) components of acceleration. From the heel and metatarsals filtered acceleration signals, the duration between the heel peak acceleration and the metatarsals peak acceleration is calculated considering the heel peak as t_0 (THM) (Fig. 9.2). Each step can be then identified as RFS, MFS or FFS using the criteria proposed by Giandolini et al., i.e. FFS < −5.49 ms <MFS < 15.2 ms < RFS (Fig. 9.3) (Giandolini et al. 2014). These criteria were obtained by extracting from the validation protocol's database all individual data corresponding to a MFS according to the gold standard video criteria of Altman and Davis, i.e. with a foot strike angle ranging from −1.6° and 8° (Altman and Davis 2012). The mean THM and standard deviation were calculated for this subsample and lower and upper THM limits were obtained as follows:

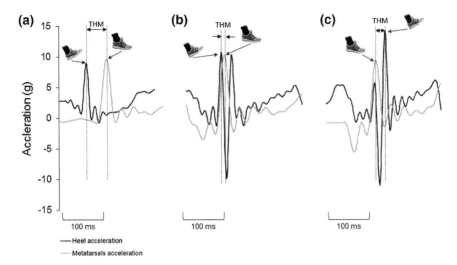

Fig. 9.2 Typical heel (black line) and metatarsal (gray line) acceleration signals with a RFS pattern (raw data obtained in level–PRS–RFS condition of *validation* for a typical subject, panel **a**), a MFS pattern (raw data obtained in level–PRS–MFS condition of *validation* for a typical subject, panel **b**), and an FFS pattern (raw data obtained in level–PRS–FFS condition of *validation* for a typical subject, panel **c**)

$$\text{THM limits} = \text{mean}\left(\text{THM}_{\text{MFSsubsample}}\right) \pm \text{SD}\left(\text{THM}_{\text{MFSsubsample}}\right)$$

9.3.2 Reliability

The temporal parameter measured (THM) has been shown to be highly correlated to the foot strike angle (R = 0.916, $P < 0.0001$, n = 288, Fig. 9.3) for a wide range of slopes and speeds whatever the foot strike pattern used, the shoes worn, the shoe size or the state of fatigue (Table 9.1) (Giandolini et al. 2014). However, the method was not validated when running at 14 or 16 km h^{-1} with a FFS because with this "sprint-like" running pattern, also called "toe run", heels never or rarely strike the ground making hence impossible the calculation of THM (Table 9.1).

Regarding the reliability of the classification criteria, 86.3% of matching results were found during the validation process when comparing the video-based classification proposed by Altman and Davis (Altman and Davis 2012) and the acceleration-based classification proposed in Giandolini et al. (2014). For instance, among the overall validation database (n = 288), 61.2% of RFS, 12.6% of MFS and 26.3% of FFS were identified by the video-based method (foot strike angle), *versus* 67.3% of RFS, 15.1% of MFS and 17.6% of FFS using the acceleration-based method (THM).

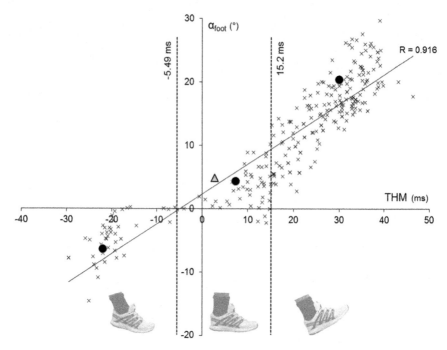

Fig. 9.3 Overall correlation between THM and strike angle (α_{foot}) computed from the validation database (n = 288, r = 0.916). Black crosses represent the mean of each participant within each condition. Black dots represent RFS, MFS and FFS condition values averaged from the two level running conditions in *protocol 1* (see Table 4.1). The black triangle represents the video-based MFS subsample. Dashed lines symbolize THM limits for pattern classification (Giandolini et al. 2014)

9.4 Field Applications

The main advantage of the acceleration-based method is to permit the analysis of an almost unlimited number of steps in situ. This field approach can allow researchers to investigate the effects of slope, speed, running duration, etc., on the foot strike pattern repartition over an entire race in order to better apprehend the influence of foot strike pattern on race performance for instance. It can also allow running footwear manufacturers to study the influence of the shoe construction on foot strike pattern in real practice. Finally, by this simple field method, clinicians can turn patients towards a footwear model better adapted to their foot strike pattern profile or towards a certain running technique better accommodated to their injury prevalence.

Hereby are presented some examples of field application of the acceleration-based method. All runners studied in these experiments were equipped with two accelerometers with air pressure sensor (Hikob Agile Fox, Hikob, Villeurbanne, France) placed at the left heel and metatarsals in fitted pockets firmly fixed by elastic

Table 9.1 Sample sizes for each correlation and Bravais-Pearson's correlation coefficients between THM and α_{foot} (foot strike angle): overall coefficients, coefficients for each condition, and maximal and minimal intra-individual coefficients for *protocol 1*

Conditions		n	r	α_{foot}	THM
Validation overall		140	0.946**	8.37 ± 11.2	9.53 ± 20.8
Natural pattern	Uphill	14	0.948**	5.08 ± 6.69	1.07 ± 21.7
	Downhill	14	0.951**	15.8 ± 12.2	17.9 ± 21.7
Level running at preferred running speed	Natural pattern	14	0.964**	12.1 ± 9.40	17.0 ± 18.4
	RFS	14	0.954**	20.4 ± 5.57	30.0 ± 9.46
	MFS	14	0.923**	4.29 ± 4.40	7.35 ± 12.5
	FFS	14	0.620*	−6.46 ± 3.44	−22.0 ± 1.51
Level running at 14 or 16 km h^{-1}	Natural pattern	14	0.908**	12.5 ± 8.77	15.8 ± 16.2
	RFS	14	0.963**	19.1 ± 6.92	25.3 ± 9.35
	MFS	14	0.856**	4.87 ± 4.46	8.56 ± 11.7
	FFS	14	0.191	−4.78 ± 2.42	−20.0 ± 2.14
Intra-individual	Max	10	0.976**		
	Min	10	0.885**		
Application overall		148	0.875**	13.0 ± 7.34	25.6 ± 12.1
Pre-ultratrail	10 km h^{-1}	34	0.897**	12.1 ± 7.97	25.2 ± 13.1
	12 km h^{-1}	34	0.885**	14.3 ± 7.96	25.5 ± 12.2
	MAS	34	0.886**	14.5 ± 8.02	26.6 ± 12.8
Post-ultratrail	10 km h^{-1}	23	0.894**	11.0 ± 5.78	24.9 ± 11.3
	MAS	23	0.870**	12.4 ± 5.30	25.4 ± 11.0

Mean ± SD values for both THM and α_{foot} parameters in *protocol 1* and *protocol 2* are also reported. Significant correlations are indicated by * ($P < 0.05$) or ** ($P < 0.0001$). The validation consists in two distinct protocols. During *protocol 1*, shoes used were standardized and the reliability of the method was tested under different conditions of speed (preferred running speed and 14 or 16 km h^{-1} depending the subject's gender), slope (level, +10%, −10%) and foot strike (natural pattern, forced RFS, forced MFS and forced FFS). During *protocol 2*, shoes were not standardized and the method was tested under different conditions of fatigue (before and after a 110-km ultratrail race) and speed (MAS, Maximal Aerobic Speed assessed with a 10% slope, 10 and 12 km h^{-1}). From Giandolini et al. (2014)

straps as presented in Figs. 9.1b and 9.4. This setting permits to protect them from rocks, water, mud, etc. Accelerometers were time-synchronized by a common acquisition system (Hikob, Villeurbanne, France). Acceleration data were sampled at 1344 Hz and pressure data at 12 Hz. They were collected on micro-SD cards. Data analysis was performed using Scilab 5.4.1 software (Scilab Entreprises, Orsay, France). Acceleration signals were 30-Hz high-pass filtered to facilitate the detection of peak accelerations. THM was calculated for all steps. Litigious steps were removed from the analysis. Using THM criteria (FFS < −5.49 ms <MFS < 15.2 ms < RFS) all analyzed steps were identified as RFS, MFS or FFS.

Fig. 9.4 Heel and metatarsals accelerometers' placements for a field experiment

The first field application of the method was performed on the trail running world leader during a trail running race (Giandolini et al. 2015). The experiment took place during the Kilian's Classik™ 2013 (Font-Romeu, France), a 45-km official trail running race with a 1627-m of positive elevation. The individual studied was the current world leader in trail and ultratrail running (26 years, 56.5 kg, 171 cm). He ended first with a finish time of 4:23:18 h. In addition to the heel and metatarsals accelerometers, a GPS was plugged into the metatarsals accelerometer (Hikob, Villeurbanne, France) to calculate running speed from altitude, latitude and longitude data. Only the first 20 km of the race were considered because the battery of the GPS unit stopped at mid-race. Since the aim of this study was to investigate the effects of speed, slope and running duration on the foot strike pattern repartition there was no interest in analyzing the last 20 km acceleration data without synchronized information about the environmental characteristics. A single acquisition of ~4.5-h was obtained (i.e. from the start to the end of the race) and ~82 min were analyzed over eleven sections presenting typical slope and speed profiles (i.e. 5530 left steps) (Fig. 9.5 and Table 9.2). From the analysis, it was observed that the runner exhibited on average $23.7 \pm 4.48\%$ of RFS steps, $26.9 \pm 11.8\%$ of MFS steps and $49.4 \pm 14.2\%$ of FFS steps over the analyzed periods (Fig. 9.5 and Table 9.3). This MFS-FFS profile is very atypical for an ultra-endurance runner. Indeed, basically the longer the race distance, the higher the percentage of rearfoot strikers (Hasegawa et al. 2007; Kasmer et al. 2013a; Larson et al. 2011). Nevertheless, these findings are based on measurements performed at one or two points of races, so do not providing an overview of the repartition of foot strike techniques over a race.

Same measurements were also performed on world class trail runners during training sessions (Fig. 9.6). Although they were not performed during an official competition, these measurements provide an overview of the foot strike pattern prevalence for these runners in real practice, i.e. considering the various environmental conditions encountered by trail runners. First, these observations confirmed

9 A Simple Method for Determining Foot Strike Pattern …

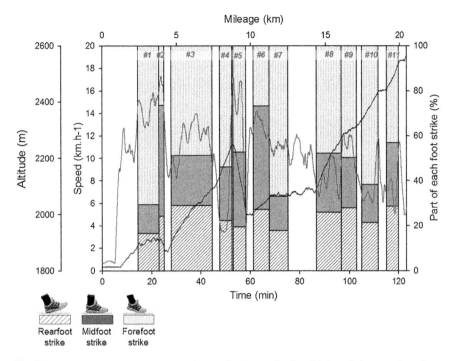

Fig. 9.5 Altitude (black line) and speed (gray line) over the first 20 km of the race. Bar charts represent the repartition of foot strikes (RFS, MFS and FFS) within the eleven analyzed sections. From Giandolini et al. (2015)

Table 9.2 Characteristics of the eleven sections analyzed

Sections #	Slope (%)	Speed (km h^{-1})	Duration (min)	Number of analyzed steps
1	1.4 ± 5.6	14.2 ± 2.1	9.4	672
2	−6.3 ± 7.1	16.3 ± 1.7	1.7	101
3	6.8 ± 4.8	11.7 ± 2.0	16.7	1102
4	34.7 ± 11.6	4.3 ± 1.4	4.9	251
5	−18.5 ± 6.8	15.8 ± 2.2	5.3	376
6	4.2 ± 1.6	13 ± 1.1	6.4	421
7	1.6 ± 6.8	11.1 ± 1.1	7.5	501
8	14.9 ± 7.0	8.2 ± 1.8	10	702
9	3.8 ± 4.8	11.6 ± 1.3	8.5	602
10	14.6 ± 4.7	6.8 ± 1.1	6.7	451
11	21.1 ± 11.7	6.8 ± 1.5	5	351
	7.1 ± 6.6	10.9 ± 1.6	82.1	5530

They were numbered according to their order of occurrence over the race. Mean ± SD for the slope and the running speed, and the overall duration of analysis and number of analyzed steps were calculated. From Giandolini et al. (2015)

Table 9.3 Mean ± SD for kinematics and foot strike parameters within the eleven sections including step frequency (SF), time between heel and metatarsal peak accelerations (THM) and percentages of each foot strike pattern (%RFS, %MFS, %FFS)

Sections #	SF (Hz)	THM (ms)	%RFS	%MFS	%FFS
1	3.05 ± 0.21	−5.1 ± 14.6	16.6	12.9	70.5
2	3.01 ± 0.26	3.6 ± 14.1	24.4	49.3	26.4
3	2.91 ± 0.13	0.3 ± 17.5	29.1	22.3	48.6
4	1.72 ± 0.12	−2.3 ± 18.8	22.4	23.7	53.9
5	3.17 ± 0.34	1.2 ± 15.2	19.4	33.3	47.3
6	3.01 ± 0.11	7.1 ± 13.1	27.3	45.9	26.8
7	2.92 ± 0.2	−4.0 ± 16.4	17.9	15.1	67.0
8	2.78 ± 0.16	1.1 ± 17.9	26.0	26.3	47.7
9	2.84 ± 0.16	2.4 ± 16.7	27.9	22.3	49.8
10	2.73 ± 0.26	−2.7 ± 17.8	21.4	16.8	61.8
11	2.72 ± 0.22	3.7 ± 17.1	28.6	28.2	43.1
	2.81 ± 0.37	0.5 ± 16.3	23.7 ± 4.5	26.9 ± 11.8	49.4 ± 14.2

From Giandolini et al. (2015)

the more or less pronounced variability in foot strike pattern in trail runners previously noticed (Giandolini et al. 2015). Second, these reports would permit to give some recommendations on shoe construction, especially in terms of drop (i.e. sole offset between heel and metatarsals), and also to enlighten on injuries prevalence. For example, the profile presented on Fig. 9.6d refer to a female athlete facing chronic shin splints, an injury partly related to the plantar flexors traction on the tibial periosteum, as mentioned above, and so indirectly to the adoption of anterior foot strike patterns (Cibulka et al. 1994; Moen et al. 2009). It is therefore interesting to notice that this athlete presents a FFS prevalence (53.1%) and only 4.2% of RFS steps over an about 1-h training session. Note also that this runner insists on wearing lightweight but low-drop shoes.

Although it cannot be affirmed that foot strike pattern influences performance in running, one can reasonably admit that it affects the load distribution applied on musculoskeletal structures, and thus the risk of certain injuries. As a consequence, identifying runners' foot strike pattern profile would allow clinicians and coaches to enlighten on the occurrence of injuries which are partly and indirectly related to the foot strike pattern adopted, and then to provide a training program specific to running technique and give some recommendation on the footwear choice.

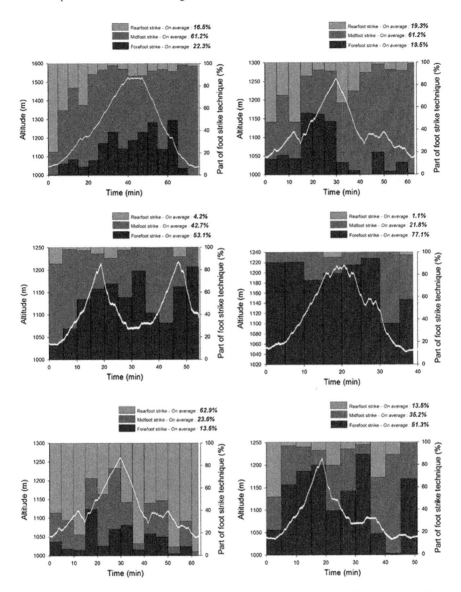

Fig. 9.6 Foot strike pattern profiles for six world class trail runners recording during a training session. Altitude (white line) was represented. Bar charts represent the repartition of foot strikes (RFS, MFS and FFS) within 5-min sections. Unpublished data

References

Almonroeder T, Willson JD, Kernozek TW (2013) The effect of foot strike pattern on achilles tendon load during running. Ann Biomed Eng 41(8):1758–1766. https://doi.org/10.1007/s10439-013-0819-1

Altman AR, Davis IS (2012) A kinematic method for footstrike pattern detection in barefoot and shod runners. Gait Posture 35(2):298–300. https://doi.org/10.1016/j.gaitpost.2011.09.104 (S0966-6362(11)00401-2 [pii])

Buczek FL, Cavanagh PR (1990) Stance phase knee and ankle kinematics and kinetics during level and downhill running. Med Sci Sports Exerc 22(5):669–677

Cavanagh PR, Lafortune MA (1980) Ground reaction forces in distance running. J Biomech 13 (5):397–406. https://doi.org/10.1016/0021-9290(80)90033-0 (0021-9290(80)90033-0 [pii])

Cibulka MT, Sinacore DR, Mueller MJ (1994) Shin splints and forefoot contact running: a case report. J Orthop Sports Phys Ther 20(2):98–102

Daoud AI, Geissler GJ, Wang F, Saretsky J, Daoud YA, Lieberman DE (2012) Foot strike and injury rates in endurance runners: a retrospective study. Med Sci Sports Exerc 44(7):1325–1334. https://doi.org/10.1249/MSS.0b013e3182465115

Davis IS (2005) Gait retraining in runners. Orthop Pract 17:8–13

Divert C, Mornieux G, Baur H, Mayer F, Belli A (2005) Mechanical comparison of barefoot and shod running. Int J Sports Med 26(7):593–598. https://doi.org/10.1055/s-2004-821327

Ewers BJ, Jayaraman VM, Banglmaier RF, Haut RC (2002) Rate of blunt impact loading affects changes in retropatellar cartilage and underlying bone in the rabbit patella. J Biomech 35 (6):747–755. https://doi.org/10.1016/S0021-9290(02)00019-2 (S0021929002000192 [pii])

Giandolini M, Arnal PJ, Millet GY, Peyrot N, Samozino P, Dubois B, Morin JB (2013) Impact reduction during running: efficiency of simple acute interventions in recreational runners. Eur J Appl Physiol 113(3):599–609. https://doi.org/10.1007/s00421-012-2465-y

Giandolini M, Poupard T, Gimenez P, Horvais N, Millet GY, Morin JB, Samozino P (2014) A simple field method to identify foot strike pattern during running. J Biomech 47(7):1588–1593. https://doi.org/10.1016/j.jbiomech.2014.03.002 (S0021-9290(14)00156-0 [pii])

Giandolini M, Pavailler S, Samozino P, Morin JB, Horvais N (2015) Foot strike pattern and impact continuous measurements during a trail running race: proof of concept in a world-class athlete. Footwear Sci. https://doi.org/10.1080/19424280.2015.1026944

Gruber AH, Umberger BR, Braun B, Hamill J (2013) Economy and rate of carbohydrate oxidation during running with rearfoot and forefoot strike patterns. J Appl Physiol. https://doi.org/10.1152/japplphysiol.01437.2012 (japplphysiol.01437.2012 [pii])

Gruber AH, Boyer KA, Derrick TR, Hamill J (2014) Impact shock frequency components and attenuation in rearfoot and forefoot running. J Sport Health Sci. https://doi.org/10.1016/j.jshs.2014.03.004

Hasegawa H, Yamauchi T, Kraemer WJ (2007) Foot strike patterns of runners at the 15-km point during an elite-level half marathon. J Strength Cond Res 21(3):888–893. https://doi.org/10.1519/R-22096.1 (R-22096 [pii])

Hayes P, Caplan N (2012) Foot strike patterns and ground contact times during high-calibre middle-distance races. J Sports Sci 30(12):1275–1283. https://doi.org/10.1080/02640414.2012.707326

Hennig EM, Lafortune MA (1988) Tibial bone and skin accelerations during running. In: Proceedings of the 5th biennial conference and human locomotion symposium of the Canadian Society for Biomechanics, London, ON

Kader D, Saxena A, Movin T, Maffulli N (2002) Achilles tendinopathy: some aspects of basic science and clinical management. Br J Sports Med 36(4):239–249

Kasmer ME, Liu XC, Roberts KG, Valadao JM (2013a) Foot-strike pattern and performance in a marathon. Int J Sports Physiol Perform 8(3):286–292. https://doi.org/10.1123/ijspp.8.3.286 (2012-0114 [pii])

Kasmer ME, Wren JJ, Hoffman MD (2013b) Foot strike pattern and gait changes during a 161-km ultramarathon. J Strength Cond Res. https://doi.org/10.1519/JSC.0000000000000282

Kulmala JP, Avela J, Pasanen K, Parkkari J (2013) Forefoot strikers exhibit lower running-induced knee loading than rearfoot strikers. Med Sci Sports Exerc 45(12):2306–2313. https://doi.org/10.1249/MSS.0b013e31829efcf7

Larson P, Higgins E, Kaminski J, Decker T, Preble J, Lyons D, McIntyre K, Normile A (2011) Foot strike patterns of recreational and sub-elite runners in a long-distance road race. J Sports Sci 29(15):1665–1673. https://doi.org/10.1080/02640414.2011.610347

Lieberman DE, Venkadesan M, Werbel WA, Daoud AI, D'Andrea S, Davis IS, Mang'eni RO, Pitsiladis Y (2010) Foot strike patterns and collision forces in habitually barefoot versus shod runners. Nature 463(7280):531–535. https://doi.org/10.1038/nature08723 (nature08723 [pii])

Lopes AD, Hespanhol Junior LC, Yeung SS, Costa LO (2012) What are the main running-related musculoskeletal injuries? A systematic review. Sports Med 42(10):891–905. https://doi.org/10.2165/11631170-000000000-00000

McCarthy C, Fleming N, Donne B, Blanksby B (2014) Barefoot running and hip kinematics: good news for the knee? Med Sci Sports Exerc. https://doi.org/10.1249/MSS.0000000000000505

McCrory JL, Martin DF, Lowery RB, Cannon DW, Curl WW, Read HM Jr, Hunter DM, Craven T, Messier SP (1999) Etiologic factors associated with Achilles tendinitis in runners. Med Sci Sports Exerc 31(10):1374–1381

Moen MH, Tol JL, Weir A, Steunebrink M, De Winter TC (2009) Medial tibial stress syndrome: a critical review. Sports Med 39(7):523–546. https://doi.org/10.2165/00007256-200939070-000022 [pii]

Morgenroth DC, Medverd JR, Seyedali M, Czerniecki JM (2014) The relationship between knee joint loading rate during walking and degenerative changes on magnetic resonance imaging. Clin Biomech (Bristol, Avon) 29(6):664–670. https://doi.org/10.1016/j.clinbiomech.2014.04.008 (S0268-0033(14)00093-X [pii])

Morin JB, Samozino P, Millet GY (2011a) Changes in running kinematics, kinetics, and spring-mass behavior over a 24-h run. Med Sci Sports Exerc 43(5):829–836. https://doi.org/10.1249/MSS.0b013e3181fec518

Morin JB, Tomazin K, Edouard P, Millet GY (2011b) Changes in running mechanics and spring-mass behavior induced by a mountain ultra-marathon race. J Biomech 44(6):1104–1107

Muller R, Siebert T, Blickhan R (2012) Muscle preactivation control: simulation of ankle joint adjustments at touchdown during running on uneven ground. J Appl Biomech 28(6):718–725. https://doi.org/10.1123/jab.28.6.718 (2010-0181 [pii])

Ogueta-Alday A, Rodriguez-Marroyo JA, Garcia-Lopez J (2013) Rearfoot striking runners are more economical than midfoot strikers. Med Sci Sports Exerc. https://doi.org/10.1249/MSS.0000000000000139

Perl DP, Daoud AI, Lieberman DE (2012) Effects of footwear and strike type on running economy. Med Sci Sports Exerc 44(7):1335–1343. https://doi.org/10.1249/MSS.0b013e318247989e

Santos-Concejero J, Tam N, Granados C, Irazusta J, Bidaurrazaga-Letona I, Zabala-Lili J, Gil SM (2014) Stride angle as a novel indicator of running economy in well-trained runners. J Strength Cond Res 28(7):1889–1895. https://doi.org/10.1519/JSC.0000000000000325

Shih Y, Lin KL, Shiang TY (2013) Is the foot striking pattern more important than barefoot or shod conditions in running? Gait Posture. https://doi.org/10.1016/j.gaitpost.2013.01.030 (S0966-6362(13)00117-3 [pii])

Shorten MR, Winslow DS (1992) Spectral analysis of impact shock during running. Int J Sports Biomech 8:288–304

Taunton JE, Ryan MB, Clement DB, McKenzie DC, Lloyd-Smith DR, Zumbo BD (2002) A retrospective case-control analysis of 2002 running injuries. Br J Sports Med 36(2):95–101

Vannatta NC, Kernozek TW (2014) Patellofemoral joint stress during running with alterations in foot strike pattern. Med Sci Sports Exerc [Epub ahead of print]

Warden SJ, Davis IS, Fredericson M (2014) Management and prevention of bone stress injuries in long-distance runners. J Orthop Sports Phys Ther 44(10):749–765. https://doi.org/10.2519/jospt.2014.5334

Zadpoor AB, Nikooyan AA (2011) The relationship between lower-extremity stress fractures and the ground reaction force: a systematic review. Clin Biomech 26:23–28

Chapter 10
The Measurement of Sprint Mechanics Using Instrumented Treadmills

Jean-Benoit Morin, Scott R. Brown and Matthew R. Cross

Abstract Since sprinting involves very fast movement velocities (up to 12 m/s in the best athletes), experimental studies in this field have always been a technical challenge. Whilst sprint kinematics and distance-time or velocity-time variables were first described by the end of the 19th century, kinetics and especially ground reaction force and mechanical power outputs have remained unexplored until the 1970s and 1980s. Cutting edge laboratory installations now allow for full-length sprint acceleration studies (single or multiple sprint protocols) with track-embedded force plates. However, a significant amount of literature and knowledge has been previously established by the use of instrumented treadmills. These were first non-motorized and not directly measuring the ground reaction force (end of the 1980s), but the most up-to-date device allows investigation of sprint mechanics and three-dimensional ground reaction force during an accelerated run (from zero to maximal velocity). In this chapter, we will present the historical development of these devices, along with their advantages and limitations, and the main experimental results obtained with the motorized accelerated treadmill. In particular, we will present the key concept of mechanical effectiveness of ground force application, and how it is related to sprint performance. Furthermore, we will discuss the muscular underpinnings of the mechanical effectiveness; specifically the role of hip extensors. Finally, we will discuss the comparison between treadmill and track

---

J.-B. Morin (✉)
Laboratory of Human Motricity, Education Sport and Health,
Université Côte D'Azur, 261 Route de Grenoble, Route de Grenoble Nice, France
e-mail: jean-benoit.morin@unice.fr

S. R. Brown · M. R. Cross
Sports Performance Research Institute New Zealand (SPRINZ),
Auckland University of Technology, Auckland, New Zealand
e-mail: scott.brown@aut.ac.nz

M. R. Cross
e-mail: matt.cross@aut.ac.nz

© Springer International Publishing AG 2018
J.-B. Morin and P. Samozino (eds.), *Biomechanics of Training and Testing*,
https://doi.org/10.1007/978-3-319-05633-3_10

sprint performance and mechanics, including data from elite sprinters, and how current and future research on this topic will allow a deeper understanding of this seemingly simple yet complex motor task.

Treadmills will get you nowhere

Anonymous

10.1 Introduction

Sprint running consists of reaching and trying to maintain one's absolute maximal running velocity. This can be done by starting from a still position like in track and field sprint events (starting-block start), or from an already ongoing movement like in soccer or rugby (flying start). Be it a direct or indirect determinant of athletic performance, sprint running is a key ability in many sports. For this reason, it is the focus of specific training programs and exercises. A deeper understanding of the mechanics of sprinting will undoubtedly help better design training exercises to improve injury prevention and/or manage rehabilitation and return-to-sport strategies. The field of sports biomechanics, like other fields of sports science, is dependent on advances in technology available to explore human locomotion. This is particularly important when studying sprinting, i.e. an all-out ballistic activity that makes the human body move at velocities ranging 8–12 m/s (30–40 km/h), making any direct biomechanical measurement rather challenging. At such maximal running velocity, a trained male athlete can cover more than 10 m in one second. Measuring any kind of biomechanical data in this context is very difficult. For instance, the first study to report ground reaction force (GRF) data on several steps of an acceleration in elite sprinters was published in 2015 (Rabita et al. 2015), on the basis of the pioneering work of Cavagna et al. (1971).

A common alternative is to have the athlete running in place on a treadmill. However, if running at sub-maximal velocity is a common exercise in both fitness and research contexts, the development of sprint treadmills has required constant improvements over the last 40 years. Recently-developed instrumented motorized treadmills have allowed a significant increase in the body of knowledge about the mechanical determinants of accelerated running performance.

In this chapter, we will briefly review the history of sprint treadmill developments, and present the devices, the concepts and some of the main results obtained. Then, we will specifically focus on an instrumented treadmill presented in 2009 that allows for both subject-driven accelerations and three-dimensional kinetics measurements, which is to our knowledge the most advanced technology for such a purpose. We will detail and discuss the mechanical concepts derived from the measurements (especially the mechanical effectiveness of ground force application) and how they allowed a more detailed understanding of sprint performance. Finally, we will review the limitations inherent to this device and the current and future developments in this area.

10.2 Devices

The understanding of running mechanics had a major breakthrough in the end of the 1970s and 1980s with the use of ground-embedded force platforms to study human locomotion including jumping, running (Cavagna 1975; Blickhan 1989) and sprinting (Cavagna et al. 1971). This has allowed researchers to investigate sprint running mechanics in real overground running conditions, but the major limitation was that only one or two steps at most could be studied. Two alternatives have been proposed to study sprint mechanics over longer distances: using a multiple sprint protocol and aggregate ground-embedded force platform data of several sprints to reconstruct a virtual acceleration mechanical profile (Cavagna et al. 1971; Rabita et al. 2015), or have subjects run on an instrumented treadmill. The advantage of the latter method is the entire sprint data could be recorded (some 400-m sprints have been previously studied Tomazin et al. 2012), but from the early developments of this kind of technology in the late 1980s, several limitations have been discussed. Because this situation is almost the opposite of real-world walking and running locomotion (in which subjects' center-of-mass moves relative to the supporting ground), authors have investigated the differences between field and treadmill running in the past 40 years or so.

In studies comparing these modalities at steady low speeds, van Ingen Schenau (1980) showed that the mechanics of running were basically similar provided that the belt speed was constant and a coordinate system moving with the belt was used. This overall similarity was also supported by other studies (e.g. Kram et al. 1998; Schache et al. 2001; Riley et al. 2008). However, in several studies authors reported significant differences between the two conditions (Nelson et al. 1972; Dal Monte et al. 1973; Elliott and Blanksby 1976; Nigg et al. 1995).

When focusing on sprint running, contradicting results have also been reported. Frishberg (1983) and Kivi et al. (2002) showed biomechanical differences between field and treadmill sprint running, whereas McKenna and Riches (2007) concluded that sprinting on a treadmill and overground was similar for the majority of the kinematic variables they studied, but specified that a motorized treadmill was necessary to reach this similarity. For this reason, we will detail the main devices used and their advantages and limitations with this non-motorized versus motorized distinction in mind.

10.2.1 Early Non-motorized Treadmills

Modern attempts at experimentally determining mechanical characteristics of the body during load-bearing sprinting occur most frequently using specialized sprint treadmill ergometry. This method requires subjects to propel a treadmill belt while tethered around the waist to an immovable stationary point at the rear of the machine (Fig. 10.1). The majority of models calculate power output in a two-dimensional nature, as the interaction of velocity of treadmill belt, measured via rotary encoders, and horizontal-force via various loading-cells and goniometers mounted on a non-elastic tether. Although published as an abstract in 1984, the first international publication we know of implementing direct measurements of sprint running kinetics was by Lakomy (1987) who used an early non-motorized treadmill (Woodway model AB, Germany) to show that power (P), estimated horizontal force (F_H), and velocity (v) could be accurately measured during a 7-s maximal sprint.

This major breakthrough came with some limitations inherent to treadmill assessment. First, instantaneous power values were sampled over large (0.5–1 s or more) brackets of time and at low sampling frequencies. This resulted in the incorrect estimation of power and underestimation of velocity. This error has been avoided in recent studies (sampling frequencies up to 1000 Hz), which improved the accuracy of averaging methods.

The main limitation of treadmill sprints using force transducers is that the recorded horizontal force is an approximation, as the tether moves up and down with each step (movement of <4°; ~7% contribution of vertical force to horizontal readings (Lakomy 1987). While some authors have used goniometers to account for this bias (Jaskolska et al. 1998; Chelly and Denis 2001), this is not a common

Fig. 10.1 Original (left) and modern (right) non-motorized treadmills. The pulling force is recorded via a load cell (white circles) and the running velocity is recorded via optical encoders installed in the rolling belt system

practice. Moreover, although the output of power results from horizontally-directed force and velocity outputs, these variables are recorded in disparate locations: along the tether (force) and at the treadmill belt under subjects' feet (velocity). Because of this, a substantial amount of force is recorded during the strides aerial phases, due to a component of body mass acting on the tether and/or some elasticity in the system (Brughelli et al. 2015).

Furthermore, significantly lower maximal running velocity (about 3 m/s, 10.8 km/h (Lakomy 1987) and acceleration have been reported on these ergometers due to friction and inertia of the overall rolling belt and components. This is an issue that persists even with motorized instrumented treadmills, with the exception of feedback controlled models (Bowtell et al. 2009), as will be discussed in the following section.

10.2.2 Modern Non-motorized Treadmills

Since these original studies, modernized sprint treadmills (Fig. 10.2) have been shown to provide reliable and accurate assessment of horizontal-power in various population groups (Brughelli et al. 2011; Cross et al. 2014; Brown et al. 2016). In addition, it has been used to accurately describe the changes in sprint running mechanics during a return-to-sport rehabilitation process in rugby players (Brown and Brughelli 2014).

Although these devices bring significant value to the study of sprint mechanics, in a safe and comfortable environment, care should be taken when processing and

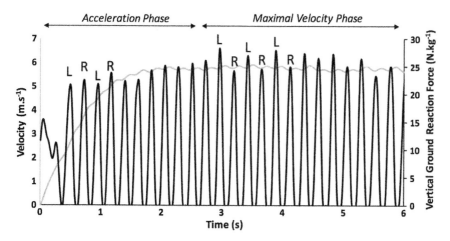

Fig. 10.2 Running velocity (grey) and vertical ground reaction force relative to body mass (black) during a sprint on a non-motorized treadmill. Vertical ground reaction force data were collected via four load cells underneath the user-driven belt at 200 Hz. The force trace clearly shows a left-right asymmetry during the maximal velocity phase

displaying the data, and a critical attitude should be adopted for the development and validation of assessment protocols in sport club settings. As pointed out in a recent letter to the Editor, simply relying on software for data analysis from equipment companies could lead to flawed results and mechanically inconsistent data (Brughelli et al. 2015).

10.2.3 Motorized Treadmills

The development of motorized sprint treadmills allow for a compensation of the aforementioned effects of friction/inertia of the belt and other rolling elements, and help the athletes reach higher running velocities. This kind of technology has been used to quantify mechanical power output in various types of sportsmen (e.g. Chelly and Denis 2001) and investigate the force-velocity relationships in sprinting (Jaskolska et al. 1998). These authors showed that a multiple-trial method could be used to accurately profile force-velocity relationships on a sprint-specialized motorized treadmill ergometer (Gymrol Sprint 1800, France). To provide resistance for each loading condition, the motors were set to apply an increasing resistance, across a set of 6 sprints (68, 108, 135, 176, 203 and 270N), during which F_H was estimated via a tether-mounted force-transducer and goniometer, and velocity via a sensor system attached to the rear drum of the belt. A major limitation remaining was that the force output measurements were made along the attachment tether, and not at the ground level. This limitation was then answered with the development of force-sensor mounted treadmills.

10.2.4 Motorized Treadmills Equipped with Force Sensors

A first attempt was made to obtain GRF by installing a force platform first directly under the rolling belt (Kram and Powell 1989). The vertical component of the GRF could be accurately measured, but due to the direct contact between the belt and the force platform upper frame at each step, the measurement of the horizontal component was altered by friction and cross-talk phenomena. A few years later, the same research group, presented a treadmill entirely mounted (frame, belt, rolling components and motor) on a force platform system (Kram et al. 1998). Following the same procedure, a three-dimension GRF motorized treadmill was also presented by Belli and his colleagues for walking purposes (Belli et al. 2001) with a split-belt system allowing for the measurement of separate limbs. This treadmill was then adapted for running (e.g. Avogadro et al. 2004; Divert et al. 2005; Morin et al. 2011c) and mounted on four rigid piezo-electric force sensors positioned under the treadmill (Kislter, Winthertür, Switzerland). The entire frame and motor are positioned on the sensors allowing measurements of the vertical, antero-posterior and medio-lateral components of the GRF. This was done after appropriate "tare" of the force sensors when the treadmill was mounted on the force sensor, and static-force calibration (Belli et al. 2001).

Both of these devices have been shown to give typical vertical and horizontal GRF-time traces that are very consistent with those reported using ground-embedded force platforms. This was a major improvement in measuring running mechanics compared to tethered systems, but regarding sprint running conditions, these treadmills still had two main limitations: they did not allow for sprint running conditions at very high velocity >6 to 7 m/s (>25 to 26 km/h), and they only allowed for steady-state running (i.e. at constant velocity), and not accelerated running, which is typical in sprinting.

The non-maximal speed limitation was lifted in studies where the treadmill motor could set high-to-maximal constant speed imposed on the subjects either to match the one previously measured in the field. An example of the first type of setting is given in the study of Frishberg (1983), who recorded a typical sprint performance (a 91.44-m sprint) through 9.14-m sections in order to set the speed of the treadmill belt with corresponding increasing values (one mean 9.14-m constant speed value for every span corresponding to those covered in field conditions). The second type of setting typically requires subjects to run at constant velocities either corresponding to their maximal speed recorded beforehand during field measurements (e.g. Frishberg 1983; Kivi et al. 2002) or to the maximal speed they are able to maintain for a few steps performed after a lowering movement from the handrails onto the moving treadmill belt (e.g. Weyand et al. 2000, 2010; Kivi et al. 2002; Bundle et al. 2003) Both of these settings do not exactly reproduce "free" sprint running on the field during which the runner's speed constantly changes as a function of the amount of force applied onto the supporting ground, and over a typical acceleration phase. In particular, during acceleration, the speed increases from null to maximal speed, and thus by definition with no constant speed phase in between. Bowtell et al. (2009) proposed that such a fundamental basic of field sprinting mechanics should be respected in treadmill conditions to better simulate field sprint. Basically, subjects (through their muscular actions) should "command" the belt speed, not the other way around: *"to measure peak performance there is a requirement for a runner to speedup and slowdown as they want and the speed of the treadmill belt should be capable of responding adequately and consistently"* (Bowtell et al. 2009). Although high-speed motorized instrumented treadmills (Fig. 10.3) allowed for major insights into sprint mechanics (e.g. Weyand et al. 2000, 2010; Clark and Weyand 2014; Clark et al. 2016), the above limitation of acceleration measurements still remained an issue, especially for investigating the role of muscle groups involved in accelerated running, as pointed out by Schache (Schache et al. 2014, 2015).

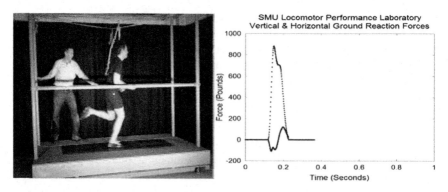

Fig. 10.3 High-speed instrumented sprint treadmill allows the athlete to drop on the rolling belt and perform a few steps at his top running velocity while anteroposterior and vertical ground reaction force is measured. *Credit* P. G. Weyand, South Methodist University Locomotor Performance Laboratory

10.3 Sprint Acceleration Mechanics on a Motorized Instrumented Treadmill

The aforementioned devices meet many of the concerns raised by previous authors; during a sprint acceleration, GRF is averaged over each single foot contact, corresponding to a single ballistic event of one push at a high sampling rate (1000 Hz) and the power output in the horizontal direction is calculated instantaneously as the product of horizontal component of the GRF and velocity collected at the same location (i.e. treadmill belt). But the most challenging technological improvement was to have the treadmill belt speed increase under the force output of the athletes, in order to simulate sprint acceleration at best, from zero to maximal running velocity.

This was made possible with a new instrumented sprint-treadmill technology (ADAL3D-WR; Medical Development, HEF Tecmachine, Andrézieux-Bouthéon, France)[1] based on the existing model used for constant-speed walking (Belli et al. 2001) or running (Avogadro et al. 2004; Morin et al. 2007). The device (Fig. 10.4) is mounted on a highly rigid metal frame fixed to the ground through four piezoelectric force transducers (KI 9077b, Kistler, Winterthur, Switzerland), and installed on a specially engineered concrete slab to ensure maximal rigidity of the supporting ground. The main technical details of the treadmill are: a 4 kW brushless motor; a force range of 60 kN in the vertical direction and 10 kN in the mediolateral and

---

[1] See this video of a typical sprint acceleration on this treadmill: https://www.youtube.com/watch?v=NkGNOPSIJus.

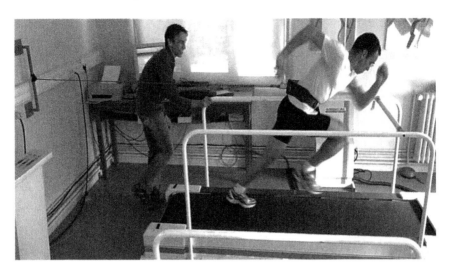

Fig. 10.4 Sprint instrumented motorized treadmill allowing direct and immediate changes in belt velocity following subject's actions during sprint acceleration, from null to maximal running velocity

anteroposterior directions; cross-talk effects between directions <2%; a natural frequency of ∼140 Hz in all three directions; and a belt surface available for running of 2.53 × 0.54 m. To our knowledge, this treadmill is currently used for research and athletes monitoring at the University of Saint-Etienne (France) and at the ASPETAR Medical Center in Doha (Qatar).

Beyond the classical constant velocity mode, a "constant driving torque" mode was added, with the torque set accurately so that, for each subject, the default motor torque could overcome the friction on the belt due to subject's body-mass. This default torque setting as a function of belt friction is in line with previous motorized-treadmill studies (Falk et al. 1996; Jaskolska et al. 1998; Chelly and Denis 2001), and with the detailed discussion by McKenna and Riches in their recent study comparing "torque treadmill" sprint to overground sprint (McKenna and Riches 2007).

As seen in Fig. 10.5, the raw vertical and horizontal force signals are sampled simultaneously with running velocity, and then averaged for each support phase (vertical force above 20N) for a step-by-step analysis of the main running kinematics (spatio-temporal variables) and kinetics (Fig. 10.6).

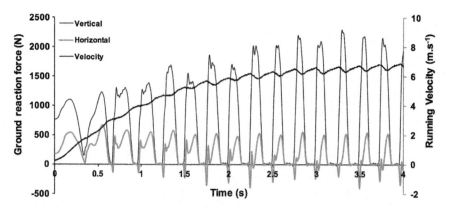

Fig. 10.5 Running velocity (black trace) and vertical (blue) and horizontal (red) components of the ground reaction force during a sprint acceleration (sampling frequency of 1000 Hz). The subject here is a physical education student not specialized in sprinting (body mass of 80 kg)

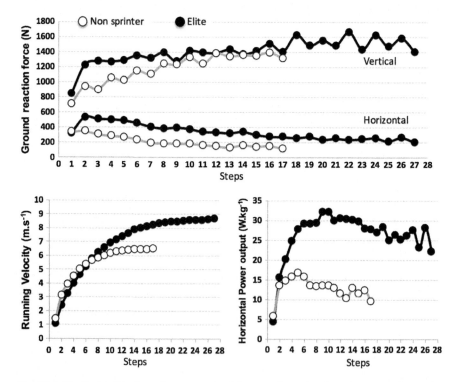

Fig. 10.6 Vertical and horizontal components of the ground reaction force (upper panel), running velocity (lower left) and power output in the horizontal direction (lower right) for all steps of the acceleration until top speed for a world-class sprinter (100-m best time: 9.92 s) and a non-specialist

Table 10.1 Mean values of contact time (*tc*), aerial time (*ta*), step frequency (*Fq*), vertical (F_V), horizontal (F_H) and resultant (F_{Tot}) GRF, vertical displacement of the center of mass (Δy) and vertical (k_{vert}) stiffness (k_{leg})

	Variable	tc (ms)	ta (ms)	Fq (Hz)	F_V (N)	F_H (N)	F_{Tot} (N)	Δy (cm)	k_{vert} (kN m^{-1})
1st half of the acceleration	Elite sprinter	134	88	4.47	1287	412	1354	–	–
	Non sprinter	208	177	2.74	1036	260	1073	–	–
2nd half of the acceleration	Elite sprinter	120	78	4.89	1506	254	1528	1.94	107
	Non sprinter	141	231	2.77	1335	151	1343	5.56	57

The values are averaged for all steps separated into equal number of steps over the acceleration (1st and 2nd half) for each athlete. For example, the elite sprinter reached his top speed in 26 steps, so averaged values for steps 1–13 are used to describe the 1st half of the acceleration, and data from steps 14–26 are used to describe the 2nd half of the acceleration. Spring-mass model data are shown for the 2nd phase only since the basis postulates of the spring-mass model for running require a constant or near-constant running speed (see Chap. 8)

10.3.1 Kinematics and Kinetics

The main running spatio-temporal, GRF and spring-mass variables (see Chap. 8) may then be obtained and compared within and between athletes. For instance, Table 10.1 shows the averaged mechanical data for the first and second half of a sprint acceleration from standing start to top-speed in an elite sprinter (best 100-m time of 9.92 s) versus a healthy young subject of similar body mass (81 kg), not specialized in sprinting.

Beyond this classical descriptive analysis of the main sprint running pattern mechanics, the instrumented accelerated treadmill allowed us to compute and study individual force-velocity and power-velocity relationships.

10.3.2 Force-Velocity and Power-Velocity Relationships

On the basis of the simultaneous, high rate GRF and running velocity measurements described above, F_H and velocity may be averaged for each support phase (dynamic action onto the running surface) and plotted throughout the course of a sprint for each stance phase (Morin et al. 2010), as in sprint-cycling studies (e.g. Dorel et al. 2010). As shown in Fig. 10.7, the entire force-velocity relationship is described by the maximal theoretical horizontal force that the lower limbs could produce over one contact at a null velocity (F_{H0}), and the theoretical maximum velocity that could be produced during a support phase in the absence of mechanical constraints (v_0). A higher v_0 value represents a greater ability to develop

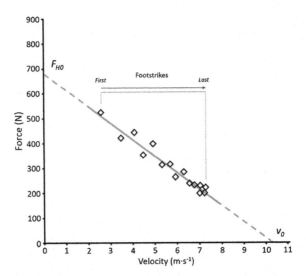

Fig. 10.7 Graphical representation of force-velocity relationship determined via treadmill ergometry. The data points represent values averaged across each support phase, from the first to the last at peak velocity. F_{H0} and v_0 represent the y and x intercepts, and the theoretical maximum of force and velocity able to be produced in absence of their opposing unit

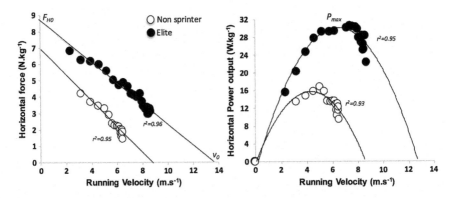

Fig. 10.8 Linear force-velocity (left panel) and 2nd degree polynomial power-velocity (right panel) relationships during treadmill sprint acceleration for a world-class sprinter (100-m best time: 9.92 s) and a non-specialist

horizontal force at high velocities, as shown in Fig. 10.8. This is a clear mechanical difference between elite and lower-level sprinters (Morin et al. 2012; Rabita et al. 2015). The criterion of averaging force over a single contact time is important from a biomechanical perspective, as the force-velocity relationship directly relates to lower limb mechanical capabilities. When multiplying F_H and v values for each support phase, the equivalent of a mechanical power in the horizontal direction is obtained (P), and the power-velocity may be computed (Fig. 10.8). Values of P_{max} obtained via this method (when converted to similar time-periods) and mechanical variables (F_{H0} and v_0) are congruent with results from comparable subject pools and loading parameters in earlier studies (Cheetham et al. 1985; Jaskolska et al. 1998; Funato et al. 2001; Morin and Belli 2004) and are highly reliable for test-retest measurement.

10.3.3 Effectiveness of Ground Force Application

Definitions In addition to the aforementioned force-velocity-power relationships, updated treadmill ergometry allows mechanical effectiveness to be quantified throughout a sprint acceleration phase. In the cycling literature (Davis and Hull 1981; Dorel et al. 2010), pedal-mounted force sensors have allowed the computation of force application indices that indicate, for each pedal down stroke, the ratio of the effective component of the force applied to the pedal by the cyclist (i.e. the component oriented perpendicularly to the crank arm) to the resultant (total) force applied to the pedal (Fig. 10.9). The higher the perpendicular component of the resultant force, the higher the mechanical effectiveness, and the more force that will actually drive the pedal for a given muscular force output. Note that the mechanical effectiveness may be equally expressed as the angle between the effective component and the resultant force (α in Fig. 10.9), or as a percentage. For example, a mechanical effectiveness of 80% means that over a pedal downstroke, the average effective force will be 80% of the resultant force applied to the pedal.

Following the same idea, GRF output from modern treadmills can be expressed as a ratio of the 'effective' horizontal component of GRF (i.e F_H) to total force (i.e. F_{Tot}) averaged across each contact phase (i.e. 'ratio of forces': RF) as defined by Morin et al. (2011a). While it is possible (and encouraged) to maximize RF (i.e. up

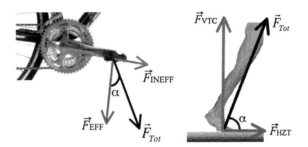

Fig. 10.9 Schematic representation of the concept of mechanical effectiveness of force application, from cycling to sprint running. In cycling (left), effectiveness is computed as the ratio between the effective component (F_{EFF} which will cause the rotation) and the total i.e. resultant force produced by the active muscles (F_{Tot}). The other component is ineffective. During a sprint running acceleration (right), the analogy we propose gives effectiveness as the ratio $RF = F_{Hzt}/F_{Tot}$

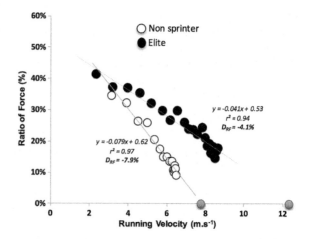

Fig. 10.10 Ratio of Force as a function of running velocity during a sprint acceleration on a motorized instrumented treadmill for a world-class sprinter (100-m best time: 9.92 s) and a non-specialist. The blue circles represent the extrapolation of the linear relationship between ratio of force and running velocity to the theoretical value of 0% reached at maximal velocity

to 100%) in cycling, the requirement of a vertical component in sprint running means that it is impossible to present a maximal RF value without falling. Instead, in sprint running we observed sub-maximal values of RF (30–45%) at the beginning an accelerated sprint on the treadmill, which systematically decreases with a clear linear trend with increasing velocity (Fig. 10.10). This is a logical outcome as when the athletes move faster during an acceleration (or propel the treadmill belt faster) their overall body posture tends to orient upright, which is related to an overall progressive vertical orientation (i.e. less effective) of the GRF vector at each support phase. Interestingly, we systematically observed that this decrease in RF with increasing speed was linear, in human runners ranging from elite sprinters to non-sprinters. In theory, and as confirmed by track-embedded force plate GRF measurements (see below), the value of RF decreases down to approximately 0% as the maximal velocity is reached. At this moment of the sprint, the athlete's acceleration is null, and F_H averaged over the support phase is null (equal braking and propulsive impulses cancel out), or quasi null in real-life sprinting since air friction force is low (between 30 and 40N).

Furthermore, the slope of this linear decrease in RF ('decrement in ratio of forces': D_{RF}), with velocity is described as an index of force application technique. This slope is negative (decrease in RF as velocity increases) and a typical D_{RF} value of −0.079 as in Fig. 10.10 may be expressed as −7.9%, which describes the fact that over this sprint acceleration, the RF as decreased on average by 7.9% for each new meter per second of running velocity generated. It follows that, in theory, the best sprinters are able to accelerate more and over a longer period, and have therefore a lower decrease in RF as velocity increases, i.e. a less negative D_{RF}. Vice versa, we hypothesized that more negative D_{RF} values would indicate a fast decrease in RF over acceleration, with runners reaching an upright body posture prematurely and not being able to accelerate anymore (RF of 0%) very early into the acceleration. These hypotheses of a relationship between mechanical effectiveness of force application and sprint acceleration and overall performance have been tested in two treadmill protocols and a study performed on a sprint track.

Relationship with sprint performance. In a first study (Morin et al. 2011a), Morin et al. tested whether mechanical effectiveness of force application (especially the D_{RF} index) was correlated to 100-m performance, i.e. were the athletes with a more effective propulsion as measured on the treadmill the fastest on the track and vice versa. The maximal power output measured in the laboratory was highly and positively correlated, but more importantly, the best performance on the track (100-m time and maximal velocity and acceleration capability over 4 s) was observed in subjects who had a "good" D_{RF}, i.e. a value of −6 to −5%. Conversely, subjects who had a "poor" D_{RF} value on the treadmill (−7 to −10%) were the slowest at the track 100-m test. An interesting additional result was that the horizontal component of the GRF (F_H) was significantly correlated to 100-m performance which makes sense in the biomechanical context of human locomotion aiming at propelling the body mass in the forward direction. However, the resultant (total) GRF averaged over the 6-s acceleration on the treadmill was not correlated to any sprint performance variable. This means that the total force output measured on the treadmill (taken regardless of its orientation in space) was not a performance indicator. At first sight, this appears to contradict previous work by Weyand et al. (2000) showing that faster athletes were able to produce more GRF per unit body mass. However, it is important to note that only the vertical component of the GRF was studied by Weyand et al., and only at the very specific moment of maximal velocity, and in a very heterogeneous population mixing untrained females and Olympian male sprinters. When the GRF output (i.e. the resulting, net, mechanical output from the entire propelling effort of the neuromuscular system) was considered over the entire acceleration phase, in a homogeneous group of athletes, it was not correlated to sprint performance. This study gave a first insight into our "force production and transmission to the ground" general view of sprint acceleration. That said, an issue with this first study was that the best athlete that participated had a 100-m best time of 10.9 s, which does not allow to extend these conclusions to elite athletes.

In a second study (Morin et al. 2012), the same group of athletes repeated the very same protocol (treadmill sprint to measure mechanical GRF output capability and mechanical effectiveness of force application and track 100-m test of sprint performance), but tested this time 3 national level sprinters (best 100-m times ranging from 10.3 to 10.6 s) and a world-class sprinter (best 100-m time of 9.92 s at the time of the study). The results confirmed those found in the first study, with the elite athlete showing a very effective force application (D_{RF} of −4%), despite a resultant GRF per unit body weight within the range of that of much slower individuals (no difference with his peers, very small difference with non-specialists). One biomechanical explanation of this world-class sprinter's ability to run fast was not his outstanding force production capability (at the time he ran 9.92 s, he had never practiced heavy-weight strength training) but his extraordinary capability of orienting the GRF vector forward when pushing the ground, especially at high running velocities (Figs. 10.10 and 10.11).

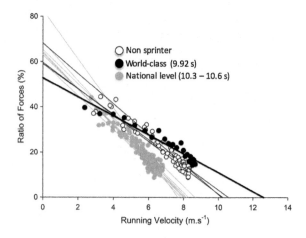

Fig. 10.11 Individual linear relationship between the ratio of force and the running velocity during a sprint acceleration for non-specialists (physical education students, in grey), national-level sprinters (in white) and a world-class sprinter (100-m best time: 9.92 s, in black)

This D_{RF} value of −4.2% is the most effective value reported in the scientific literature to our knowledge (Table 10.2).

Practically, these variables demonstrate the ability to maintain effective orientation of global force production throughout a sprint (independently from the magnitude of the resultant GRF output) under increasing velocity, and add another level of analysis to sprint running acceleration mechanics. That being said, a correct interpretation of Newton's fundamental law of dynamics would have led to the same conclusions: the first part of the law states that the alteration of motion (acceleration) is proportional to the motive force impressed: more net external force applied to a given mass (e.g. GRF and body mass) means more acceleration. But the second, often overlooked, part of the law states that this acceleration is made in the direction of the right line in which that force is impressed: more net force in the fore-aft direction means more forward acceleration.

Muscular determinants: the hip extensors hypothesis. The aforementioned studies presented the concept of mechanical effectiveness of force application, its main indices RF and D_{RF}, and showed that they were significantly correlated to track sprint acceleration performance, from non-specialists to elite sprinters. In order to transfer these laboratory conclusions into sports practice, the main questions that needed to be answered were then: (1) what are the neuromuscular determinants of the mechanical effectiveness? (2) are they trainable, and if so, how? (3) since all the results and conclusions drawn above are based on instrumented treadmill measurements, do they still old true in "real" track sprint conditions? The latter question will be addressed in the limitation section of this chapter.

The second question is still under study, but a recent work has brought an answer to the first question as to the muscular determinants of an effective forward propulsion. We termed this set of research the "hip extensors hypothesis". The three most important arguments supporting this hypothesis were:

First, several experimental and/or modeling studies showed the key role of hip extensors muscles (gluteal and hamstring) in running performance (e.g. Schache

Table 10.2 Main sprint acceleration mechanical outputs averaged over the 6-s acceleration on an instrumented motorized treadmill (±Standard deviation)

Variable	World-class sprinter	National-level sprinters	% difference with WCS	Non-specialists	% difference with non-specialists
Horizontal component of GRF (F_H in N/kg)	3.90	3.44 ± 0.29	−11.8	3.04 ± 0.51	−22.1[a]
Vertical component of GRF (F_V in N/kg)	18.1	17.6 ± 0.58	−3.24	15.7 ± 1.17	−13.5[a]
Resultant (total) GRF (F_{Tot} in N/kg)	18.6	17.9 ± 0.59	−3.68	16.0 ± 1.19	−14.2[a]
Mechanical effectiveness of force application D_{RF} (%)	−4.21	−6.00 ± 0.6	−42.9[b]	−8.21 ± 2.9	−95.2[b]

Comparison between a world-class sprinter (WCS, personal 100-m best time of 9.92 s), national-level sprinters (personal best times ranging from 10.3 and 10.6 s) and active physical education students, non-specialists of sprinting
[a]Difference greater than 2 Standard deviations
[b]Difference greater than 3 standard deviations

et al. 2014, 2015). Based on experimental and modeling approaches used in various types of subjects including elite sprinters, research has consistently showed that the hip extensors played an increasing role as running speed increased up to maximal sprint values (typically above 7 m/s, 25.2 km/h) (Mann and Sprague 1980; Simonsen et al. 1985; Belli et al. 2002; Kyröläinen et al. 2005; Dorn et al. 2012; Schache et al. 2015). However, these studies mostly considered high but constant running velocity (no acceleration) and did not relate this hip extensor activity to direct measurements of propulsive GRF or mechanical effectiveness.

Second, to produce high amounts of horizontal GRF and impulse (Morin et al. 2015b), especially at high velocity (i.e. when the body posture is upright), intense backward actions of the lower limb are necessary during both swing and support phases. It is therefore anatomically and functionally reasonable to expect the hip extensors to produce very high forces during both phases (Sun et al. 2015). Because sprinting is associated with great limb velocities prior to ground impact at each step, this swing-support transition moment is crucial for hamstrings, which counteract very high external hip flexion and knee extension moments in the context of support forces as high as eight times body weight (Sun et al. 2015). In addition, it is interesting to note that, as Fig. 10.11 clearly shows, the difference in mechanical effectiveness between non-specialists, high-level sprinters and the elite athlete tested is visible at high running velocities, not at the beginning of the sprint acceleration. This tends to suggest that the ability to produce high RF and orient the GRF with a good effectiveness is particularly higher in best sprinters at high running velocity (Morin et al. 2012), i.e. when the overall body orientation is vertical, and lower limbs joint angular velocity and the associated muscle mechanical stress are at their highest. What distinguishes the best sprinters (Fig. 10.11) is not their effectiveness and ability to produce high amounts of F_H at slow running velocity, but their ability to do so at high to very high velocity. The key here is to produce high amounts of F_H in the very specific context of high running velocity: "strong at high velocity".

Finally, and interestingly, the fact that hamstring injuries are the most prevalent and recurrent lower limb muscle injury associated with sprinting tasks presupposes the importance of this muscle group in the context of high speed and/or accelerations (e.g. Ekstrand et al. 2011; Feddermann-Demont et al. 2014).

In order to test this hypothesis, Morin et al. (2015a) used an instrumented treadmill that allowed two novel experimental features: (1) measure simultaneously and time-synchronize the muscular activity (surface electromyography) of the main lower limb muscles and the GRF (including F_H) (2) do so over an entire sprint acceleration from zero to maximal velocity and analyze every step for one leg. To our knowledge, no equivalent measurements of muscle activity synchronized with GRF outputs have been performed, except very recent GRF measurements performed over an entire 50-m sprint, thanks to an innovative force plate measurement system (Nagahara et al. 2017). As shown in Fig. 10.12, this experimental setting was completed by isokinetic joint torque testing to provide an indirect idea of subjects' torque production capability at the hip and knee, in extension and flexion, concentric and eccentric modes. Although contraction velocities and range of

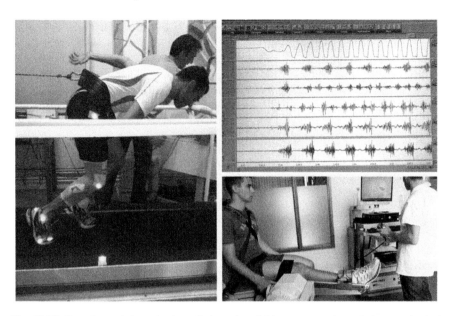

Fig. 10.12 Experimental investigation of the role of hip extensors in sprinting mechanical effectiveness. Left: subjects were equipped with surface electromyography electrodes during their sprint acceleration on the instrumented motorized treadmill. Right: the ground reaction force signal (first line of data) was synchronized with the right leg muscle activity (other lines) throughout the acceleration. In addition, subjects hip and knee extensor and flexor torque capability was tested in concentric and eccentric mode on an isokinetic dynamometer

motion strongly differ between sprint running and isokinetic testing, this technique was selected to provide an indication about subjects' overall force capability, since it is currently impossible to directly assess muscle force output during sprint running. The main results of the study showed that, as hypothesized, the highest level of F_H during the sprint acceleration was observed in subjects who had *both* the highest torque production capability of the hip extensors (especially hamstring muscles in eccentric mode) and the highest hamstring EMG activity during the end-of-swing phase. A subsequent analysis focused on the initial acceleration phase (first 10 steps) showed a significant relationship between F_H and glutei concentric torque capability and glutei EMG activity during the end-of-swing phase. In other words, hamstring EMG activity during the swing and end-of-the-swing phases and eccentric knee flexor peak torque are related to the amount of F_H produced during sprinting. These findings help the suggestion that the hip extensors (especially the hamstrings), play a significant role in sprint acceleration performance via mechanical effectiveness an F_H production.

This singular study requires confirmation, and other muscular determinants of mechanical effectiveness deserve deeper investigation, such as the capability of the ankle to resist ground impact and play a role in transmitting the power generated by the lower limb neuromuscular system to the ground. As commonly reiterated,

"A chain is only as strong as its weakest link", and although much power is generated at the hip and knee during sprint acceleration, one should keep in mind that the GRF is in fine applied onto the supporting ground at the ankle/foot. Future studies should test whether, as training anecdotal evidence and locomotion biomechanics suggest, a "strong foot" also plays a role in applying the ground force with effectiveness, i.e. with a forward orientation, during sprint acceleration.

10.4 Limitations and Future Studies

Although sprint instrumented treadmills have allowed coaches and researchers to significantly improve their level of understanding about human sprinting, this approach has several limitations that need to be discussed. Keeping these limitations in mind, this corpus of research is relatively novel and future studies will surely bring more insight into the results hitherto obtained, and a deeper understanding of the mechanisms underlying the mechanical effectiveness, its trainability and the associated training methods.

10.4.1 Main Limitations

Although the most updated motorized instrumented treadmill presented in this Chapter allows direct GRF measurements in three dimensions during sprint acceleration, an issue that persists is that the application of compensatory friction appears to restrict the subject's ability to reach levels of maximal running velocity near to overground sprint running (Morin et al. 2010; Morin and Sève 2011). Moreover, the individualized determination of the default torque is heavily time consuming, and familiarization persists as a limitation with even the most modern machines (>10 trials) (Morin et al. 2010). Although the reduction in sprinting velocity with torque compensated treadmills varies when compared to over-ground sprinting ($\sim$20% according to (Morin and Sève 2011), the main sprint performance output were significantly and highly correlated within the same subjects when compared with their track sprint performance. Thus, the overall inter-subject difference in track sprint performance is maintained during treadmill measurements, and one could argue that the ability to measure direct kinetics over a virtually unlimited time-period (for instance during a 400-m sprint Tomazin et al. 2012; Girard et al. 2016c) potentially mitigates this limitation. In addition, some unpublished data show that when extrapolating the RF-velocity linear relationship to the velocity axis (see Fig. 10.10), the theoretical maximal velocity calculated is very close to the actual maximal sprinting velocity reached by the athletes tested during a track 100-m within the same testing session. In fact, in theory, at maximal velocity, the acceleration of the athlete is null and thus F_H is quasi-null (only the force to overcome air friction during track sprinting) and $RF \approx 0\%$. Due to the inertia and friction of the treadmill system (see discussion in Morin and Sève 2011), the athletes cannot reach their track maximal velocity while sprinting on the treadmill (values of about 80% are reached). However, should the treadmill allow such a realistic end of acceleration phase, the estimated maximal velocity ($RF \approx 0\%$ in line with RF values actually measured during the treadmill acceleration, Fig. 10.10) almost perfectly matches track measurements and "real-life" sprinting conditions.

The scientific study of human sprinters began more than one century ago with pioneer biomechanist Marey (2002) and the first scientific articles by physiologist Archibald V. Hill and his colleagues (Furusawa et al. 1927; Bassett 2002), who published analyses of sprint mechanics based on time-motion and center of mass spatiotemporal variables. The analysis of sprint kinetics and GRF requires ground-embedded force plate systems over the acceleration distance. Since this technological feat has only very recently been accomplished by Japanese researchers (Nagahara et al. 2017), the only possibility with the 2- to 7-m force plate systems hitherto available was to use multiple sprint protocols and aggregate the data collected to virtually "reconstruct" a complete sprint acceleration dataset. For example, if a 7-m force plate system is used, athletes step onto the force plate system for their first 3 steps of an acceleration during their first trial, and then moving the starting blocks back for the second trial, the 4th and 5th steps are recorded, and so on. This approach was used by Cavagna et al. (1971) and more recently by Rabita et al. (2015). In the latter study, it is interesting to notice that these authors reported, in highly trained to elite athletes (personal 100-m best time below 10 s) sprint acceleration kinetics that were very close to those obtained with the instrumented treadmill presented in this chapter:

- the resultant GRF values and horizontal and vertical components measured on the track were within the range of those measured on the instrumented treadmill (for example, maximal horizontal force F_{H0} of ~ 10 N/kg)
- although maximal RF values in the starting-blocks ($\sim 60\%$) were much higher than those observed on the treadmill (which is logical given the crouched versus standing positions), RF values for the first steps on the track were within the range of those measured on the treadmill
- the linear decrease in RF with increasing running velocity observed on the treadmill (Fig. 10.10) was also found on the track (Fig. 3 in Rabita et al. 2015), and the D_{RF} values calculated on the track for highly trained sprinters ($\sim -6\%$) were in line with those found for the same population on the treadmill

Overall, the main conclusion one can draw from the comparison of published treadmill (Morin et al. 2010, 2011a, b, 2012; Morin and Sève 2011; Girard et al. 2016c) and track (Morin et al. 2015b; Rabita et al. 2015; Nagahara et al. 2017) studies of sprint acceleration kinetics is that despite the above-mentioned limitations and obvious difference in sprinting conditions, the main mechanical outputs (GRF, mechanical effectiveness) are similar when tested in athletes of similar training and performance level. This of course does not remove the limitations associated with sprint treadmill measurements, but it brings support to the fact that these limitations are outweighed by the numerous experimental possibilities offered to bring new insights into to complexity of human sprinting. Indeed, as shown in recent studies, sprint instrumented treadmills allow a step-by-step analysis of sprint mechanics, and many external conditions to be easily simulated compared to real-life track sprinting (e.g. repeated sprints, hypoxia, heat).

10.4.2 Latest Studies and Future Research

In 2015, Morin and his colleagues were able to synchronize muscle activity measurements with GRF during entire sprint accelerations to better investigate the role of lower limb muscles (hip extensors in particular) in sprint acceleration performance and mechanical effectiveness. In this study, and as shown in most Figures from this Chapter, a step-by-step analysis is possible since all the steps of an entire sprint acceleration are recorded. Based on this key advantage of treadmill sprinting, it has been possible to investigate fatigue and changes in running mechanics during short (100-m) to long (400-m sprint events) (Tomazin et al. 2012; Girard et al. 2016c), but also during repeated sprint series. This context is close to the demands of many team sports such as rugby, soccer or hockey, and has been the recent focus of several investigations during which it was experimentally easy to require subjects to repeat sprint accelerations on the treadmill and to set effort-recovery sequences. This has led to several recent publications by leading author Olivier Girard about the methodological issues with studying multiple steps and repeated sprints mechanics (Girard et al. 2015a, 2016a), or the changes in running mechanics associated with interval training (Girard et al. 2017b).

In addition, some external conditions such as hypoxia and heat may be studied in laboratory conditions, with the use of an instrumented treadmill within a specific laboratory environment allowing the simulation of hypoxia and heat conditions. This has led to a series of recent works from which a deeper knowledge has been obtained as to the mechanical consequences of performing sprints under heat stress (Girard et al. 2016b, 2017a) or in hypoxic conditions (Girard et al. 2015b, 2016b; Brocherie et al. 2016).

The above-mentioned possibility to record and study each step of a sprint allows the investigation of inter-limb asymmetry during sprint running. This area of research has recently been advanced[2] using biomechanical modeling (Clark et al. 2017), but also more direct experimental measurements during sprinting (e.g. Girard et al. 2017c). In particular, a recent work by Brown et al. (2017) compared the asymmetry between the vertical and horizontal components of the resultant GRF in rugby players. This study showed that interestingly, the magnitude and

---

[2]See this comment on Usain Bolt's running mechanics: https://phys.org/news/2017-06-symmetry-usain-asymmetrical-gait.html.

range of asymmetry computed on the horizontal GRF component during acceleration and maximal velocity phases, where much greater than those found in the vertical component. All these results may have significant implications for better-designed sports training and injury prevention/rehabilitation programs.

This transition towards applications to sport performance and training is another expected research perspective. In particular, training methods aiming at improving the mechanical effectiveness of ground force application should be studied, in both unfatigued and fatigued conditions. Cross-sectional studies have shown that the mechanical effectiveness of ground force application was a key factor of sprint performance and largely impaired due to fatigue associated with repeated sprints (Morin et al. 2011b). It is now time for controlled, longitudinal training studies to guide sports practitioners toward the most effective way(s) of developing this ability in athletes. For example, although sprint mechanics were not tested on an instrumented treadmill, a pilot training study has recently shown that the use of very heavy sleds (80% of body mass) was an effective way to improve maximal F_H and RF in soccer players (Morin et al. 2016). In the future, the continued use of instrumented sprint treadmills and additional laboratory-based methods will undoubtedly improve our understanding of human locomotion and sports performance in this area.

In conclusion, the recent developments of instrumented sprint treadmills, in particular those allowing three-dimensional ground reaction force measurements during acceleration (from zero to maximal velocity), have led to significant insights into the mechanics of this key sports ability. Despite some inevitable limitations associated to the device itself, and hoping for more detailed and numerous studies performed on track-embedded force plate systems, treadmill sprinting has and will continue to allow researchers and practitioners to gain more direct experimental insights into the mechanical underpinnings of this seemingly simple yet in fact so complicated motor task.

References

Avogadro P, Chaux C, Bourdin M et al (2004) The use of treadmill ergometers for extensive calculation of external work and leg stiffness during running. Eur J Appl Physiol 92:182–185
Bassett DR (2002) Scientific contributions of A. V. Hill: exercise physiology pioneer. J Appl Physiol 93:1567–1582
Belli A, Bui P, Berger A et al (2001) A treadmill ergometer for three-dimensional ground reaction forces measurement during walking. J Biomech 34:105–112
Belli A, Kyröläinen H, Komi PV (2002) Moment and power of lower limb joints in running. Int J Sports Med 23:136–141
Blickhan R (1989) The spring-mass model for running and hopping. J Biomech 22:1217–1227
Bowtell MV, Tan H, Wilson AM (2009) The consistency of maximum running speed measurements in humans using a feedback-controlled treadmill, and a comparison with maximum attainable speed during overground locomotion. J Biomech 42:2569–2574
Brocherie F, Millet GP, Morin J-B, Girard O (2016) Mechanical alterations to repeated treadmill sprints in normobaric hypoxia. Med Sci Sport Exerc 48:1570–1579

Brown SR, Brughelli M (2014) Determining return-to-sport status with a multi-component assessment strategy: a case study in rugby. Phys Ther Sport 15:211–215

Brown SR, Brughelli M, Cross MR (2016) Profiling sprint mechanics by leg preference and position in rugby union athletes. Int J Sports Med 37:890–897

Brown SR, Cross MR, Girard O et al (2017) Kinetic sprint asymmetries on a non-motorised treadmill in rugby union athletes. Int J Sport Med (In press)

Brughelli M, Cronin J, Chaouachi A (2011) Effects of running velocity on running kinetics and kinematics. J Strength Cond Res 25:933–939

Brughelli M, Morin J-B, Mendiguchia J (2015) Asymmetry after hamstring injury in English Premier League: issue resolved, or perhaps not? Int J Sports Med 36:603

Bundle MW, Hoyt RW, Weyand PG (2003) High-speed running performance: a new approach to assessment and prediction. J Appl Physiol 95:1955–1962

Cavagna GA (1975) Force platforms as ergometers. J Appl Physiol 39:174–179

Cavagna GA, Komarek L, Mazzoleni S (1971) The mechanics of sprint running. J Physiol 217:709–721

Cheetham ME, Williams C, Lakomy HK (1985) A laboratory running test: metabolic responses of sprint and endurance trained athletes. Br J Sports Med 19:81–84

Chelly SM, Denis C (2001) Leg power and hopping stiffness: relationship with sprint running performance. Med Sci Sports Exerc 33:326–333

Clark KP, Ryan LJ, Weyand PG (2016) A general relationship links gait mechanics and running ground reaction forces. J Exp Biol (jeb.138057)

Clark KP, Ryan LJ, Weyand PG (2017) A general relationship links gait mechanics and running ground reaction forces. J Exp Biol 220:247–258

Clark KP, Weyand PG (2014) Are running speeds maximized with simple-spring stance mechanics? J Appl Physiol 117:604–615

Cross MR, Brughelli ME, Cronin JB (2014) Effects of vest loading on sprint kinetics and kinematics. J Strength Cond Res 28:1867–1874

Dal Monte A, Fucci S, Manoni A (1973) The treadmill used as a training and simulator instrument in middle- and long-distance running. In: Krager (ed) Medicine and sport, Basel, pp 359–363

Davis RR, Hull ML (1981) Measurement of pedal loading in bicycling: II. Analysis and results. J Biomech 14:857–872

Divert C, Mornieux G, Baur H et al (2005) Mechanical comparison of barefoot and shod running. Int J Sports Med 26:593–598

Dorel S, Couturier A, Lacour JR et al (2010) Force-velocity relationship in cycling revisited: benefit of two-dimensional pedal forces analysis. Med Sci Sports Exerc 42:1174–1183

Dorn TW, Schache AG, Pandy MG (2012) Muscular strategy shift in human running: dependence of running speed on hip and ankle muscle performance. J Exp Biol 215:1944–1956

Ekstrand J, Hägglund M, Walden M (2011) Epidemiology of muscle injuries in professional football (Soccer). Am J Sports Med 39:1–7

Elliott BC, Blanksby BA (1976) A cinematographic analysis of overground and treadmill running by males and females. Med Sci Sport Exerc 8:84–87

Falk B, Weinstein Y, Dotan R et al (1996) A treadmill test of sprint running. Scand J Med Sci Sports 6:259–264

Feddermann-Demont N, Junge A, Edouard P et al (2014) Injuries in 13 international Athletics championships between 2007-2012. Br J Sports Med 48:513–522

Frishberg BA (1983) An analysis of overground and treadmill sprinting. Med Sci Sports Exerc 15:478–485

Funato K, Yanagiya T, Fukunaga T (2001) Ergometry for estimation of mechanical power output in sprinting in humans using a newly developed self-driven treadmill. Eur J Appl Physiol 84:169–173

Furusawa K, Hill AV, Parkinson JL (1927) The dynamics of "Sprint" running. Proc R Soc B Biol Sci 102:29–42

Girard O, Brocherie F, Morin J-B et al (2015a) Comparison of four sections for analyzing running mechanics alterations during repeated treadmill sprints. J Appl Biomech 31:389–395

Girard O, Brocherie F, Morin J-B et al (2017a) Mechanical alterations associated with repeated treadmill sprinting under heat stress. PLoS ONE 12:e0170679

Girard O, Brocherie F, Morin J-B, Millet GP (2016a) Intrasession and intersession reliability of running mechanics during treadmill sprints. Int J Sports Physiol Perform 11:432–439

Girard O, Brocherie F, Morin J-B, Millet GP (2017b) Mechanical alterations during interval-training treadmill runs in high-level male team-sport players. J Sci Med Sport 20:87–91

Girard O, Brocherie F, Morin J-B, Millet GP (2016b) Running mechanical alterations during repeated treadmill sprints in hot versus hypoxic environments. A pilot study. J Sports Sci 34:1190–1198

Girard O, Brocherie F, Morin J-B, Millet GP (2015b) Neuro-mechanical determinants of repeated treadmill sprints—usefulness of an "hypoxic to normoxic recovery" approach. Front Physiol 6:260

Girard O, Brocherie F, Morin J-B, Millet GP (2017c) Lower limb mechanical asymmetry during repeated treadmill sprints. Hum Mov Sci 52:203–214

Girard O, Brocherie F, Tomazin K et al (2016c) Changes in running mechanics over 100-m, 200-m and 400-m treadmill sprints. J Biomech 49:1490–1497

Jaskolska A, Goossens P, Veentra B et al (1998) Treadmill measurement of the force-velocity relationship and power output in subjects with different maximal running velocities. Sports Med 8:347–358

Kivi DMR, Maraj BKV, Gervais P (2002) A kinematic analysis of high-speed treadmill sprinting over a range of velocities. Med Sci Sports Exerc 34:662–666

Kram R, Griffin TM, Donelan JM, Chang YH (1998) Force treadmill for measuring vertical and horizontal ground reaction forces. J Appl Physiol 85:764–769

Kram R, Powell AJ (1989) A treadmill-mounted force platform. J Appl Physiol 67:1692–1698

Kyröläinen H, Avela J, Komi PV (2005) Changes in muscle activity with increasing running speed. J Sports Sci 23:1101–1109

Lakomy HKA (1987) The use of a non-motorized treadmill for analysing sprint performance. Ergonomics 30:627–637

Mann R, Sprague P (1980) A kinetic analysis of the ground leg during sprint running. Res Q Exerc Sport 51:334–348

Marey EJ (2002) Le mouvement, Editions J. Chambon

McKenna M, Riches PE (2007) A comparison of sprinting kinematics on two types of treadmill and over-ground. Scand J Med Sci Sport 17:649–655

Morin J-B, Petrakos G, Jimenez-Reyes P et al (2016) Very-heavy sled training for improving horizontal force output in soccer players. Int J Sports Physiol Perform 1–13

Morin J-B, Belli A (2004) A simple method for measurement of maximal downstroke power on friction-loaded cycle ergometer. J Biomech 37:141–145

Morin J-B, Bourdin M, Edouard P et al (2012) Mechanical determinants of 100-m sprint running performance. Eur J Appl Physiol 112:3921–3930

Morin J-B, Edouard P, Samozino P (2011a) Technical ability of force application as a determinant factor of sprint performance. Med Sci Sport Exerc 43:1680–1688

Morin J-B, Gimenez P, Edouard P et al (2015a) Sprint acceleration mechanics: the major role of hamstrings in horizontal force production. Front Physiol 6:e404

Morin JB, Samozino P, Bonnefoy R et al (2010) Direct measurement of power during one single sprint on treadmill. J Biomech 43:1970–1975. https://doi.org/10.1016/j.jbiomech.2010.03.012

Morin J-B, Samozino P, Edouard P, Tomazin K (2011b) Effect of fatigue on force production and force application technique during repeated sprints. J Biomech 44:2719–2723

Morin J-B, Samozino P, Millet GY (2011c) Changes in running kinematics, kinetics, and spring-mass behavior over a 24-h run. Med Sci Sports Exerc 43:829–836

Morin J-B, Samozino P, Zameziati K, Belli A (2007) Effects of altered stride frequency and contact time on leg-spring behavior in human running. J Biomech 40:3341–3348

Morin J-B, Sève P (2011) Sprint running performance: comparison between treadmill and field conditions. Eur J Appl Physiol 111:1695–1703

Morin J-B, Slawinski J, Dorel S et al (2015b) Acceleration capability in elite sprinters and ground impulse: push more, brake less? J Biomech 48:3149–3154

Nagahara R, Mizutani M, Matsuo A et al (2017) Association of step width with accelerated sprinting performance and ground reaction force. Int J Sports Med 38:534–540

Nelson RC, Dillman CJ, Lagasse P, Bickett P (1972) Biomechanics of overground versus treadmill running. Med Sci Sports 4:233–240

Nigg BM, De Boer RW, Fisher V (1995) A kinematic comparison of overground and treadmill running. Med Sci Sports Exerc 27:98–105

Rabita G, Dorel S, Slawinski J et al (2015) Sprint mechanics in world-class athletes: a new insight into the limits of human locomotion. Scand J Med Sci Sports 25:583–594

Riley PO, Dicharry J, Franz J et al (2008) A kinematics and kinetic comparison of overground and treadmill running. Med Sci Sports Exerc 40:1093–1100

Schache AG, Blanch PD, Rath DA et al (2001) A comparison of overground and treadmill running for measuring the three-dimensional kinematics of the lumbo-pelvic-hip complex. Clin Biomech 16:667–680

Schache AG, Brown NAT, Pandy MG (2015) Modulation of work and power by the human lower-limb joints with increasing steady-state locomotion speed. J Exp Biol 218:2472–2481

Schache AG, Dorn TW, Williams GP et al (2014) Lower-limb muscular strategies for increasing running speed. J Orthop Sport Phys Ther 44:813–824

Simonsen EB, Thomsen L, Klausen K (1985) Activity of mono- and biarticular leg muscles during sprint running. Eur J Appl Physiol Occup Physiol 54:524–532

Sun Y, Wei S, Zhong Y et al (2015) How joint torques affect hamstring injury risk in sprinting swing-stance transition. Med Sci Sports Exerc 47:373–380

Tomazin K, Morin J-B, Strojnik V et al (2012) Fatigue after short (100-m), medium (200-m) and long (400-m) treadmill sprints. Eur J Appl Physiol 112:1027–1036

van Ingen Schenau GJ (1980) Some fundamental aspects of the biomechanics of overground versus treadmill locomotion. Med Sci Sports Exerc 12:257–261

Weyand PG, Sandell RF, Prime DNL, Bundle MW (2010) The biological limits to running speed are imposed from the ground up. J Appl Physiol 108:950–961

Weyand PG, Sternlight DB, Bellizzi MJ, Wright S (2000) Faster top running speeds are achieved with greater ground forces not more rapid leg movements. J Appl Physiol 89:1991–1999

Chapter 11
A Simple Method for Measuring Force, Velocity and Power Capabilities and Mechanical Effectiveness During Sprint Running

Pierre Samozino

Abstract A macroscopic view of sprint mechanics during an acceleration phase, and notably athlete's propulsion capacities, can be given by Force-velocity (F-v) and Power-velocity (P-v) relationships. They characterize the change in athlete's maximal horizontal force and power production capabilities when running speed increases and directly determine sprint acceleration performance. This chapter presents an accurate and reliable simple method to determine these mechanical capabilities during sprinting. This method, based on a macroscopic biomechanical model and validated in laboratory conditions in comparison to force plate measurements, is very convenient for field use since it only requires anthropometric (body mass and stature) and spatio-temporal (split times or instantaneous velocity) input variables. It provides different information on athlete's horizontal force production capabilities: maximal power output, maximal horizontal force, maximal velocity until which horizontal force can be produced and mechanical effectiveness of force application onto the ground. This information presents interesting practical applications for sport practitioners to individualize training focusing on sprint acceleration performance, but also perspectives in injury management. This chapter presents different examples of such applications. Moreover, this simple method can also help to bring new insight into the limits of human locomotion since it makes possible to estimate sprinting mechanical properties of the fastest men and women without testing them in a laboratory.

P. Samozino (✉)
Laboratoire Inter-universitaire de Biologie de la Motricité,
Université de Savoie Mont Blanc, Campus Scientifique,
73000 Le Bourget du Lac, Chambéry, France
e-mail: pierre.samozino@univ-smb.fr

11.1 Introduction

The previous chapter presented recent innovative concepts and measurement methodologies to explore, evaluate and better understand mechanics of sprinting. A macroscopic view of the sprint mechanics during an acceleration phase, and notably the athlete's propulsion capacities, can be given by Force-velocity (F-v) and Power-velocity (P-v) relationships. These relationships characterize the change in athlete's maximal horizontal force and power production capabilities when running speed increases. These capabilities directly determine athlete's forward acceleration, which is a key factor of performance in many sport activities, not only to reach the highest top velocity, but also and most importantly to cover a given distance in the shortest time possible, be it in track and field events or in team sports (see previous chapter).

As previously described for other movements (pedaling, squat jumps, bench press), these relationships allow the determination of key mechanical variables being complex integration of the different physiological, neural and biomechanical mechanisms involved in the total external force production and characterizing different athlete's muscle abilities. However, in contrast to acyclic ballistic push-offs (e.g. squat jumps, Chaps. 4 and 5), F-v and P-v relationships in sprint running are specific to running acceleration propulsion and in turn also integrate the ability to apply the external force effectively (i.e. horizontally in the antero-posterior direction) onto the ground (Morin et al. 2011a, b, 2012; Rabita et al. 2015). The technical ability of force application during sprint running and its implication in sprinting performance has been well presented and detailed in the previous chapter (see Chap. 10, Sect. 10.3.3). F-v and P-v relationships provide thus an objective quantification of the maximal power output an athlete can develop in the horizontal direction (P_{max}, power capabilities), the theoretical maximal horizontal force an athlete could produce onto the ground (F_0, force capabilities) and the theoretical maximal velocity at which he/she could run if there is no external constraints to overcome (v_0, velocity capabilities). The latter can be also interpreted as the maximal running velocity until which the athlete is still able to produce positive net horizontal force, which well represents his/her ability to produce horizontal force at high running velocities. As for the other movements, force and velocity capabilities in sprint are independent and do not refer to the same physical and technical abilities. The ratio between both corresponds to athlete's F-v profile (S_{Fv}). These different mechanical variables integrate athlete's "physical" qualities (lower limb muscle force production capacities) and "technical" abilities (mechanical effectiveness of force application). The latter can be computed at each step by the ratio of horizontal to total force (RF). Mechanical effectiveness can so be well described by the maximal value of RF observed at the first step ($RFmax$) and its rate of decrease when velocity increases (D_{RF}). Quantifying individually the mechanical effectiveness can help to distinguish the physical and technical origins of inter or intra-individual differences in both Force-velocity-Power (FvP) mechanical profiles and sprint performances, which can be useful to more appropriately orient the training process towards the specific mechanical qualities to develop.

As presented in the previous chapter, different methodologies have been proposed to assess all these mechanical variables characterizing sprint propulsion capabilities. Briefly, the first ones in the nineties used motorized or non-motorized specific treadmills of whom the belt was accelerated by the athlete and required several 6-s sprints against different loads to determine FvP relationships from peak velocity (Jaskolska et al. 1999; Jaskolski et al. 1996; Chelly and Denis 2001). Then, in 2010, a single sprint method was validated in our lab using a dynamometric treadmill and foot-ground contact phase averaged values to compute the different relationships and mechanical variables (Morin et al. 2010, 2011a). Few years later, to respond to the criticisms on treadmill measurements for sprinting evaluations (non-natural movement due to waist attachment, a belt narrower than a typical track lane, the impossibility to use starting block, the need to set a default torque), we collaborated with colleagues of the French National Institute of Sport to propose and validate a method to measure ground reaction force over an entire sprint acceleration phase from 5 sprints overground and a 6.6-m force plate system, since to date no 30- to 60-m long force plate systems exist (Rabita et al. 2015). This allowed, for the first time, to provide the data to entirely characterize the mechanics of overground sprint acceleration. These different laboratory methodologies, each of them presenting advantage and inconvenient, present very accurate and reliable measurements of the different mechanical propulsion qualities (force, velocity and power capabilities). However, sport practitioners do not have easy access to such expensive and rare devices, and often do not have the technical expertise to process the raw force data measured. In the best cases, this forces athletes to report to a laboratory. This explains that, although very accurate and potentially useful for training purposes, this kind of evaluation has almost been never performed.

Consequently, sport scientists investigating sprint mechanics and performance usually assess, at best, only very few steps of a sprint (e.g. Kawamori et al. 2014; Lockie et al. 2013). Sport practitioners do not explore kinetic variables, i.e. athlete's force and power production, but only their consequences on movement, i.e. kinematic parameters (e.g. split times or distances covered in a given time). The exploration of sprint performance in field conditions through kinematic analyses of the horizontal displacement of the body center of mass has already been proposed in 1920s by Furusawa, Hill and colleagues. They used a magnet carried by the athlete which induced an electrical current each time the athlete runs in front of coils of wire connected to a galvanometer placed at given distances in parallel to the track (Fig. 11.1, Furusawa et al. 1927). Another ingenious device was used later in 1954 by Henry who equipped a track (on the roof of a building) with several timing contact gates allowing to measure times with an accuracy of 0.01 s (Fig. 11.1, Henry 1954). These were the ancestors of current photocells or high speed cameras giving the different split times during a sprint acceleration phase. In parallel to time measurements, instantaneous speed assessments can be done out of labs using radar or laser guns which measure athlete positions at very high sample rates

Fig. 11.1 Main devices used in the past and at the moment to analyse sprint kinematics out of labs

(from 30 to 100 Hz). Even if kinematic variables provide very interesting information on sprint performance, they do not give insights about the athlete's force and power production capabilities nor about the distinction between "physical" and "technical" abilities. Typical examples will be presented at the end of this chapter to show the interest of kinetic assessments in addition to kinematic ones.

A simple method for determining F-v and P-v relationships and force application effectiveness during sprint running in overground realistic conditions, out of the lab, and from only one single sprint, seems to be therefore very interesting to generalize sprint mechanics evaluations for training or scientific purposes. This chapter will present a simple field method we proposed to compute accurately force, velocity and power lower limb capabilities from few data inputs that are easy to obtain in typical training practice. The validation protocol and results, limitations and practical applications will then be presented and discussed.

11.2 Theoretical Bases and Equations

As for the other simple methods presented in this book, the simple sprint method for measuring force, velocity and power capabilities during sprint running is based on the fundamental principles of dynamics applied to the body center of mass (CM) (for more details, see Samozino et al. 2016). The biomechanical model used here is an analysis of kinematics and kinetics of the runner's CM during sprint acceleration using a macroscopic inverse dynamics approach aiming to be the

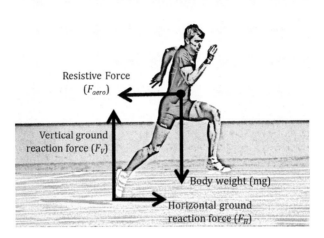

Fig. 11.2 Schematic representation of the external forces applied to a sprinter during his acceleration phase: body weight (mg), aerodynamic resistive force (F_{aero}) and vertical (F_V) and horizontal (F_H) components of the ground reaction force

simplest possible (Helene and Yamashita 2010; Furusawa et al. 1927; di Prampero et al. 2015). In this approach, as for the previous ones, mechanical variables are modeled over time, without considering intra-step changes, and thus corresponds to step-averaged values (contact plus aerial times).

During a running maximal acceleration, let us consider the different external forces applied to the athlete's CM: the body weight, the aerodynamic resistive forces and the ground reaction force (GRF) with its horizontal and vertical components (Fig. 11.2). Applying the fundamental laws of dynamics in the horizontal direction, the net horizontal antero-posterior GRF (F_H) applied to the body CM can be modeled over time as:

$$F_H(t) = m.a_H(t) + F_{aero}(t) \qquad (11.1)$$

with m the runner's body mass (in kg), $a_H(t)$ the CM horizontal acceleration and $F_{aero}(t)$ the aerodynamic drag to overcome during sprint running. Basic computational fluid dynamics principles show that $F_{aero}(t)$ is proportional to the square of the velocity of air relative to the runner and can be modelled as:

$$F_{aero}(t) = k.(v_H(t) - v_w)^2 \qquad (11.2)$$

with v_w the wind velocity (if any) and k the runner's aerodynamic friction coefficient. The latter can be estimated from values of air density (ρ, in kg m^{-3}), frontal area of the runner (Af; in m^2), and drag coefficient ($Cd = 0.9$, van Ingen Schenau et al. 1991) as proposed by Arsac and Locatelli (2002):

$$k = 0.5.\rho.Af.Cd \qquad (11.3)$$

with

$$\rho = \rho_0 . \frac{Pb}{760} . \frac{273}{273 + T°} \qquad (11.4)$$

$$Af = \left(0.2025.h^{0.725}.m^{0.425}\right).0.266 \qquad (11.5)$$

where $\rho_0 = 1.293$ kg m^{-1} is the ρ at 760 Torr and 273K, Pb is the barometric pressure (in Torr), $T°$ is the air temperature (in °C) and h is the runner's stature (in m).

To sum up, beyond to know the athlete's stature and body mass, as well as approximations of ambient temperature and barometric pressure (which have only negligible incidences on mechanical output, see Sect. 11.3), $a_H(t)$ is the only required kinematic input variable to model F_H.

During a running maximal acceleration, horizontal velocity (v_H) systematically increases following a mono-exponential function. This can be observed for recreational athletes, team sport players, the best sprinter of all time, a young 4-year old boy or a 95-year old man and have been shown many times in scientific literature (Fig. 11.3, (e.g. Furusawa et al. 1927; di Prampero et al. 2005; Morin et al. 2006; Chelly and Denis 2001):

$$v_H(t) = v_{H_{max}} . \left(1 - e^{-t/\tau}\right) \qquad (11.6)$$

with $v_{H_{max}}$ the maximal velocity reached at the end of the acceleration and τ the acceleration time constant. After integration and derivation of $v_H(t)$ over time, the

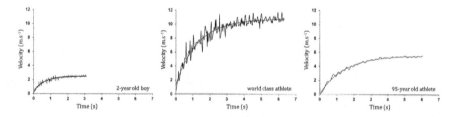

Fig. 11.3 Velocity-time curves obtained during sprinting acceleration of a 2-year old young boy, a world class athlete and a 95-year old man. The noisy lines correspond to the radar gun data and the solid smoothed lines represent the mono-exponential model function. The differences in magnitudes of noise in the radar signals are independent from the age or level of the individual, but is associated to different filters applied on these typical raw data

horizontal position (x_H) and acceleration (a_H) of the body CM as a function of time during the acceleration phase can be expressed as follows:

$$x(t) = v_{H_{max}} \cdot \left(t + \tau \cdot e^{-\frac{t}{\tau}}\right) - v_{H_{max}} \cdot \tau \tag{11.7}$$

$$a_H(t) = \left(\frac{v_{H_{max}}}{\tau}\right) \cdot e^{-\frac{t}{\tau}} \tag{11.8}$$

Consequently, $v_{H_{max}}$ and τ values can be determined from velocity or position/time measurements and using least square regression method and Eq. 11.6 or 11.7. Then, $a_H(t)$ can be computed at each instant (for instance, every 0.1 or 0.05 s) using Eq. 11.8, F_H can be obtained using Eq. 11.1 and v_H using Eq. 11.6 [if the initial measurement is $x(t)$]. The mean net horizontal antero-posterior power output applied to the body CM (P_H in W) can then be modelled at each instant as the product of F_H and v_H. Plotting F_H versus v_H and modeling this relationship by a linear equation gives the F-v relationship which can be extrapolated to obtain the maximal force (F_0) and velocity (v_0) values as the intercept with the force- and velocity-axis, respectively (Fig. 11.4). The P-v relationship can be obtained using a 2nd order polynomial regression on P_H-v_H plot, the apex of the latter corresponding to the maximal power output ($P_{H_{max}}$).

Then, applying the fundamental laws of dynamics in the vertical direction, the mean net vertical component of GRF (F_V) applied to the body CM over each complete step can be modeled over time as equal to body weight (di Prampero et al. 2015):

$$F_V(t) = m \cdot g \tag{11.9}$$

where g is the gravitational acceleration (9.81 m s^{-2}).

The ratio of force (*RF* in %) can be modeled at each instant by:

$$RF = \frac{F_H}{\sqrt{F_H^2 + F_V^2}} \cdot 100 \tag{11.10}$$

After plotting *RF* versus v_H with a linear regression, the slope of this relationship corresponds to the rate of decrease in *RF* when velocity increases over the entire acceleration phase (D_{RF}, in % s m^{-1}). It is worth noting that, since the starting block phase (push-off and following aerial time) lasts between 0.5 and 0.6 s (Slawinski et al. 2010; Rabita et al. 2015) and so occurs at an averaged time of $\sim$0.3 s, *RF* and D_{RF} can be reasonably computed from F_H and F_V values modeled for t > 0.3 s.

The biomechanical model and associated equations presented here allows the estimation of GRFs in the sagittal plane of motion during one single sprint running acceleration from simple inputs: anthropometric (body mass and stature) and spatiotemporal (split times or instantaneous running velocity) data. This model can then be used as a simple method to evaluate force, velocity and power capabilities and mechanical effectiveness during sprint running.

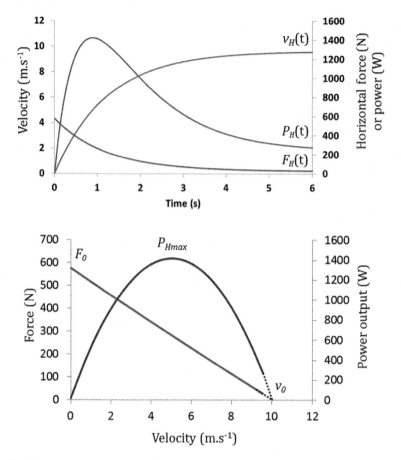

Fig. 11.4 Modelled changes in horizontal velocity (v_H), horizontal antero-posterior force (F_H) and power (P_H) over time during the acceleration phase of a sprint, and associated modelled force- and power-velocity relationships with maximal force (F_0), velocity (v_0) and power ($P_{H_{max}}$)

11.3 Limits of the Method

The biomechanical model presented in the previous section is a simple application of the basic principles of dynamics. However, as for all models, some simplifying assumptions have been required to model the horizontal force developed by the runner onto the ground during the entire sprint acceleration phase. Moreover, the macroscopic level of the model (kinematics and kinetics of the body CM) induces some limits in the level of analysis possible with this approach, which has to be considered for the interpretation of output variables. These different points are following discussed.

- The dynamics principles were applied to a whole body considered as a system and represented by its centre of mass (e.g. Samozino et al. 2008, 2010, 2012;

Cavagna et al. 1971; van Ingen Schenau et al. 1991; Helene and Yamashita 2010; Rabita et al. 2015). Only the forces aiming at moving the CM were considered.

- The biomechanical model only focuses on GRF components in the sagittal plane of motion (i.e. vertical and antero-posterior) and neglects the medio-lateral component which was shown to constitute a negligible part in the total force produced by athletes (Rabita et al. 2015).
- The horizontal aerodynamic friction coefficient (k) was proposed to be estimated from only stature, body mass and a fixed drag coefficient (Arsac and Locatelli 2002). Even if it does not represent the "gold standard" process, this represents a very simple method to estimate the aerodynamic resistive force out of laboratory. Moreover, the sensibility of the mechanical output variables to a likely error in the estimation of frontal area, air density or drag coefficient is low: a $\sim 10\%$ error in these input estimations leads to an error lower than 0.5% in the output variables.
- In contrast with previous kinetic measurements during sprint running which averaged mechanical variables over each support phase (Morin et al. 2010, 2011a, 2012; Lockie et al. 2013; Rabita et al. 2015; Kawamori et al. 2014), computations lead here to modelled values over complete steps, i.e. contact plus aerial times, which induces lower values of force or power output. Step-averaged variables characterize more the mechanics of the overall sprint running propulsion than specifically the mechanical capabilities of lower limb neuromuscular system during each contact phase. However, this does not affect RF (and in turn D_{RF}) values since it is a ratio between two force components averaged over the same duration. Note that such values modeled over complete steps cannot bring information about inter-step variability, intra-step analyses or explorations of contact and aerial times, step length/frequency and force impulse and rate of development during sprint running.
- Mechanical effectiveness computations required to model GRF vertical component over each step as equal to the runner's body weight. This required the assumption of a quasi-null CM vertical acceleration over the acceleration phase of the sprint. However, be it with or without using starting blocks (but largely more pronounced with starting blocks), the runner's body CM goes up during this phase from the starting crouched position to the standing running position, and then does not change from one complete step to another. Since the initial upward movement of the CM is overall smoothed through a relative long time/distance (~ 20–40 m, Cavagna et al. 1971; Slawinski et al. 2010), we can consider that it does not require any large vertical acceleration, and so that the mean net vertical acceleration of the CM over each step is quasi null throughout the sprint acceleration phase. This is even more correct for standing sprint starts which represents the most of the cases in sport other than track and field sprint events using starting blocks.

The above-mentioned simplifying assumptions are those inherent of all biomechanical models. The important thing is to quantify the errors induced by these simplifications. For that purpose, the simple method based on these computations and on simple anthropometric and spatiotemporal measurements was validated in comparison to the reference force plate measurements.

11.4 Validation of the Method

The validity and the reliability of the simple sprint method presented in this chapter were tested through two different experimental protocols reported in details in Samozino et al. (2016).

11.4.1 Concurrent Validity Compared to Force Plate Measurements

The concurrent validity of the simple method was tested by comparison of modelled mechanical values obtained using the proposed computations to reference force plate measurements during an entire acceleration phase of a sprint.

Reference method. Antero-posterior and vertical GRF components, F-v, P-v and *RF*-v relationships and associated variables $(F_0, v_0, P_{H_{max}}, S_{FV}, D_{RF})$ were determined for nine elite or sub-elite sprinters using both the simple method and the method using a 6.60-m long force platform system (KI 9067; Kistler, Winterthur, Switzerland, sampling rate of 1000 Hz) recently proposed by Rabita et al. (2015), more details in Chap. 10). Briefly, since no 30- to 60-m long force plate systems existed, this method consisted in virtually reconstruct for each athlete the GRF signal of an entire single 40-m sprint acceleration by setting differently for each sprint the position of the starting blocks relatively to the 6.60-m long force platform system. So each athlete performed, after a standardized 45 min warm-up, 7 maximal sprints in an indoor stadium (2 × 10 m, 2 × 15 m, 20 m, 30 m and 40 m with 4 min rest between each trial). During each sprint, GRF data were collected over the 6.60 m section covered by the force plate system, the latter being placed at different positions of the acceleration phase for each sprint. Instantaneous data of vertical (F_V) and horizontal antero-posterior (F_H) GRF components were averaged for each step (contact + aerial phase) to compute the above-mentioned variables.

Simple method. In parallel to force plate measurements, sprint times were measured with a pair of photocells (Microgate, Bolzano, Italy) located at the finish line of the different sprints. The 5 split times at 10 (best one of the two trials), 15 (idem), 20, 30 and 40 m were then used to determine $v_{H_{max}}$ and τ using Eq. 11.7 and least square regression method. From these two parameters, $v_H(t)$ and $a_H(t)$ were modelled over time (every 0.1 s) using Eqs. 11.6 and 11.8, respectively. From $a_H(t), F_H(t), P_H(t)$ and $RF(t)$ were computed using equations and data processing presented in Sect. 11.2 in order to determine all the mechanical variables for each subject.

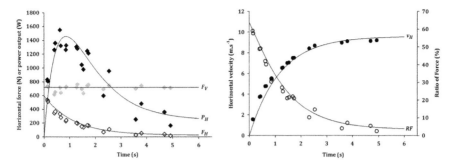

Fig. 11.5 Changes over the acceleration phase in horizontal velocity (v_H, black points), horizontal (F_H, open diamonds) and vertical (F_V, grey diamonds) force components, horizontal power output ($P_{H_{max}}$, black diamonds), and ratio of force (RF, open circles) for a typical sprinter. Points represent averaged values over each step obtained from force plate method (from five sprints) and lines represent modelled values computed by the simple sprint method from split times

Results showed that modeled force (F_H, F_V, resultant), power and RF values were very close to values measured by force plates at each step with low standard errors of estimate of ~30–50N, ~230W and 3.7%, respectively (as shown for a typical subject in Fig. 11.5). Over the first 20–30 m, F_V values measured with force plates were not particularly higher than body weight (Fig. 11.5) and were very close to body weight when averaged over the entire acceleration phase (difference lower than ~2.40% in average). This supports the assumption of a quasi-null vertical acceleration of the CM over this phase (despite the use of starting blocks), and in turn supports the validity of step averaged F_V modelled values as equal to body weight. Beyond the very good agreement in the modeled GRF in the sagittal plane of motion (horizontal, vertical and resultant) during sprint running acceleration, low bias in the determination of F_0, v_0, $P_{H_{max}}$, S_{FV} and D_{RF} were observed: lower than ~5% for F_0, v_0 and $P_{H_{max}}$; and lower than 8% for S_{FV} and D_{RF} (Table 11.1). These low bias were associated to narrowed 95% agreement limits crossing 0, which support the high accuracy and validity of the proposed simple method to determine F-v, P-v and RF-v relationships and their associated mechanical variables $\left(P_{H_{max}}, F_0, v_0, S_{FV}, D_{RF}\right)$ in sprint running (Fig. 11.6).

Table 11.1 Mean ± SD values of variables attesting the concurrent validity of the simple sprint method

	Reference method	Proposed method	Bias	95% agreement limits	Absolute bias
F_0 (N)	654 ± 80	638 ± 84	−15.9 ± 25.7	[−66.3;34.5]	3.74 ± 2.69
v_0 (m.s^{-1})	10.20 ± 0.36	10.51 ± 0.74	0.32 ± 0.52	[−0.7;1.3]	4.77 ± 3.26
P_{Hmax} (W)	1669 ± 253	1680 ± 280	10.56 ± 45.01	[−77.7;98.8]	1.88 ± 1.88
S_{FV} (N s m^{-1})	−64.06 ± 6.30	−60.8 ± 7.71	3.26 ± 5.22	[−6.97;13.49]	7.93 ± 5.32
D_{RF} (% s m^{-1})	−6.80 ± 0.28	−6.80 ± 0.74	−0.002 ± 0.58	[−1.139;1.135]	6.04 ± 5.70

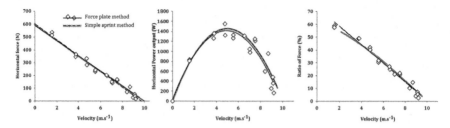

Fig. 11.6 Force- power- and RF-velocity relationships obtained by both methods for a typical athlete. Filled diamonds represent averaged values over each step obtained from force plate method, solid lines the associated regressions, and grey lines the modelled values computed by the proposed simple method confounded with the associated regressions (dashed lines)

The kinematic input variables of the present simple method are spatio-temporal data: split times (as used in the validation protocol) or instantaneous running velocity measurements (as could be obtained from radar guns, (e.g. di Prampero et al. 2005; Morin et al. 2006), or laser beams, (e.g. Bezodis et al. 2012) during one single sprint. So, the simple sprint method was also tested during the same protocol from data measured using a radar (Stalker ATS System, Radar Sales, Minneapolis MN, USA, 46.875 Hz) during the best sprints of the 30- and 40-m trials. Results were very similar to those obtained from split times, with slightly higher bias values (absolute bias from 3 to 7%) due the fact that the comparison was made between mechanical variables obtained from one single sprint using the simple method and data obtained from 5 sprints with the reference method. This could have added a bias that would have only been associated to the validation protocol itself, and not to the method.

11.4.2 Reliability

The inter-trial reliability of the simple method was tested in a second protocol during which six high-level sprinters performed three maximal 50-m sprints with 10 min of rest between each trial. The different mechanical variables $F_0, v_0, S_{FV}, P_{H_{max}}$ and D_{RF} were obtained with the same data processing as presented before for the proposed computation method, except that $v_{H_{max}}$ and τ were determined using Eq. 11.6, least square regression method and $v_H(t)$ measured by the radar system (sample rate 46.875 Hz). The latter was placed on a tripod 10 m behind the subjects at a height of 1 m corresponding approximately to the height of subjects' CM (di Prampero et al. 2005; Morin et al. 2012). For all the mechanical variables, low coefficients of variation between the two best trials and standard errors of measurement were observed (<5%, Table 11.2) and were associated to change in the mean close to 0. This showed low systematic and random errors, and in turn high test-to-test reliability.

Table 11.2 Mean ± SD of the coefficient of variation (CV), change in the mean and standard error of measurement between the 2 best trials

	CV (%)	Change in the mean	Standard error of measurement (%)
F_0 (N)	2.93 ± 2.00	−1.53 ± 32.2	3.57
v_0 (m s^{-1})	1.11 ± 0.86	−0.171 ± 0.776	1.40
P_{max} (W)	1.87 ± 1.36	−0.167 ± 0.66	2.33
S_{FV} (N s m^{-1})	4.04 ± 2.72	−0.20 ± 4.18	4.94
D_{RF} (% s m^{-1})	3.99 ± 2.80	−0.110 ± 0.45	4.86

Through this study (Samozino et al. 2016), we clearly show that the simple method based on a macroscopic biomechanical model and only anthropometric (body mass and stature) and spatio-temporal (split times or instantaneous velocity) variables easy to obtain out of laboratory, is accurate, reliable and valid to evaluate force, velocity and power capabilities, as well as mechanical effectiveness, during sprint running.

11.5 Technologies and Input Measurements

As for the different simple methods presented in this book, the accuracy and reliability of the present simple sprint method depends on the accuracy of the devices used to obtain the mechanical inputs of the model, i.e. the position-time or velocity-time data here. Note that body mass has to be measured with shoes and clothes used during the tests.

11.5.1 Split Times

When using split times as input data of the simple sprint method (using Eq. 11.7), as it was done in the above-described validation protocol, at least 4 or 5 split times are required over the acceleration phase to obtain reliable mechanical output variables. According to the level of the athletes, the distances associated to the different split times would be different in order to cover all the acceleration phase: from start line to 30 m (for non-expert sprinters), until to 50–60 m (for track and field sprinters). These distances have to be shorter at the beginning than at the end of the acceleration phase since the higher the acceleration magnitude (and so in the first meters of the sprint), the higher the number of split time needed to well describe the change in motion velocity. For instance, for soccer or rugby players, split times should be at 5, 10, 15, 20 and 30 m. For 100-m sprinters, they could be at 5, 10, 15, 30 and 40 m. Different devices can be used in field conditions to measure split times during a sprint acceleration, the main of them are presented here.

Fully automatic timing systems. The gold standard device to measure accurately and reliably split times during sprints is the fully automatic timing systems including silent gun and photo-finish camera (Haugen and Buchheit 2016). They present timing resolution up to 0.0005 s and are mostly and quasi only used in international athletics competition. They are too expensive and impractical for sport practitioners and scientists. However, since the 1987 World Championships in Rome, the International Association of Athletics Federations (IAAF) has provided biomechanics reports presenting split times for each 10- or 20-m sections for the 100-m race in both men and women. Analysing these data, combined those obtained during individual World Record races or Olympic Games, with the above-presented simple computations has recently allowed us to explore the mechanical determinants of 100-m sprint running performance in the world's fastest men and women, and so the limits of human sprinting performance (Slawinski et al. 2017, see Sect. 11.6.3). Since a sprint acceleration actually starts when the force production on the ground firstly rises, the reaction time has to be removed from the different split times to consider only the CM kinematics (and not the global 100-m sprint performance) to estimate kinetics variables.

Photocells. In sprint testing or training, photocells are the most used system to measure split times usually with a resolution of 1 ms. Different photocell timing systems exist with different methods to start the timer: a pair of cells placed just (20–50 cm) in front of the athlete positioned on the starting line, a finger pod on the floor under the thumb for three-point starts or foot pod under the back foot (Haugen and Buchheit 2016). As mentioned before, to use split times as input in the simple sprint method, they have to be measured from the first force production on the ground. We determined that the best method to estimate these split times is to add 0.1 s to split times measured with athletes using a three-point crouching starting position and with photocells using a finger pod on the floor under the thumb to start the timer. This 0.1 s time delay was quantified on several sprinters using force plate and high speed camera (Samozino et al. 2016). Other methods, for instance using a pair of photocells placed just in front of the starting line, would overestimate force and power output computed by the simple method, even if reliability remains very good if photocells are always set at the same place over the different testing sessions.

High speed cameras. With the multiplication and generalization of high-speed cameras presenting high pixel resolutions and high frame rates (up to 240 frames per second in the very recent smartphone and tab), measuring split times using video provides accurate enough information to compute reliably mechanical variables from the simple sprint method. The higher the frame rate, the higher the accuracy. We think that the frame rate has to be at least 100 frames per second to obtain relevant outcomes. To obtain split times from video analysis, it requires to position several markers on the track at the 4–5 different distances and determine times when athletes cross the markers with their hip or shoulders. This can be done using a travelling camera moving at the same velocity as the athlete to film the sprint from the side or using a fixed camera placed at a given distance from the track at the marker positioned at the half of the acceleration targeted distance (i.e. at 15 or

20 m for a 30- or 40-m acceleration, respectively). In the latter case, video parallax was corrected to ensure the different split times are measured properly when athletes cross the different targeted distances. A simple methods for parallax correction is proposed, detailed and illustrated in Romero-Franco et al. (Fig. 1 in Romero-Franco et al. 2016). As for photocells, the critical point is the criterion considered to start the timer, i.e. to determine the frame corresponding to the start of the sprint which corresponds, from a mechanical point of view, of the beginning of the force production. We proposed to make athletes start with a three-point starting position, to consider the frame at which the thumb leaves the ground as the frame 0, and then add 0.1 s at each split time. Recently, a smartphone application (MySprint) was designed by Pedro Jimenez-Reyes to compute all the sprint mechanical variables using the computations of present simple sprint method and split times measured with the 240-fps camera of recent iPhones or Ipad (Fig. 11.7). Performance inputs (split times and velocity-time curves) and mechanical outputs (horizontal force and power, mechanical effectiveness) showed very high reliability (ICC > 0.99) and concurrent validity (standard error of estimate <1.3%) compared to reference devices (photocells and radar) (Romero-Franco et al. 2016). In our opinion, this low cost system (Iphone/Ipad + MySprint App) represents nowadays the best compromise between cost, validity/accuracy/reliability, direct feedback data, and ease to use and to carry on field.

11.5.2 Instantaneous Velocity

Instantaneous velocity can also be used as input data of the simple sprint method and lead to similar concurrent validity in the estimation of mechanical variables characterizing force production during sprint acceleration (Samozino et al. 2016). Here are the main used devices to obtain velocity signal over time during such an all-out effort.

Laser and radar guns. Instantaneous velocity can be obtained using laser (e.g. LAVEG Sport laser speed gun, Jenoptik, Jena, Germany) or radar (e.g. Stalker ATS radar gun, Radar Sales, Minneapolis, MN, USA) systems which presents sampling rates of 100 and 46.875 Hz, respectively, high reliability and validity (Haugen and Buchheit 2016). Both of these guns are typically positioned 3–10 m behind the athletes at the starting line at a height of 1 m (corresponding approximately to the height of subjects' centre of mass). For data analysis, only the acceleration phase is required to compute $v_{H_{max}}$ and τ using an exponential regression (Eq. 11.6). So, all velocity values measured before the actual sprint start and after the maximal velocity plateau have to be deleted (Fig. 11.8). Contrary to split times for which each data depends on measurement triggering, instantaneous velocity values are independent from the previous ones, which avoids any bias due to the instant or the way the measurement is started. However, the detection of the actual sprint start on velocity data is often compromised by noise on the velocity signal due to

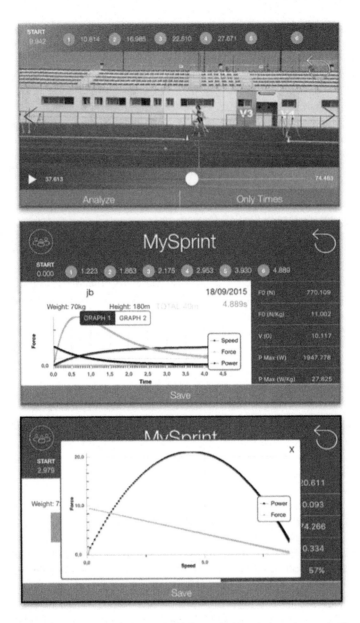

Fig. 11.7 The iPhone app "Mysprint" uses the high-speed video slow motion mode (frame rate of 240 fps) to measure the split times of a 30-m sprint acceleration, and thus compute all Force-velocity-Power profile and mechanical effectiveness variables

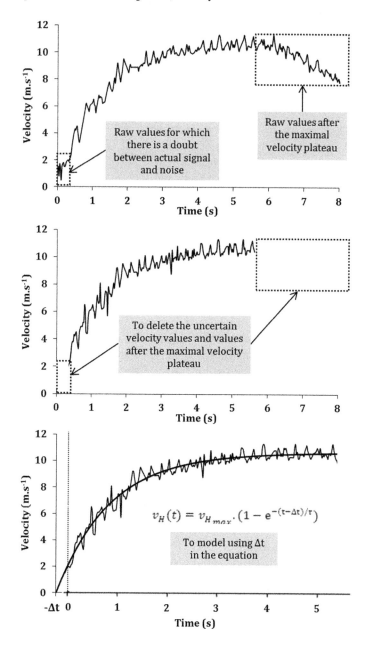

Fig. 11.8 Typical data analysis when instantaneous velocity is measured by radar gun during a sprint acceleration. The first step is to delete uncertain raw velocity values for which there is a doubt between actual signal and noise at the start of the sprint and values after the maximal velocity plateau at the end of the sprint. Then, the raw values can be modelled using the mono-exponential regression including a time delay (Δt) in the equation

movements of other things in the gun field of measurement just before the sprint start. If the first non-null velocity values considered is not the actual first one (i.e. athlete's velocity), this could lead to overestimations of force and power variables. To face to this issue, we suggested to delete all the values for which there is a doubt between actual signal and noise, and then add a third parameter (a time delay, Δt, in s) to the mathematical exponential function in order to associate the first velocity value considered to the true time (Fig. 11.8). Including Δt in the Eq. 11.6 gives:

$$v_H(t) = v_{H_{max}} \cdot \left(1 - e^{-(t-\Delta t)/\tau}\right) \qquad (11.11)$$

This data analysis method allows to improve the reliability and validity of velocity values, and associated computed distance and acceleration data, measured in the first 5–10 m of the acceleration phase (see Sect. 11.4), which was previously shown to be an issue for these devices (Haugen and Buchheit 2016; Bezodis et al. 2012). For instance, this time delay was not included in our computations for one of our first studies using this method (Buchheit et al. 2014b): F_0 showed acceptable absolute (CV 7.8%), but not relative reliability (ICC, 0.64) (Simperingham et al. 2016; Buchheit et al. 2014b). After using this time delay, reliability was highly improved (Samozino et al. 2016).

Global positioning systems. Global positioning system (GPS) devices have become popular in large-field ball sports for assessing players' physical activity during competition or training (Aughey 2011). Their validity and reliability have been tested for measuring player kinematics on the field, notably distance and speed (Aughey 2011; Barbero-Alvarez et al. 2010; Jennings et al. 2010; Rampinini et al. 2015), accelerations being acceptable only with high-sampling-rate (>10 Hz) GPS devices (Buchheit et al. 2014a). Recently, the concurrent validity (radar-based method was the reference) of two distinct GPS units (5 and 20 Hz sampling rates) to obtain the sprint mechanical properties using the simple method was investigated (Nagahara et al. 2017). The results showed that GPS devices made possible to obtain consistent FvP profiles, but the percentage bias showed a wide range of overestimation or underestimation for both systems (−5.1 to 2.9% and −7.9 to 9.7% for 5- and 20-Hz GPS), even if the 90% confidence intervals of errors were smaller for the 20-Hz GPS than those for 5-Hz GPS. Consequently, the validity of computing mechanical outputs during sprint acceleration with GPS units (even at high sample rate) came short of acceptable level, although the accuracy with the higher-sampling-rate system was markedly higher than that with lower-sampling-rate system. When improved global or local positioning system technology becomes available, this opens interesting possibilities for practical measurements of sprint-acceleration mechanical features directly during training exercises or competition games, without setting specific tests.

11.6 Practical Applications

11.6.1 Testing Considerations

In addition to the accuracy of the devices used to measure velocity- or position-time data, the accuracy and reliability of the present simple sprint method depends also on the rigor with which the testing protocol is set and the data analyzed. Here, we will detail the different practical points of a typical testing session using the simple sprint method that contributes to decrease the measurement errors.

Warm-up. Since we aim to assess the individual maximal sprint capabilities, athletes have to perform a sprint specific warm-up allowing him to reach his maximal performances during the following test. For instance, warm-up should comprise ~5 to 10 min of low-pace running, followed by 3 min of lower limb muscle stretching, 5 min of sprint-specific drills, and 3–5 progressive 30–40 m sprints separated by 2 min of passive rest.

Number of trials and sprint distances. After warm-up, each athlete has to perform at least two sprints with maximal effort. More trials (3–5) is better to be sure to measure the athlete's maximal capabilities, while rest between trials is sufficient to be sure that no fatigue occurs. Only the best performance (e.g. time at 30 m) will be then considered for analysis. Note that if the performance or mechanical outputs are too different between trials, the relevance of the data is altered. As mentioned before, the sprint distances should be different according to the level of the athletes in order to cover all the acceleration phase: ~30 m for non-expert sprinters (e.g. soccer or rugby players) until to 50–60 m for track and field sprinters.

Starting position. As mentioned in the previous section, when using split times (from photocells or high speed camera) as input measurement, the starting position should be a three-point crouching position with the start of the timer when the thumb leaves the ground and split times have to be corrected adding 0.1 s. When velocity is measured (radar or laser guns), the starting position can be set according to the specificity of the sport activity: starting-blocks, three-point crouching position, stand-up or others. The choice has to be done in order to evaluate the athlete's force production capabilities in similar body configuration as during competition.

Shoes and running surfaces. Since the type of the ground surface (tarmac, wet/dry grass, tartan®, wood floor) and the properties of the shoe outsole (classical running shoes, spikes, cleats) influence the friction between feet and ground during sprint acceleration, the horizontal force production can be affected by both of them. Obviously, sprint testing should be performed in surface and shoe conditions similar as those used in competition. So, for routine and follow-up testing, it is important that sprints are always performed on the same surface using the same kind of shoes.

11.6.2 Data Interpretation

Beyond its accuracy and reliability, the interest of a simple field method rests on the good interpretation of the mechanical outputs, each of them presenting a very specific meaning, and the transfer to practical information for training purposes. Definition and practical interpretation of the main variables of interest when using FvP profiling in sprint have been well presented and discussed in Morin and Samozino (2016) and in previous section/chapter of the present book. Here, we will briefly sum-up the main indexes characterizing the athlete's force production capabilities, in a logical order to well understand, and then improve, sprint mechanics (an example of such indexes for a typical athlete is presented in Table 11.3). Note that these several variables can be obtained after having determined τ and $v_{H_{max}}$ from experimental data (velocity, split times) using least square regression method and Eq. 11.6 or 11.7, and computing at each instant (every 0.1 s or lower) a_H, F_H, v_H, x, P_H and RF (see Sect. 11.2 for more details).

Index of sprint acceleration performance. The variable representing the athlete's sprint acceleration performance is the most important since it is what athletes and coaches want to improve. This index should be chosen regarding the sport activity and the position of the player on the field for team sports. For instance, the performance can be characterized by the time at 50 m for 100-m sprinters, the time at 20 or 30 m for soccer players or rugby backs and the time at 10 or 15 m for a soccer goal keeper or rugby forwards. In some cases, the distance cover during a given time (for instance 2 or 4 s) also makes lots of sense. For the following variables, the distance or duration to consider for the acceleration phase have also to be determined according to the specificity of the performance to improve and the distance over which sprint acceleration should be optimized (short or long sprint accelerations).

Index of horizontal power and force produced during acceleration. Sprint acceleration performance is directly related to the mean horizontal power output developed over the acceleration phase (Morin et al. 2011a, 2012), and also in turn to the mean horizontal force produced since running speed is the consequence of the horizontal force production. So, mean horizontal power and mean horizontal force (expressed or not relatively to body mass) over the targeted distance/time are two macroscopic indexes of the athlete's actual mechanical production during the test, even if the associated information are very close of those brought by the acceleration performance indexes.

Index of horizontal power and force production capabilities. The horizontal power and force actually produced during the sprint acceleration phase depends on the maximal horizontal force production capability. The latter can be dissociated in two different and independent abilities. First, the ability to produce very high level of horizontal force at low velocities is characterized by the theoretical maximal F_H athlete can produced (F_0) and mainly refers to the initial pushes of the athlete onto the ground during sprint acceleration. Second, the ability to keep developing horizontal force at very high velocities is quantified by the theoretical maximal velocity (v_0) which represents the maximal velocity until which the athlete is able to produced horizontal force. The weight of F_0 and v_0 in mean force production, and so in sprint acceleration performance, depends on the acceleration distance to cover. The maximal power output $(P_{H_{max}})$ is the combination of F_0 and v_0, and reflects the overall mechanical output capabilities of the athlete.

Indexes of mechanical effectiveness of force application. The capability to produce horizontal force during sprinting depends on lower limb capability to produce force and on the ability to orient it effectively (i.e. horizontally) onto the ground. The overall mechanical effectiveness of an athlete during sprinting acceleration is well quantified by the mean *RF* over the targeted distance/time. As for the force production capabilities, the overall mechanical effectiveness can be dissociated in two different abilities. First, the ability to orient effectively the force produced at low velocities, i.e. in the first steps of the acceleration phase, is well characterized by the maximal *RF* value (*RFmax*). Second, the ability to maintain high level of effectiveness despite the increase in velocity, well quantified by the rate of decrease in *RF* with velocity (D_{RF}). The analysis of these different mechanical effectiveness indexes in parallel with the previous ones related to force production capabilities allows coaches to distinguish what is associated to athlete's "physical" qualities (lower limb muscle force production capabilities) and "technical" abilities (mechanical effectiveness of force application), which can help them to orient training on each athlete's strengths and/or weaknesses in the aims to improve performance or prevent from some muscular injuries. This will be addressed in the two following sections.

11.6.3 Optimization of Sprint Acceleration Performance

When a training program is designed to improve sprint acceleration performance, the simple sprint method and all the above-mentioned indexes can be used to compare athletes to others or to the rest of the team (e.g. Cross et al. 2015), and to follow each athlete within the season or between seasons. The training content can be individualized to mainly focus on weaknesses, while trying to keep strengths at similar level, and planned regarding the distance over which sprint acceleration should be optimized.

Comparison between athletes. The interest of FvP profiling to optimize sprint acceleration performance has been well illustrated through the case report presented in Morin and Samozino (2016) with two rugby players of an elite union team (Fig. 11.9). They have similar sprint acceleration performance over 20 m, but with opposite force production capabilities. Player #1 presents higher horizontal force production capabilities in the first slow steps of the sprint (i.e. higher F_0) but lower ones at high velocities (i.e. v_0) than Player #2. These differences in horizontal force production are mainly due to differences in mechanical effectiveness: Player #1 has a higher mechanical effectiveness at low running speeds than Player #2 (*RFmax*), but he is less able to maintain this effectiveness when speed increases ($D_{RF} = -7.7\%$ s m^{-1}, i.e. loss of 7.7% of effectiveness at each increase of 1 m s^{-1} in speed) compared to Player #2 for who the effectiveness is only altered by 5.8% for the same speed increment. If the training program for these two players is designed to improve sprint performance (e.g., here 20-m time), it should target different capabilities. A similar program given to these players, based on the fact

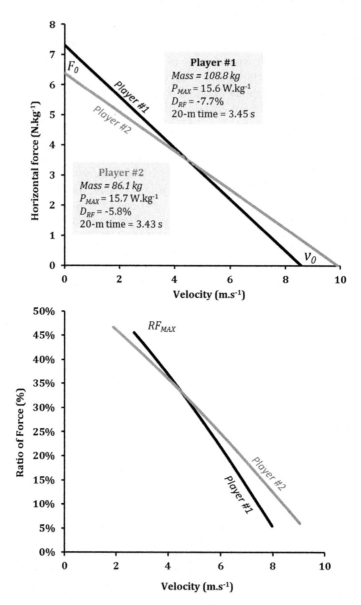

Fig. 11.9 Horizontal force-velocity profiles of 2 elite rugby union players obtained using the simple sprint method from maximal 30-m sprints. Both players reached their maximal running speed before the 30-m mark

that they have similar 20-m sprint times, will very likely result in suboptimal adaptations for both of them. Player #2 should develop F_0 through increasing *RFmax* and/or his lower limb muscle power (to increase the amount of total force produced). The latter can be assessed using FvP profiling in jumping described in

Chap. 4. Contrastingly, Player #1 has to increase v_0, notably by improving his mechanical effectiveness at high velocities in order to decrease D_{RF}.

Effect of training. Training programs focusing on F_0 or v_0 are very different since they refer to opposite training modalities associated to different movement velocities, force to produce, body positions or segment configurations. For instance, resisted sled training represents a specific means of providing overload to horizontal-force capacities which is practical and cost-effective training modality, and which can be used very easily by soccer players of all levels (Petrakos et al. 2016). We recently showed that very-heavy sled training using much greater loads than traditionally recommended (sled with a load of ~80% of body mass, Cross et al. 2017) clearly increased F_0 and *RFmax*, with trivial effect on v_0 (Morin et al. 2017). Contrastingly, training horizontal force production specifically at very high velocities, as during over speed conditions, should improve D_{RF} and v_0 (work currently in progress). Recently, a case report from Cameron Josse, an US strength and conditioning coach, brought supports to the sensibility of v_0 to specific training. He used FvP profile in sprinting (using MySprint App) with a National Football League Linebacker during an 8 week training focusing on high velocity sprinting. This training included long accelerations (from 20 to 50 m), most of them being performed in an upright posture, technical drills aiming at emphasizing frontside mechanics when running upright and some horizontal plyometrics (e.g. power skips for maximal distance). This training induced a decrease in F_0, *RFmax* and $P_{H_{max}}$, but an increase in v_0, 10-m mean *RF* and D_{RF} (Table 11.2). This leads to improvements in the actual maximal running speed reached (~+6%) and time to cover 30 m (~+3%), while no gain in 10- and 20-m split times (Table 11.2). This case report well supports the positive effect of training exercises focusing on improving the

Table 11.3 Changes in sprint acceleration performance, horizontal power and force production capabilities and mechanical effectiveness for a NFL linebacker following 8 weeks of specific high-velocity sprint training

		Pre-training	Post-training	% change
Indexes of sprint acceleration performance	10-m time (s)	1.93	1.96	1.6
	20-m time (s)	3.22	3.21	−0.3
	30-m time (s)	4.42	4.33	−2.0
	Actual maximal velocity (m s^{-1})	8.22	8.7	5.8
Indexes of horizontal power and force production capabilities	F_0 (N kg^{-1})	10.2	9.2	−9.8
	v_0 (m s^{-1})	8.39	8.91	6.2
	P_{Hmax} (W kg^{-1})	21.5	20.5	−4.7
	S_{FV} (N s m^{-1} kg^{-1})	−1.22	−1.03	−15.1
Indexes of mechanical effectiveness of force application	RF_{mean} over 10 m (%)	31.9	33.1	3.8
	RF_{MAX} (%)	59	57	−3.4
	D_{RF} (% s m^{-1})	−11.1	−9.4	−15.3

ability to keep producing horizontal force at high velocities (and so with an upright position) on v_0 and D_{RF}, and in turn on long sprint acceleration performance.

Force-velocity profile versus Split times. Since 5-m splits and 30–40-m flying splits are well correlated with F_0 or v_0, respectively (personal data), some strength and conditioning coaches have used these split times or their ratio (5-m/30–40 m) to have information about force and velocity capabilities of their athletes/players. This could give an overall good view of the individual F-v profile in sprinting, but limited and approximate. Indeed, considering a performance (here sprint split times) as an interchangeable information for the underlying neuromuscular properties or physiological mechanisms (even if there are good correlations between them) can lead to inaccuracies, which can be acceptable for recreational athletes but risked for high level population. This is a bit like considering jump height is a good index of lower limbs maximal power, which is not exactly the case as discussed in the previous chapter on the optimal F-v profile (Chap. 5). Personal simulations performed using the published equations show that two players may have the same 5-m splits and different F-v profiles characterized by different F_0 (up to 10–15%). The F-v profile is interesting since it brings information about what causes the performance, and not the performance per se (Buchheit et al. 2014b). The latter depends on other factors than only the mechanical properties, as the body mass (same split times with two different body mass does not correspond to the same F-v profile) or the distance targeted for the split times. Moreover, split times do not differentiate between force production capability and the effectiveness of force application, which brings very interesting additional information, as illustrated in the previous case report. Finally, we have no idea towards which value should tend a short/long split ratio (5-m/30–40 m), while the optimization of the F-v profile in sprint is conceivable, as it has been done in jumping with the optimal F-v profile (current work in progress).

11.6.4 Hamstring Injury Prevention and Monitoring of the Return to Sport

Given (i) the role of hip extensors, and notably hamstrings, in horizontal force production during sprinting (see section about "the hip extensors hypothesis" in previous chapter) and (ii) the high occurrence of hamstring muscle injuries during high-speed and power actions such as sprinting, notably in soccer (Woods et al. 2004), it could have been expected that the alteration of hamstring muscle function (as before or after an injury) could affect FvP profile in sprinting. So, through collaborations leaded by Jurdan Mendiguchia, we have recently studied FvP profile in sprint in the context of hamstring injuries.

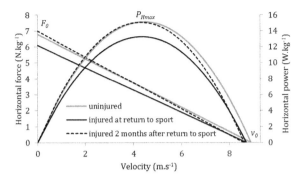

Fig. 11.10 Force- and power-velocity relationships (and associated F_0, v_0 and P_{Hmax}) of soccer players at the moment of return to sport after an hamstring muscle injury and 2 months after return to sport, and of uninjured soccer players. Lines represents averaged curves of the groups (standard deviations are not shown for clarity reasons) (from Mendiguchia et al. 2014)

FvP profile in sprint and return to sport after a hamstring injury. First, we showed that soccer players returning from a recent hamstring injury and being cleared to play, had substantial lower sprinting speed performance and reduced mechanical horizontal properties, notably F_0, compared with the uninjured players (Mendiguchia et al. 2014). Approximately two months of regular soccer training after return to sport, substantial improvements in sprinting speed (acceleration) were observed concomitantly with an increase in F_0 until similar levels as uninjured players, whereas v_0 remained unaltered (Fig. 11.10). Therefore, practitioners should consider assessing and training horizontal force production during sprint running after acute hamstring injuries in soccer players before return to sport. Second, the sensibility of F_0 to weakness or alteration of hamstring muscles were supported by the changes in FvP properties of sprinting in two injury case studies related to hamstring strain management (Mendiguchia et al. 2016). The Case #1 concerned a professional rugby player during a repeated sprint task (10 sprints of 40 m) when an injury occurred (5th sprint). The Case #2 refers to a professional soccer player prior to (8 days) and after (33 days) an acute hamstring injury. The results showed that F_0 was altered both before and after return to sport from a hamstring injury in these two elite athletes with little or no change in v_0. They also underlined that the simple on-field sprinting method was sensitive enough to indicate specific changes in horizontal mechanical properties pre- or pro-ceding an acute hamstring injury (the delay between the change observed in mechanical outputs and the injury might be as short as one or two sprints in the rugby player's case). So, practitioners should consider regularly monitoring horizontal force production during sprint running both from a performance and injury prevention perspective.

FvP profile in sprint and hamstring injury prevention. Given the previous results about association between horizontal force capabilities and hamstring muscles alterations, we hypothesized that lower and/or decrease in the horizontal force propulsion could reveal a functional weakness of hamstring muscles which could predispose them to an upcoming injury (Edouard et al. in submission process). This was tested on a cohort of 93 collegiate Japanese soccer players over an entire season with tests of sprint FvP profile every three months. Results showed that, in addition to the fact that previous hamstring injury was associated with higher risk of new hamstring injury (as previously reported), a lower F_0 was statistically related to a higher risk of sustaining a new hamstring injury. These findings, based only on tests over one season and 8 hamstring injuries, have to be confirmed by further prospective cohort study, which is currently in progress.

11.6.5 Better Understanding of the Limit of Human Sprinting Performance

Besides to be a practical method to be used for performance optimization and injury prevention, the sprint simple method can also help to bring new insights into the limits of human locomotion since it makes possible to estimate sprinting mechanical properties of the fastest men and women on earth of all time, without having testing them in the lab. In 2017, we compared the FvP profiles of women and men during 100-m finals of international events (World championships and Olympic Games) of the past 30 years based on split times provided by the IAAF Biomechanics reports after most of these competitions[1] (Slawinski et al. 2017). This comparison allowed to better understand the origins of differences in sprint running performances between men and women, notably during the acceleration phase. All the sprinting FvP profile variables were greater in men than in women. The $\sim 20\%$ higher $P_{H_{max}}$ values for men were explained by both $\sim 10\%$ higher F_0 (normalized to body mass) and v_0 values (Fig. 11.11). However, when standardized to inter-individual variability, the difference in v_0 between men and women is extremely large (effect size of 5.5) while the difference in F_0 is only moderate (effect size = 0.88). Moreover, only v_0 was correlated to 100-m performance, which means that the higher 100-m sprinting performances in men compared to women are mainly explained by a higher capability to keep producing horizontal force onto the ground at very high velocity, and thus to keep accelerating.

When we focus on the acceleration phase of some historical 100-m World records, we can have a roughly view of the evolution over time in the limit of human sprinting mechanical properties. Based on split and reaction times provided by the IAAF Biomechanics reports, sprinting FvP profiles has been estimated for the world records of Carl Lewis in 1991 (9.86 s, Tokyo), Maurice Greene in 1999 (9.79 s, Athens) and Usain Bolt in 2009 (9.58 s, Berlin), each of them representing a specific era in the sprint story. An approximate estimation of the FvP profile of Jesse Owens when he obtained the World record in 1936 (10.2 s, Chicago) has also been done, but with less reliability than the previous ones. Figure 11.12 presented their FvP profiles, and shows that the increase in 100-m performance is associated

---

[1] http://www.iaaf.org/.

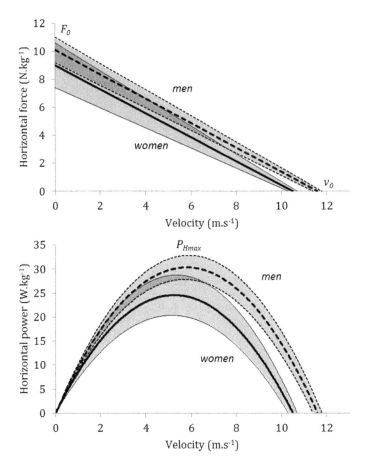

Fig. 11.11 Mean ± SD of force- and power-velocity relationships of world class women (solid lines) and men (dashed lines) athletes during 100-m finals of international events (World championships and Olympic Games) of the past 30 years based on split times provided by the International Association of Athletics Federations Biomechanics reports (from Slawinski et al. 2017)

to an increase in power capabilities and an overall shift to the top and the left of the F-v relationship. These changes were mainly due to changes in equipment (notably from forties to nineties with for instance spikes and Tartan[R]), training methods and loads or athlete's professionalization. It is worth noting that Maurice Green and Usain Bolt presented similar maximal power output during their World record, but with a different Fv profile: a Fv profile more oriented towards velocity capabilities for Bolt. Even if Bolt ran the 10 first meters slower than Green (1.74 s vs. 1.73 s, respectively, reaction time not included), he would have virtually been in front of Green from ∼20 m (2.76 s vs. 2.73 s, respectively, reaction time not included) to the finish line. This underlines the higher importance, in long sprint performance, of

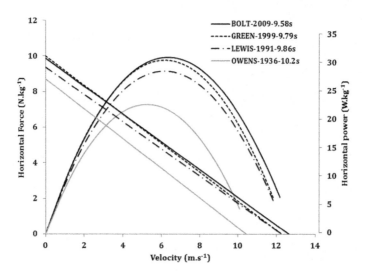

Fig. 11.12 Estimations of force- and power-velocity relationships of four historic 100-m World record holders: Jesse Owens in 1936 (10.2 s, Chicago), Carl Lewis in 1991 (9.86 s, Tokyo), Maurice Greene in 1999 (9.79 s, Athens), Usain Bolt in 2009 (9.58 s, Berlin). These sprinting mechanical properties have been estimated from split times of the acceleration phase of their respective World record run, which was provided by the International Association of Athletics Federations Biomechanics reports

the capability to keep producing horizontal force at high velocities than the capability to produce high level of horizontal force at low ones. This specific capability seems to be the main limit of human high speed bipedal locomotion.

11.7 Conclusion

This chapter presents an accurate and reliable simple method to determine the mechanical properties of the force production during sprinting. This method, based on a macroscopic biomechanical model and validated in laboratory conditions in comparison to force plate measurements, is very convenient for field use since it only requires anthropometric (body mass and stature) and spatio-temporal (split times or instantaneous velocity) input variables. It provides different information on the athlete's horizontal force production capabilities: maximal power output, maximal horizontal force, maximal velocity until which horizontal force can be produced and mechanical effectiveness of force application onto the ground. These information present interesting practical applications for sport practitioners to individualize training focusing on sprint acceleration performance, but also in hamstring injury prevention perspectives.

References

Arsac LM, Locatelli E (2002) Modeling the energetics of 100-m running by using speed curves of world champions. J Appl Physiol 92(5):1781–1788. https://doi.org/10.1152/japplphysiol.00754.2001

Aughey RJ (2011) Applications of GPS technologies to field sports. Int J Sports Physiol Perform 6(3):295–310

Barbero-Alvarez JC, Coutts A, Granda J, Barbero-Alvarez V, Castagna C (2010) The validity and reliability of a global positioning satellite system device to assess speed and repeated sprint ability (RSA) in athletes. J Sci Med Sport 13(2):232–235. https://doi.org/10.1016/j.jsams.2009.02.005

Bezodis NE, Salo AI, Trewartha G (2012) Measurement error in estimates of sprint velocity from a laser displacement measurement device. Int J Sports Med 33(6):439–444. https://doi.org/10.1055/s-0031-1301313

Buchheit M, Al Haddad H, Simpson BM, Palazzi D, Bourdon PC, Di Salvo V, Mendez-Villanueva A (2014a) Monitoring accelerations with GPS in football: time to slow down? Int J Sports Physiol Perform 9(3):442–445. https://doi.org/10.1123/ijspp.2013-0187

Buchheit M, Samozino P, Glynn JA, Michael BS, Al Haddad H, Mendez-Villanueva A, Morin JB (2014b) Mechanical determinants of acceleration and maximal sprinting speed in highly trained young soccer players. J Sports Sci 32(20):1906–1913. https://doi.org/10.1080/02640414.2014.965191

Cavagna GA, Komarek L, Mazzoleni S (1971) The mechanics of sprint running. J Physiol 217(3):709–721

Chelly SM, Denis C (2001) Leg power and hopping stiffness: relationship with sprint running performance. Med Sci Sports Exerc 33(2):326–333

Cross MR, Brughelli M, Brown SR, Samozino P, Gill ND, Cronin JB, Morin JB (2015) Mechanical properties of sprinting in elite rugby union and rugby league. Int J Sports Physiol Perform 10(6):695–702. https://doi.org/10.1123/ijspp.2014-0151

Cross MR, Brughelli M, Samozino P, Brown SR, Morin JB (2017) Optimal loading for maximising power during sled-resisted sprinting. Int J Sports Physiol Perform 1–25. https://doi.org/10.1123/ijspp.2016-0362

di Prampero PE, Botter A, Osgnach C (2015) The energy cost of sprint running and the role of metabolic power in setting top performances. Eur J Appl Physiol 115(3):451–469. https://doi.org/10.1007/s00421-014-3086-4

di Prampero PE, Fusi S, Sepulcri L, Morin JB, Belli A, Antonutto G (2005) Sprint running: a new energetic approach. J Exp Biol 208:2809–2816

Edouard P, Nagahara R, Samozino P, Rossi J, Brughelli M, Mendiguchia J, Morin J (under review) Is maximal horizontal force output during sprint acceleration associated with increased risk of hamstring muscle injuries in soccer: a pilot prospective study?

Furusawa K, Hill AV, Parkinson JL (1927) The dynamics of "sprint" running. Proc R Soc B 102:29–42

Haugen T, Buchheit M (2016) Sprint running performance monitoring: methodological and practical considerations. Sports Med 46(5):641–656. https://doi.org/10.1007/s40279-015-0446-0

Helene O, Yamashita MT (2010) The force, power and energy of the 100 meter sprint. Am J Phys 78:307–309

Henry FM (1954) Time-velocity equations and oxygen requirements of "all-out" and "steady-pace" running. Res Q 25:164–177

Jaskolska A, Goossens P, Veenstra B, Jaskolski A, Skinner JS (1999) Treadmill measurement of the force-velocity relationship and power output in subjects with different maximal running velocities. Sports Med Train Rehab 8:347–358

Jaskolski A, Veenstra B, Goossens P, Jaskolska A, Skinner JS (1996) Optimal resistance for maximal power during treadmill running. Sports Med Train Rehabil 7:17–30

Jennings D, Cormack S, Coutts AJ, Boyd LJ, Aughey RJ (2010) Variability of GPS units for measuring distance in team sport movements. Int J Sports Physiol Perform 5(4):565–569

Kawamori N, Newton R, Nosaka K (2014) Effects of weighted sled towing on ground reaction force during the acceleration phase of sprint running. J Sports Sci 32(12):1139–1145. https://doi.org/10.1080/02640414.2014.886129

Lockie RG, Murphy AJ, Schultz AB, Jeffriess MD, Callaghan SJ (2013) Influence of sprint acceleration stance kinetics on velocity and step kinematics in field sport athletes. J Strength Cond Res 27(9):2494–2503. https://doi.org/10.1519/JSC.0b013e31827f5103

Mendiguchia J, Edouard P, Samozino P, Brughelli M, Cross M, Ross A, Gill N, Morin JB (2016) Field monitoring of sprinting power-force-velocity profile before, during and after hamstring injury: two case reports. J Sports Sci 34(6):535–541. https://doi.org/10.1080/02640414.2015.1122207

Mendiguchia J, Samozino P, Martinez-Ruiz E, Brughelli M, Schmikli S, Morin JB, Mendez-Villanueva A (2014) Progression of mechanical properties during on-field sprint running after returning to sports from a hamstring muscle injury in soccer players. Int J Sports Med 35(8):690–695. https://doi.org/10.1055/s-0033-1363192

Morin JB, Bourdin M, Edouard P, Peyrot N, Samozino P, Lacour JR (2012) Mechanical determinants of 100-m sprint running performance. Eur J Appl Physiol 112(11):3921–3930. https://doi.org/10.1007/s00421-012-2379-8

Morin JB, Edouard P, Samozino P (2011a) Technical ability of force application as a determinant factor of sprint performance. Med Sci Sports Exerc 43(9):1680–1688

Morin JB, Jeannin T, Chevallier B, Belli A (2006) Spring-mass model characteristics during sprint running: correlation with performance and fatigue-induced changes. Int J Sports Med 27(2):158–165. https://doi.org/10.1055/s-2005-837569

Morin JB, Petrakos G, Jimenez-Reyes P, Brown SR, Samozino P, Cross MR (2017) Very-heavy sled training for improving horizontal-force output in soccer players. Int J Sports Physiol Perform 12(6):840–844. https://doi.org/10.1123/ijspp.2016-0444

Morin JB, Samozino P (2016) Interpreting power-force-velocity profiles for individualized and specific training. Int J Sports Physiol Perform 11(2):267–272

Morin JB, Samozino P, Bonnefoy R, Edouard P, Belli A (2010) Direct measurement of power during one single sprint on treadmill. J Biomech 43(10):1970–1975

Morin JB, Samozino P, Edouard P, Tomazin K (2011b) Effect of fatigue on force production and force application technique during repeated sprints. J Biomech 44(15):2719–2723. https://doi.org/doi:10.1016/j.jbiomech.2011.07.020 (S0021-9290(11)00526-4 [pii])

Nagahara R, Botter A, Rejc E, Koido M, Shimizu T, Samozino P, Morin JB (2017) Concurrent validity of GPS for deriving mechanical properties of sprint acceleration. Int J Sports Physiol Perform 12(1):129–132. https://doi.org/10.1123/ijspp.2015-0566

Petrakos G, Morin JB, Egan B (2016) Resisted sled sprint training to improve sprint performance: a systematic review. Sports Med 46(3):381–400. https://doi.org/10.1007/s40279-015-0422-8

Rabita G, Dorel S, Slawinski J, Saez de villarreal E, Couturier A, Samozino P, Morin JB (2015) Sprint mechanics in world-class athletes: a new insight into the limits of human locomotion. Scand J Med Sci Sports. https://doi.org/10.1111/sms.12389

Rampinini E, Alberti G, Fiorenza M, Riggio M, Sassi R, Borges TO, Coutts AJ (2015) Accuracy of GPS devices for measuring high-intensity running in field-based team sports. Int J Sports Med 36(1):49–53. https://doi.org/10.1055/s-0034-1385866

Romero-Franco N, Jimenez-Reyes P, Castano-Zambudio A, Capelo-Ramirez F, Rodriguez-Juan JJ, Gonzalez-Hernandez J, Toscano-Bendala FJ, Cuadrado-Penafiel V, Balsalobre-Fernandez C (2016) Sprint performance and mechanical outputs computed with an iPhone app: comparison with existing reference methods. Eur J Sport Sci:1–7. https://doi.org/10.1080/17461391.2016.1249031

Samozino P, Morin JB, Hintzy F, Belli A (2008) A simple method for measuring force, velocity and power output during squat jump. J Biomech 41(14):2940–2945

Samozino P, Morin JB, Hintzy F, Belli A (2010) Jumping ability: a theoretical integrative approach. J Theor Biol 264(1):11–18

Samozino P, Rabita G, Dorel S, Slawinski J, Peyrot N, Saez de Villarreal E, Morin JB (2016) A simple method for measuring power, force, velocity properties, and mechanical effectiveness in sprint running. Scand J Med Sci Sports 26(6):648–658. https://doi.org/10.1111/sms.12490

Samozino P, Rejc E, Di Prampero PE, Belli A, Morin JB (2012) Optimal force-velocity profile in ballistic movements. Altius: citius or fortius? Med Sci Sports Exerc 44(2):313–322

Simperingham KD, Cronin JB, Ross A (2016) Advances in sprint acceleration profiling for field-based team-sport athletes: utility, reliability, validity and limitations. Sports Med 46 (11):1619–1645. https://doi.org/10.1007/s40279-016-0508-y

Slawinski J, Bonnefoy A, Ontanon G, Leveque JM, Miller C, Riquet A, Cheze L, Dumas R (2010) Segment-interaction in sprint start: analysis of 3D angular velocity and kinetic energy in elite sprinters. J Biomech 43(8):1494–1502. https://doi.org/10.1016/j.jbiomech.2010.01.044

Slawinski J, Termoz N, Rabita G, Guilhem G, Dorel S, Morin JB, Samozino P (2017) How 100-m event analyses improve our understanding of world-class men's and women's sprint performance. Scand J Med Sci Sports 27(1):45–54. https://doi.org/10.1111/sms.12627

van Ingen Schenau GJ, Jacobs R, de Koning JJ (1991) Can cycle power predict sprint running performance? Eur J Appl Physiol Occup Physiol 63(3–4):255–260

Woods C, Hawkins RD, Maltby S, Hulse M, Thomas A, Hodson A, Football Association Medical Research P (2004) The football association medical research programme: an audit of injuries in professional football-analysis of hamstring injuries. Br J Sports Med 38(1):36–41

Chapter 12
The Energy Cost of Sprint Running and the Energy Balance of Current World Records from 100 to 5000 m

Pietro E. di Prampero and Cristian Osgnach

Abstract The time course of metabolic power during 100–400 m top running performances in world class athletes was estimated assuming that accelerated running on flat terrain is biomechanically equivalent to uphill running at constant speed, the slope being dictated by the forward acceleration. Hence, since the energy cost of running uphill is known, energy cost and metabolic power of accelerated running can be obtained, provided that the time course of the speed is determined. Peak metabolic power during the 100 and 200 m current world records (9.58 and 19.19 s) and during a 400 m top performance (44.06 s) amounted to 163, 99 and 75 W kg^{-1}, respectively. Average metabolic power and overall energy expenditure during 100–5000 m current world records in running were also estimated as follows. The energy spent in the acceleration phase, as calculated from mechanical kinetic energy (obtained from average speed) and assuming 25% efficiency for the transformation of metabolic into mechanical energy, was added to the energy spent for constant speed running (air resistance included). In turn, this was estimated as: $(3.8 + k' v^2) \cdot d$, where 3.8 J kg^{-1} m^{-1} is the energy cost of treadmill running, k′ = 0.01 J s^2 kg^{-1} m^{-3}, v is the average speed (m s^{-1}) and d (m) the overall distance. Average metabolic power decreased from 73.8 to 28.1 W kg^{-1} with increasing distance from 100 to 5000 m. For the three shorter distances (100, 200 and 400 m), this approach yielded results rather close to mean metabolic power values obtained from the more refined analysis described above. For distances between 1000 and 5000 m the overall energy expenditure increases linearly with the corresponding world record time. The slope and intercept of the regression are assumed to yield maximal aerobic power and maximal amount of energy derived from anaerobic stores in current world records holders; they amount to 27 W kg^{-1} (corresponding to a maximal O$_2$ consumption of 77.5 ml O$_2$ kg^{-1} min^{-1} above resting) and 1.6 kJ kg^{-1} (76.5 ml O$_2$ kg^{-1}). This last value is on the same order of the maximal amount of energy that can be derived from complete utilisation of

---

P. E. di Prampero (✉)
Department of Medical Sciences, University of Udine, 33100 Udine, Italy
e-mail: pietro.prampero@uniud.it

C. Osgnach
Department of Sport Sciences, Exelio srl, 33100 Udine, Italy

© Springer International Publishing AG 2018
J.-B. Morin and P. Samozino (eds.), *Biomechanics of Training and Testing*,
https://doi.org/10.1007/978-3-319-05633-3_12

phosphocreatine in the active muscle mass and from maximal tolerable blood lactate accumulation. The anaerobic energy yield has also been estimated, throughout the overall set of distances (100–5000 m), assuming that, at work onset, the rate of O_2 consumption increases with a time constant of 20 s tending to the appropriate metabolic power, but stops increasing once the maximal O_2 consumption is attained. Hence the overall energy expenditure can be partitioned into its aerobic and anaerobic components. This last increases from about 0.6 kJ kg^{-1} for the shortest distance (100 m) to a maximum close to that estimated above (1.6 kJ kg^{-1}) for distances of 1500 m or longer.

12.1 Introduction

Since the beginning of the last century, the analysis of world records has fascinated scientists dealing with muscle and exercise physiology (e.g. Hill 1925) because, on one side, world records represent the best possible performances in any given sport event in any specific point in time; on the other, the assessment of world record times and speeds is by several order of magnitude more accurate than any possible laboratory measurement. Following this time honoured tradition, we will here present an energetic analysis of the current world records in running from 100 to 5000 m. Necessary prerequisite of the matter at stake will be the discussion of a recently developed approach to estimate the energy cost of sprint running, an obviously crucial requirement when dealing with the shorter distances.

Hence, the present chapter is organised as follows.

We will first describe a model proposed by di Prampero et al. (2005) wherein accelerated/decelerated running on flat terrain is considered biomechanically equivalent to uphill/downhill running at constant speed, the slope being dictated by the forward acceleration. If this is so, since the energy cost of uphill/downhill running at constant speed is fairly well know, it is a rather straightforward matter to estimate the energy cost of accelerated/decelerated running, once the acceleration is determined. This will allow us to estimate peak metabolic power and overall energy expenditure during top running performances over 100–400 m distances.

We will then discuss a model to estimate the overall energy spent over 100–5000 m covered at current world record speed, based on a simplified assessment of the energy spent to accelerate the runner's body in the initial part of the race. We will show that, for the three shorter distances (100, 200 and 400 m), wherein, thanks to the availability of the time course of the speed throughout the race, the more refined analysis described above can be applied, the two approaches yield rather close estimates of overall energy expenditure and mean metabolic power.

The so obtained overall energy expenditure over distances between 100 and 5000 m will then be plotted as a function of the corresponding world record time. The slope and intercept of the resulting straight line regression, as obtained for the five longer distances (1000–5000 m) will be interpreted to yield an estimate of the maximal aerobic power and of the maximal amount of energy derived from anaerobic stores in the current world records holders over these distances. The resulting values turn out to be 27 W kg^{-1} (corresponding to a maximal O$_2$ consumption of 77.5 ml O$_2$ kg^{-1} min^{-1} above resting) and to 1.6 kJ kg^{-1} (76.5 ml O$_2$ kg^{-1}). It is fair to point out here that this approach is conceptually similar to that proposed by B. B. Lloyd in 1966 for assessing the maximal aerobic speed and the maximal distance covered at the expense of anaerobic energy.

The anaerobic energy yield will also be estimated, throughout the overall set of investigated distances (100–5000 m), assuming that, at work onset, the rate of O$_2$ consumption increases with a time constant of 20 s tending to the appropriate metabolic power, but stops increasing once the maximal O$_2$ consumption is attained. This will allow us to estimate the overall amount of energy derived from aerobic sources wherefrom, since the overall energy spent over the corresponding distance is known, the energy yield from the anaerobic sources is easily obtained; it increases from about 0.6 kJ kg^{-1} for the shortest distance (100 m) to attain a maximal value close to that estimated above (1.6 kJ kg^{-1}) for distances of 1500 m or longer (1 mile, 2000, 3000 and 5000 m).

It can therefore be concluded that the maximal capacity of the anaerobic stores in world class athletes is on the order of 1.6 kJ kg^{-1}. This is compatible with the maximal amount of energy that can be derived from complete utilisation of phosphocreatine in the active muscle mass and from maximal tolerable blood lactate accumulation.

12.2 The Energy Cost of Sprint Running

Direct measurements of energy expenditure during sprint running are rather problematic because of the massive utilization of anaerobic sources and because of the resulting short duration of any such events that prevents the attainment of a steady state. Indeed, so far the energetics of sprint running has mainly been estimated indirectly from biomechanical analyses, on the bases of assumed overall efficiencies of metabolic to mechanical energy transformation (Cavagna et al. 1971; Fenn 1930a, b; Kersting 1998; Mero et al. 1992; Murase et al. 1976; Plamondon and Roy 1984), or else assessed by means of rather indirect procedures (Arsac 2002; Arsac and Locatelli 2002; di Prampero et al. 1993; Summers 1997; van Ingen Schenau et al. 1991, 1994; Ward-Smith and Radford 2000).

An alternative approach is to assume that sprint running on flat terrain, during the acceleration phase is biomechanically equivalent to uphill running at constant speed, the slope being dictated by the forward acceleration and that, conversely, during the deceleration phase it is biomechanically equivalent to running downhill. If this is so, since the energy cost of uphill (downhill) running at constant speed is rather well know over a fairly large range of inclines, it is a rather straightforward matter to estimate the energy cost of accelerated (decelerated) running, once the acceleration (deceleration) is known.

12.3 Theory

The theory underlying the analogy mentioned above is summarised graphically in Fig. 12.1, as from the original paper by di Prampero et al. (2005); it has been recently reviewed by di Prampero et al. (2015) and is briefly sketched below, the interested reader being referred to the original papers.

Figure 12.1 (left panel) shows that a runner accelerating on flat terrain must lean forward in such a way that the angle α between his/her mean body axis and the terrain is smaller the greater the forward acceleration (a_f). This state of affairs is analogous to running uphill at constant speed, provided that the angle α between mean body axis and terrain is unchanged (Fig. 12.1, right panel). It necessarily follows that the complement of α, i.e. the angle between the terrain and the horizontal (90 − α), increases with a_f.

The incline of the terrain is generally expressed by the tangent of the angle between the terrain and the horizontal (90 − α). As shown in Fig. 12.1 (left panel), this is the ratio of the segment AB to the acceleration of gravity (g):

$$\tan(90-\alpha) = AB/g \tag{12.1}$$

In turn since the length of the segment AB is equal to a_f, Eq. (12.1) can be rewritten as:

$$\tan(90-\alpha) = a_f/g = ES \tag{12.2}$$

where ES is the tangent of the angle which makes accelerated running (Fig. 12.1, left panel) biomechanically equivalent to running at constant speed up a corresponding slope (Fig. 12.1, right panel), hence the definition Equivalent Slope (ES).

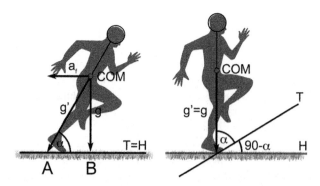

Fig. 12.1 The subject is accelerating forward while running on flat terrain (left) or running uphill at constant speed (right). COM, subject's centre of mass; a_f, forward acceleration; g, acceleration of gravity; $g' = \sqrt{(a_f^2 + g^2)}$, vectorial sum of a_f plus g; T, terrain; H, horizontal; α, angle between the runner's mean body axis throughout the stride and T; 90 − α, angle between T and H. See text for details

12 The Energy Balance of Current World Records from 100 to 5000 m

Inspection of Fig. 12.1 also shows that, in addition to being equivalent to running uphill, accelerated running is characterised by yet another difference, as compared to constant speed running. Indeed, the force that the runner must develop (average throughout a whole stride), as given by the product of the body mass and the acceleration, is greater in the former case (= M · g') as compared to the latter one (= M · g), because g' > g (Fig. 12.1, left panel). Thus, accelerated running is equivalent to uphill running wherein, however, the body mass is increased in direct proportion to the ratio g'/g. Since $g' = \sqrt{(a_f^2 + g^2)}$, this ratio, which will here be defined "equivalent body mass" (EM) is described by:

$$EM = M \cdot g'/M \cdot g = \sqrt{(a_f^2 + g^2)}/g = \sqrt{[(a_f^2/g^2) + 1]} \qquad (12.3)$$

Substituting Eq. (12.2) into Eq. (12.3), one obtains:

$$EM = \sqrt{[(a_f^2/g^2) + 1]} = \sqrt{(ES^2 + 1)} \qquad (12.4)$$

It must also be pointed out that during decelerated running, which is equivalent to downhill running, and in which case the equivalent slope (ES) is negative, EM will nevertheless assume a positive value because ES in Eq. (12.4) is raised to the power of 2. It can be concluded that, if the time course of the velocity during accelerated/decelerated running is determined, and the corresponding instantaneous accelerations/decelerations calculated, Eqs. (12.2) and (12.4) allow one to obtain the appropriate ES and EM values, thus converting accelerated/decelerated running into the equivalent constant speed uphill/downhill running. Hence, if the energy cost of this last is also known, the corresponding energy cost of accelerated/decelerated running can be easily obtained.

The energetics of running at constant speed, on the level, uphill or downhill, has been extensively investigated since the second half of the XIX century (for references see Margaria 1938; di Prampero 1986). Minetti et al. (2002) determined the energy cost of running at constant speed over the widest range of inclines studied so far (at least to our knowledge): from −0.45 to +0.45 and showed that throughout this whole range of inclines the energy cost of running per unit body mass and distance (C_r) is independent of the speed and is described by the polynomial equation that follows (Fig. 12.2):

$$C_r = 155.4 \cdot i^5 - 30.4 \cdot i^4 - 43.3 \cdot i^3 + 46.3 \cdot i^2 + 19.5 \cdot i + 3.6 \qquad (12.5)$$

where C_r is expressed in J kg^{-1} m^{-1}, i is the incline of the terrain, i.e. the tangent of the angle between the terrain and the horizontal, and the last term (3.6 J kg^{-1} m^{-1}) is the energy cost of constant speed running on compact flat terrain. Thus, substituting i with the equivalent slope (ES), denoting C_0 the energy cost of constant speed level running, and multiplying by the equivalent body mass (EM), Eq. (12.5) can be rewritten as:

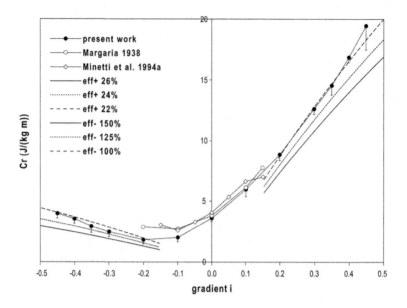

Fig. 12.2 Energy cost of running at constant speed, C_r (J kg^{-1} m^{-1}), as a function of the incline (i) of the terrain. The function interpolating the full dots is described by Eq. (12.5). Straight lines irradiating from the origin indicate net efficiency of work against gravity, the values of which are indicated in the inset. Open dots and diamonds are data from previous studies (from Minetti et al. 2002)

$$C_r = (155.4 \cdot ES^5 - 30.4 \cdot ES^4 - 43.3 \cdot ES^3 + 46.3 \cdot ES^2 + 19.5 \cdot ES + C_0) \cdot EM$$
(12.6)

Equation (12.6) allows one to estimate the energy cost of accelerated running provided that the instantaneous velocity, the corresponding acceleration values, and hence ES and EM, are known. Strictly speaking the individual value of C_0 should also be known; however, in view of its relatively minor interindividual variation (Lacour and Bourdin 2015), it is often convenient to utilise an average value from the literature, a fact that it is not likely to greatly affect the final outcome.

This approach was originally proposed by di Prampero et al. (2005) and applied to 12 medium level sprinters [best performance time over 100 m: 11.30 s (±0.35, SD)]. It was then utilised by Osgnach et al. (2010) to determine metabolic power and energy expenditure in elite soccer players during official games and it was recently extended to evaluate Usain Bolt's 100 m world record performance (9.58 s, Berlin, 2009) (di Prampero et al. 2015). Throughout these studies, the energy cost of running was obtained by means of Eq. (12.6), wherefrom the metabolic power, as given by the product of C_r and speed, was obtained.

However, it should be noted that Eq. (12.6) applies only within the range of inclines actually studied by Minetti et al. (2002). Indeed, whenever the calculated ES is far from the maximal (+0.45) or minimal (−0.45) inclines on which the equation was based the obtained values of C_r become unreasonably high, or even negative (in the case of large down-slopes). This state of affairs is not likely to affect greatly the data obtained on medium level sprinters or in soccer players in which cases the maximal values of ES very rarely exceed 0.50 and even when they do the corresponding duration is substantially less than 1 s. In the case of Bolt's 100 m record, however, the maximal value of ES attained at the very beginning of the race was on the order of 0.85 and, even if it reached the canonical values within the first second (di Prampero et al. 2015), as pointed out in the paper, it led to an astonishingly high value of C_r and of maximal metabolic power.

In view of these considerations, the aim of the section that follows is to recalculate the values of C_r and of metabolic power throughout a 100 m dash on medium level sprinter and on Usain Bolt, utilising a more conservative approach, but still based on the data obtained by Minetti et al. (2002) during uphill running at constant speed. Specifically, Eq. (12.6), will be used to estimate C_r whenever the calculated ES lies within the range of inclines on which the equation was based ($-0.45 \geq$ ES $\leq$ +0.45); for ES > 0.45 C_r will be estimated on the basis of the equation that follow:

$$C_r = (55.65 \cdot ES - 5.61) \cdot EM \qquad (12.7)$$

In turn this equation describes the tangent to the C_r versus incline data reported in Fig. 12.2 for i = +0.45.

This same approach will also be applied to calculate the time course of the metabolic power for the 200 m world record of Usain Bolt (19.19 s, Berlin, 2009) and of the top 400 m performance of L. S. Merritt (44.06 s) at the 2009 IAAF World Championship in Berlin.

12.4 Methods

In all cases, the velocity was assumed to increase exponentially during the acceleration phase, as described by:

$$v(t) = v(f) \cdot (1 - e^{-t/\tau}) \qquad (12.8)$$

where v(t) is velocity at time t, v(f) the peak velocity, and τ the appropriate time constant, the values of which are reported in Table 12.1, together with the corresponding v(f).

In turn the values of τ reported in Table 12.1 for the 100 m (MLS and U. Bolt) were calculated interpolating the velocity data obtained by a radar system, (di Prampero et al. 2005; Hernandez Gomez et al. 2013). For the 200 and 400 m (U. Bolt and L. S. Merritt) the velocity values utilised for estimating τ were those reported over 50 m intervals by Graubner and Nixdorf (2011), up to the attainment of the highest speed. The speed values, as calculated from Eq. (12.8) and Table 12.1, are reported in Fig. 12.3 for Bolt's and Merritt's performances, together with the actually measured ones.

Table 12.1 Peak velocity (v(f), m s^{-1}) over the indicated distances (m), together with the corresponding time constants (τ) are reported for U. Bolt's 100 and 200 m current world records, L. S. Merritt's 400 m top performance (44.04 s, Berlin, 2009) and for 12 medium level sprinters (MLS) (best time over 100 m = 11.30 s ± 0.30)

Distance (m)	Athlete	τ (s)	v(f) (m s^{-1})
100	MLS	1.42	9.46
100	Bolt	1.25	12.35
200	Bolt	1.60	11.57
400	Merritt	1.51	9.88

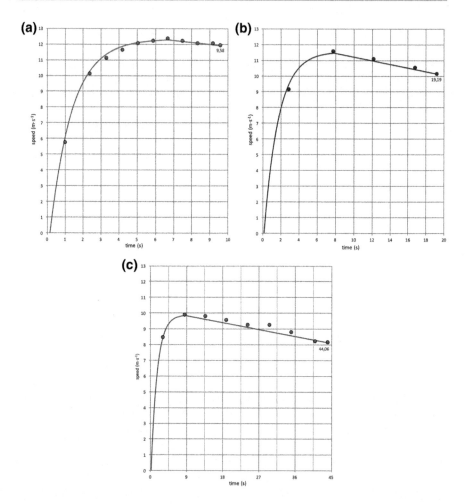

Fig. 12.3 Time course of the speed during the 100 and 200 m world records by U. Bolt (panels A and B) and the 400 m top performance by L. S. Merritt (panel C) (Berlin, 2009), as calculated with the aid of Eq. (12.8) from the τ and v(f) values reported in Table 12.1 (continuous line). Full dots: actually measured speed. After the attainment of v(f) actual speed data were linearly interpolated. See text for details and references

The time course of the acceleration was then obtained from the first derivative of Eq. (12.8):

$$a(t) = (v(f) - v(t))/\tau = [v(f) - v(f) \cdot (1 - e^{-t/\tau})]/\tau \qquad (12.9)$$

where a(t) is the acceleration at time t and all other terms have been defined above.

Equations (12.8) and (12.9) allowed us to estimate energy cost of running [as from Eqs. (12.6) and (12.7)], and hence the instantaneous metabolic power (as given by the product of energy cost and speed) during the acceleration phase. In the case of the 200 and 400 m performances, for times greater than that corresponding to peak speed, the energy cost of running was assumed to be equal to C_0, that, in this latter case as well as in Eq. (12.6), was assumed to be 3.8 J kg^{-1} m^{-1}. In all cases the so obtained values of C_r were corrected for the energy spent against the air resistance (per unit body mass and distance, J kg^{-1} m^{-1}), as given by: $0.01 \cdot v(t)^2$, where v(t) is expressed in m s^{-1} (Pugh 1970; di Prampero 1986).

12.5 Metabolic Power and Overall Energy Expenditure

The time course of the metabolic power is reported in Fig. 12.4 for Bolt's (100 and 200 m) current world records and Merritt's 400 m top performance, together with the average for Medium Level Sprinters (MLS) over 100 m, as estimated from the data by di Prampero et al. (2005). The time integral of the so obtained metabolic power reported in Fig. 12.4 allowed us to estimate the overall energy expenditure (air resistance included) to cover the distances in question as well as the corresponding average power; they are reported in Table 12.2, together with peak power values.

The peak metabolic power reported in Table 12.2 for U. Bolt 100 m world record is essentially equal to the calculated for Bolt himself by Beneke and Taylor (2010) and by di Prampero et al. (2005) for C. Lewis, winner of the gold medal over same distance, with 9.92 s, at the 1988 Olympic Games in Seoul. It is, however, substantially less than that estimated in a previous study by di Prampero et al. (2015) which amounted to 197 W kg^{-1}. Indeed, this last value was obtained on the basis of Minetti et al.'s polynomial Eq. (12.6) which, as mentioned in the paper itself and briefly discussed above, overestimates the energy cost of accelerated running for equivalent slopes (ES) greater than 0.5, whereas in this study, for ES > 0.45 the energy cost of running was obtained by means of Eq. (12.7). It is also interesting to note that the peak power of medium level sprinters at the onset of a 100 m dash is about 50% of Bolt's value (see Fig. 12.4) for an average performance time over the same distance of 11.3 s, corresponding to an average speed only about 15% slower than Bolt's record. This highlights the fact that top performances over short distances require large accelerations in the initial phase of the run, a prerequisite for achieving this feat, in addition to the appropriate anthropometric and biomechanical characteristics (Charles and Bejan 2009; Maćkała and

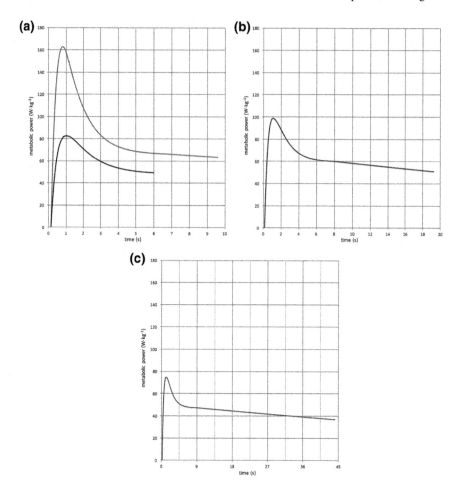

Fig. 12.4 Time course of metabolic power during the 100 and 200 m world records by U. Bolt (panels **a** and **b**) and the 400 m top performance by L. S. Merritt (panel **c**). The lower curve in Panel a is the average time course of metabolic power during the first 6 s of a 100 m dash in 12 medium level sprinters. See text for details and references

Mero 2013; Taylor and Beneke 2012), being the capability to develop very large metabolic power outputs in the shortest possible time.

The procedure to estimate total energy expenditure and average metabolic power values reported in Table 12.2 requires knowledge of the time course of the speed and hence of the acceleration. These data are not always available, nor are they easy to obtain. The paragraphs that follow are therefore devoted to utilise an alternative approach based on the average speed only, as originally proposed by di Prampero et al. (1993) (see also Hautier et al. 2010; Rittweger et al. 2009). It will be shown that this approach yields total energy expenditure and average metabolic power values rather close to those estimated (for the 100–400 m distances) on the basis of the more rigorous procedure described above.

Table 12.2 Overall energy expenditure (E_{tot}*, kJ kg^{-1}), peak and average metabolic power (P_{met}, W kg^{-1} or ml O$_2$ kg^{-1} min^{-1} over the indicated distances covered at (or close to) world record speed, as obtained from the time integral of the metabolic power curves reported in Fig. 12.4)

Distance (m)	Time (s)	Athlete	E_{tot}* (kJ kg^{-1})	Metabolic power[a]			
				Average		Peak	
				(W kg^{-1})	(ml O$_2$ kg^{-1} min^{-1})	(W kg^{-1})	(ml O$_2$ kg^{-1} min^{-1})
100	9.58	Bolt	0.757	79.0	226.8	163.0	467.9
200	19.19	Bolt	1.143	59.6	171.1	99.9	286.8
400	44.06	Merritt	1.885	43.6	125.2	74.9	215.0

[a]Above resting

Table 12.3 Overall energy expenditure (E_{tot}, kJ kg^{-1}) over the indicated distances covered at (or close to) current world record velocity in running as from Table 12.2 (E_{tot}*) is indicated together with the corresponding values obtained via the simplified procedure described in the text [$E_{tot}°$, as from Eq. (12.12); $E_{tot}^\wedge$, as from Eq. (12.12')]

Distance (m)	Athlete	E_{tot}* (kJ kg^{-1})	$E_{tot}°$ (kJ kg^{-1})	$E_{tot}^\wedge$ (Eq. 12.12') (kJ kg^{-1})
100	Bolt	0.757	0.707	0.794
200	Bolt	1.143	1.195	1.246
400	Merritt	1.885	2.035	2.059

*Time integral of metabolic power; ° from Eq. (12.12); ^ from Eq. (12.12')

It will be assumed that the additional energy spent in the acceleration phase (E_{acc}), over and above that for constant speed running, is described by:

$$E_{acc} = M v_{mean}^2 / (2 \cdot \eta) \qquad (12.10)$$

where M is the subject's body mass, v_{mean} the average speed and η the efficiency of converting metabolic into mechanical energy. Thus, expressing E_{acc} per unit body mass and assuming $\eta = 0.25$ (Cavagna and Kaneko 1977; Cavagna et al. 1971), Eq. (12.10) can be rewritten as:

$$E_{acc}/M = v_{mean}^2 / 0.5 = E_{acc}sp \qquad (12.11)$$

If this is so, the overall energy spent to cover any given distance (d) from a still start, per unit body mass, is given by:

$$E_{tot} = (C_0 + k' v_{mean}^2) \cdot d + E_{acc}sp = (C_0 + k' v_{mean}^2) \cdot d + v_{mean}^2 / 0.5 \qquad (12.12)$$

The so obtained values (above resting), for $C_0 = 3.8$ (J kg^{-1} m^{-1}) and $k' = 0.01$ (J s^2 kg^{-1} m^{-3}), are shown in Table 12.3, where the last column reports the E_{tot} values obtained replacing, in the last term of Eq. (12.12), average with peak speed (v_{peak}):

$$E_{tot} = (C_0 + k' v_{mean}^2) \cdot d + E_{acc}sp = (C_0 + k' v_{mean}^2) \cdot d + v_{peak}^2 / 0.5 \qquad (12.12')$$

Table 12.3 shows that the ratio between the E_{tot} values obtained from the time integral of the metabolic power curve and those estimated from the simplified procedures described above for the 100–400 m range from 1.07 to 0.93 if E_{tot} is calculated from the mean speed (Eq. 12.12) or from 0.89 to 0.99 if it is calculated from the peak speed (Eq. 12.12'). In view of the fact that the mean speed is easily available, whereas this is not always the case for peak speed, we will use the simplified procedure described by Eq. (12.12) to estimate the overall energy spent, as well as the mean metabolic power requirement, for the current world records over distances from 100 to 5000 m; they are reported in Table 12.4.

Table 12.4 Overall energy expenditure (E_{tot}, kJ kg^{-1}), as obtained according to the simplified procedure (Eq. 12.12) for the indicated distances covered in current world record (WR) times (s) are reported together with corresponding mean metabolic power (P_{met}, W kg^{-1})

Distance (m)	E_{tot} (kJ kg^{-1})	WR time (s)	Mean power (W kg^{-1})	Mean $\dot{V}O_2$ (W kg^{-1})	
				For $\dot{V}O_2$max 22 W kg^{-1}*	For $\dot{V}O_2$max 27 W kg^{-1}**
100	0.707	9.58	73.8	14.5	15.1
200	1.195	19.19	62.2	17.3	19.8
400	2.035	43.18	47.1	19.1	22.4
800	3.669	100.91	36.3	20.3	24.2
1000	4.489	131.96	34.0	20.6	24.6
1500	6.599	206.00	32.0	21.0	25.3
1 mile	7.057	223.13	31.6	21.1	25.4
2000	8.685	284.79	30.5	21.2	25.6
3000	12.883	446.67	29.2	21.5	26.0
5000	21.266	757.35	28.1	21.7	26.4

Last two columns are the mean O_2 consumption above resting (W kg^{-1}) estimated as described in the text for the indicated $\dot{V}O_2$max values
*63.2 ml O_2 kg^{-1} min^{-1} above resting; **77.5 ml O_2 kg^{-1} min^{-1} above resting.
1 mile = 1609.35 m

12.6 Aerobic Versus Anaerobic Energy Expenditure

The last two columns of Table 12.4 report the mean $\dot{V}O_2$ throughout the entire competition calculated as described in detail elsewhere (di Prampero et al. 2015). Suffice it here to say that $\dot{V}O_2$ at the start of the race was assumed to increase exponentially with a time constant of 20 s, tending to the average power requirement, without increasing any further once $\dot{V}O_2$max had been attained (Margaria et al. 1965). This procedure allowed us to estimate the average $\dot{V}O_2$ throughout the race for any given $\dot{V}O_2$max value (see Table 12.4 and Fig. 12.5). The difference between the average values of metabolic power and of $\dot{V}O_2$ must be fulfilled by the anaerobic stores, the overall energetic contribution of which can therefore be easily obtained from the product of this difference and the time of the race; it is reported in Fig. 12.6 for the two selected $\dot{V}O_2$max values of 22 and 27 W kg^{-1}.

The average $\dot{V}O_2$ throughout the 100–400 m world records has also been calculated as described above, but replacing the average metabolic power with its actual time course, as reported in Fig. 12.4. For these three distances, therefore, the contribution of the anaerobic stores to the overall energy requirement could also be calculated from the difference between the time integrals of the metabolic power and the estimated $\dot{V}O_2$ curves; they are not substantially different than those obtained by the simplified procedure described in the preceding paragraphs.

Figure 12.6 shows that the anaerobic contribution to the overall energy requirement, for $\dot{V}O_2$max on the order of 26–27 W kg^{-1} (74.6–77.5 ml O_2 kg^{-1} min^{-1}

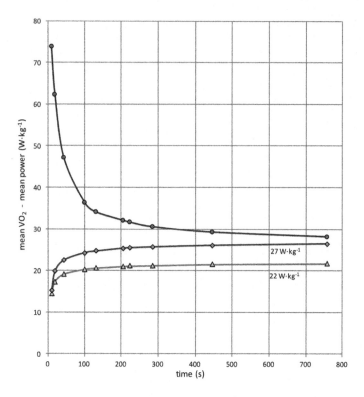

Fig. 12.5 Average metabolic power (mean power, W kg^{-1}) as a function of the current world record time (s) over 100–5000 m distance (blue line and dots). Diamonds and red line, triangles and green line are mean $\dot{V}O_2$ calculated as described in the text, for $\dot{V}O_2$ max of 27 (red line and diamonds) or 22 (green line and triangles) W kg^{-1}. See also Table 12.4 (Color figure online)

above resting) tends to a plateau of about 1.6 kJ kg^{-1}, whereas for smaller $\dot{V}O_2$max values it keeps increasing without reaching an asymptote. This suggests that the maximal capacity of the anaerobic stores in world class male runners competing over distances between 100 and 5000 m is on the order of 1.6 kJ kg^{-1} (76.5 ml O_2 kg^{-1}).

The maximal capacity of the anaerobic stores (AnSmax) can also be assessed from a different approach, essentially equal to that proposed by Lloyd (1966) for estimating "maximal aerobic speed" and "anaerobic distance" for the running world records. Indeed, Lloyd proposed to plot the distance covered as a function of the world record time and showed that, for distances greater than 1000 m the regression becomes a straight line with a positive intercept on the y axis (Fig. 12.7, left panel). He therefore suggested that the slope of the so obtained straight line yields the aerobic speed, i.e. the speed sustained solely on the basis of $\dot{V}O_2$max, whereas the y intercept of the regression represents the distance covered at the expense of the anaerobic stores. It should be pointed out that Lloyd's interpretation is correct only if the slope of the straight line is calculated within the range of distances allowing $\dot{V}O_2$max to be attained and maintained at the 100% level throughout the competition, i.e., for world class athletes between 1000 and 5000 m (131.96–757.35 s).

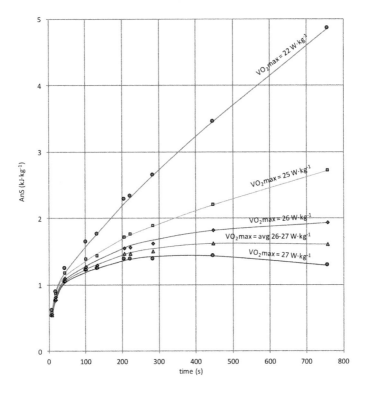

Fig. 12.6 Anaerobic contribution to overall energy requirement (AnS, kJ kg^{-1}) as a function of the current world record time (s) over 100–5000 m distance, calculated as described in the text for the indicated values of $\dot{V}O_2$max

Lloyd's approach can be applied to the overall energy expenditure, as calculated above (see Table 12.4) rather that to the distance covered. Also in this case, for distances between 1000 and 5000 m, the regression becomes a straight line with a positive intercept on the y axis (Fig. 12.7, right panel), as described by:

$$E_{tot} = 1.06 + 0.027 \cdot t \qquad (12.13)$$

for E_{tot} in kJ kg^{-1} and t in s. Along the lines propose by Lloyd, the slope of this regression can be interpreted as the average $\dot{V}O_2$max of the athletes competing in these events, that amounts then to 27 W kg^{-1}, whereas its y intercept is an index for the anaerobic stores capacity, as discussed in some detail below.

The overall energy (E_{tot}) spent during a supramaximal effort to exhaustion is the sum of the energy derived from aerobic and anaerobic stores, as described by:

$$E_{tot} = AnS + \dot{V}O_2max \cdot t_e - \dot{V}O_2max \cdot (1 - e^{-t_e/\tau}) \cdot \tau \qquad (12.14)$$

where t_e is the time to exhaustion and τ is the time constant of the $\dot{V}O_2$ kinetics at work onset (Scherrer and Monod 1960; Wilkie 1980). The third term of Eq. (12.14)

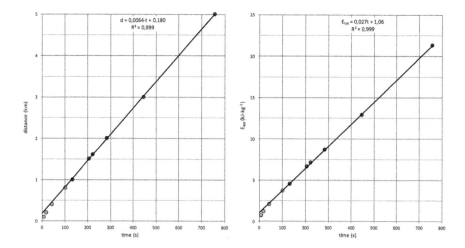

Fig. 12.7 Distance (km, *left panel*) and overall energy expenditure (kJ kg^{-1}, *right panel*) as a function of the corresponding world record time (s) in running (100–5000 m). The regressions, as calculated from 1000 to 5000 m and reported in the figures, show that, for world class athletes: **a** the maximal aerobic speed amounts to 6.4 m s^{-1} (23.04 km h^{-1}); and **b** the maximal aerobic power to 27 W kg^{-1} (77.5 ml O$_2$ kg^{-1} min^{-1}) above resting. The intercept on the y axis of the regression of the lower panel (1.06 kJ kg^{-1}) allows one to estimate the maximal anaerobic capacity in these athletes. See text for details

takes into account the fact that $\dot{V}O_2$max is not attained at the very onset of the exercise, but it is reached following an exponential function with the time constant τ. As such, to yield the correct amount of aerobic energy, the quantity $\dot{V}O_2$max · t_e must be reduced accordingly.

This equation becomes particularly useful if the effort duration to exhaustion is comprised between the minimum necessary for complete utilisation of the anaerobic stores (high energy phosphate breakdown and lactate accumulation) and the maximum allowing for $\dot{V}O_2$max to be maintained at the 100% level throughout the effort duration (50 s ≤ t_e ≤ 15 min) (Wilkie 1980; di Prampero et al. 1993; di Prampero 2003). If this is indeed the case, both AnS and $\dot{V}O_2$max can be safely assumed to be constant and maximal. Furthermore, since τ ≈ 20 s (di Prampero et al. 1993; di Prampero 2003), if t_e is greater than about 100 s the quantity $e^{-t_e/\tau}$ becomes vanishingly small; therefore, under these conditions, also the third term of Eq. (12.14) becomes constant ($\approx \dot{V}O_2$max ·τ). Thus, rearranging Eq. (12.14):

$$E_{tot} = [AnSmax - \dot{V}O_2max \cdot \tau] + \dot{V}O_2max \cdot t_e = \text{Constant} + \dot{V}O_2max \cdot t_e \quad (12.15)$$

The regression of Fig. 12.7, right panel, was calculated for world record times over distances between 1000 and 5000 m, i.e. within the above mentioned time range (131.96–757.35 s). If the additional simplifying assumptions are also made that: (1) world class athletes competing over these distances are characterised by an equal $\dot{V}O_2$max and (2) world record times can be identified with t_e, from Eqs. (12.13) to (12.15):

$$[AnSmax - \dot{V}O_2max \cdot \tau] = 1.06 \qquad (12.16)$$

Thus, setting $\dot{V}O_2$max $= 0.027$ (kW kg^{-1}) and $\tau = 20$ s, one obtains:

$$AnSmax = 1.06 + 0.027 \cdot 20 = 1.6 (\text{kJ kg}^{-1}) \qquad (12.17)$$

a value astonishingly close to that estimated from Fig. 12.6 for $\dot{V}O_2$max of 26–27 W kg^{-1}.[1]

These considerations can not be applied to shorter distances because: (i) the corresponding points in Fig. 12.7 (right panel) tend to fall below the regression calculated between 1000 and 5000 m, and (ii) it can not be reasonably assumed that $\dot{V}O_2$max in these athletes is equal to that of the athletes competing over the longer distances. Nor can they be applied to longer distances, in which case it cannot be safely assumed that $\dot{V}O_2$max is maintained to the 100% level throughout the entire duration of the race. Indeed, the slope of the regression between distance and record time, as calculated between 5000 and 10,000 m decreases to 6.09 m s^{-1}, and to 5.61 m s^{-1} for distances between 10,000 m and the marathon. The corresponding $\dot{V}O_2$ values (for $C_0 = 3.8$ J kg^{-1} m^{-1}) and k' $= 0.01$ (J s^2 kg^{-1} m^{-3}) become then 25.4 and 23.1 W kg^{-1}, to be compared to a "maximal aerobic speed" of 6.34 m s^{-1} and a $\dot{V}O_2$ of 27 W kg^{-1} (assumed to yield $\dot{V}O_2$max) for distances between 1000 and 5000 m, Fig. 12.7.

The fraction of the overall energy expenditure (E_{tot}, see Table 12.4) derived from anaerobic sources (FAnS) has been estimated over all distances between 100 and 5000 m for the current world record times for two $\dot{V}O_2$max values (22 and 27 W kg^{-1}). As mentioned above, in all cases it was assumed that, at the start of the race, $\dot{V}O_2$ increases with a time constant of 20 s tending to the appropriate mean

---

[1]The estimate of AnS, as from Eq. (12.14) is based on the simplifying assumption that the $\dot{V}O_2$ kinetics at the onset of square wave supramaximal exercise (as is necessarily the case for world record performances over the distances in question) increases exponentially with a time constant τ ($\approx$20 s) towards $\dot{V}O_2$max. However, it seems more realistic to assume that at work onset, $\dot{V}O_2$ increase exponentially with the same time constant towards the metabolic power requirement ($\dot{E}$), but stops increasing abruptly once $\dot{V}O_2$max is attained (Margaria et al. 1965). If this is the case, a more rigorous approach for estimating AnS, as derived from first principles in the Appendix (Eq. 12.32), is as follows:

$$AnS = (\dot{E} - \dot{V}O_2max) \cdot t_e + \dot{V}O_2max \cdot \tau - [-\ln(1 - \dot{V}O_2max/\dot{E}) \cdot \tau \cdot (\dot{E} - \dot{V}O_2max)] \quad (12.32)$$

The values of AnS calculated on the basis of this equation are not far from those reported in Fig. 12.6; for world record performances from 1000 to 5000 m, assuming $\dot{V}O_2$max $= 27$ W kg^{-1}, they amount on the average to 1.35 kJ kg^{-1} (range: 1.28–1.41), to be compared to a value of 1.6 kJ kg^{-1}, as calculated from Eqs. (12.14) and (12.17). Thus, on the basis of this approach the maximal capacity of the anaerobic stores would turn out to be about 15% lower than that estimated above.

Table 12.5 Fraction of the overall energy derived from anaerobic sources (FAnS) for distances of 100–5000 m covered in current world record times (s) for the two indicated $\dot{V}O_2$max values

Distance (m)	For $\dot{V}O_2$max 22 W kg^{-1}*		For $\dot{V}O_2$max 27 W kg^{-1}**	
	t @ $\dot{V}O_2$max (s)	FAnS	t @ $\dot{V}O_2$max (s)	FAnS
100	7.1	0.803	9.1	0.795
200	8.7	0.722	11.4	0.681
400	12.6	0.597	17.1	0.527
800	18.6	0.441	27.2	0.333
1000	20.8	0.394	31.5	0.276
1500	23.3	0.344	37.3	0.209
1 mile	24.0	0.332	38.5	0.196
2000	25.5	0.305	43.2	0.161
3000	28.0	0.264	52.0	0.110
5000	30.5	0.228	64.7	0.060

The time (s) necessary to reach the appropriate $\dot{V}O_2$max (t @ $\dot{V}O_2$max) is also reported. The absolute values of the anaerobic contribution to the overall energy requirement are reported in Fig. 12.6
*63.2 ml O_2 kg^{-1} min^{-1} above resting; **77.5 ml O_2 kg^{-1} min^{-1} above resting.
1 mile = 1609.35 m

power requirement (see Table 12.4), but stops increasing once $\dot{V}O_2$max is attained. FAnS is reported in Table 12.5, together with the time to reach $\dot{V}O_2$max. As shown in this Table, this is shorter the higher the metabolic power requirement, and, for a given metabolic power, is longer the higher $\dot{V}O_2$max. It should be pointed out that, whereas for distances $\geq$ 1000 m it can be reasonably assumed that $\dot{V}O_2$max amounts to 27 W kg^{-1}, for the shorter distances the FAnS values estimated for the two smaller $\dot{V}O_2$max values (22 W kg^{-1}) are probably closer to the "truth".

12.7 Discussion

The preceding sections of this chapter were devoted to an analysis of the energetics of current world records in running over distances from 100 to 5000 m. This was carried out along two different lines, as follows.

For the three shorter distances (100, 200 and 400 m) the time course of the metabolic power was estimated according to the model proposed by di Prampero et al. (2005) wherein accelerated running on flat terrain is considered analogous to uphill running at constant speed, the incline of the terrain being dictated by the forward acceleration. The time integral of the so obtained metabolic power curves allowed us to assess the overall energy spent to cover the distances in question. This approach can be performed only if the time course of the speed throughout the run is known, as such it could not be applied for distances longer that 400 m, in which case these data are not available. Indeed, even in the case of the 400 m, speed data were available only over discrete 50 m intervals, thus weakening the time resolution of the metabolic power curve.

Therefore, for the longer distances (800–5000 m), a different approach was followed, [see Sect. 12.5 and Eqs. (12.12), and (12.12')]. This, while taking care of the overall energy spent in the acceleration phase, did not allow us to estimate the time course of the metabolic power throughout the race, but only its average values. Even so, when applying this simplified procedure to the 100–400 m distances, the resulting overall energy expenditure turned out fairly close (at least for the 200 and 400 m distances) to that estimated from the time integral of the metabolic power time course (see Table 12.3).

The anaerobic contribution to the overall energy expenditure was then estimated as follows. As described in detail elsewhere (di Prampero et al. 2015), it was assumed that the rate of O_2 consumption at the onset of the run increases exponentially with a time constant of 20 s, tending to the metabolic power requirement, but stops abruptly once $\dot{V}O_2$max is attained (Margaria et al. 1965). In turn, for the distances between 1000 and 5000 m, this was assumed to be 27 W kg^{-1} above resting (see Fig. 12.7). For the shorter distances, however, this assumption does not seem reasonable; as a consequence the rate of $\dot{V}O_2$ at work onset was estimated also for a smaller $\dot{V}O_2$max value (22 W kg^{-1}).

The time integral of the so obtained $\dot{V}O_2$ kinetics was then subtracted from the total energy expenditure to estimate the overall anaerobic yield over the distances in question, for the appropriate metabolic power and $\dot{V}O_2$max values. It is reported in Table 12.5 for the investigated world records as a fraction of the overall energy expenditure.

As shown in Fig. 12.6, the maximal capacity of the anaerobic stores for world class athletes competing over distances between 1000 and 5000 m and assumed to be characterised by a $\dot{V}O_2$max of 26–27 W kg^{-1} turned out to be on the order of 1.6 kJ kg^{-1}.

The maximal capacity of the anaerobic stores was also assessed according to a different approach, essentially equal to that proposed by Lloyd (1966) for estimating "maximal aerobic speed" and "anaerobic distance" for the running world records. Briefly, the overall energy expenditure over distances between 100 and 5000 m has been plotted as a function of the current world record time (see Table 12.4). The resulting regression, for distances between 1000 and 5000 m turns out to be linear (Fig. 12.7), its slope yielding the mean maximal aerobic power of the current world record holders over these distances, which amounted to 27 W kg^{-1} (77.5 ml O_2 kg^{-1} min^{-1}) above resting. In turn, the y intercept of this same regression allowed us to estimate the maximal capacity of the anaerobic stores that, for a $\dot{V}O_2$max of 27 W kg^{-1}, turned out to be 1.6 kJ kg^{-1}, essentially equal to that estimated from the data of Fig. 12.6 for athletes whose $\dot{V}O_2$max is on the order of 26–27 W kg^{-1}.

It can therefore be concluded that, for high calibre world class athletes, the maximal capacity of the anaerobic stores is indeed on the order of 1.6 kJ kg^{-1} (76.5 ml O_2 kg^{-1}).

This is consistent with the considerations that follow. (i) The capacity of the anaerobic stores is the sum of the maximal amount of energy that can be obtained from high energy phosphates (ATP + phosphocreatine (PCr)) splitting and from lactate (La) accumulation. (ii) When competing over the distances in question at world record speed, the maximally exercising muscles in world class athletes constitute 20–25% of the body mass. (iii) In resting muscles the ATP and PCr concentrations amount to about 6 and 20 m-mol kg^{-1} fresh tissue, respectively (e.g. see Francescato et al. 2003, 2008), and since (iv) ATP can not decrease substantially without greatly compromising muscle power production (di Prampero and Piiper 2003), only about 20 m-mol kg^{-1} of PCr can be utilised from rest to exhaustion, i.e. 4–5 m-mol PCr kg^{-1} body mass (see point ii above). (v) Hence, assuming a P/O_2 ratio of 6 (mol/mol), the amount of O_2 "spared" thanks to the splitting of PCr in the transition from rest to exhaustion corresponds to 0.67–0.83 m-mol O_2 kg^{-1} body mass. (vi) Assuming further an energy equivalent of 20.9 J ml^{-1} O_2, and since 1 m-mol O_2 = 22.4 ml, the maximal amount of energy derived from PCr splitting can be estimated in the range 0.31–0.39 kJ kg^{-1} body mass. (vii) The maximal blood La concentration attained at the end of competitions at maximal speed over the distances $\geq$ 400 m is on the order of 15–20 mM above resting (Arcelli et al. 2014; Hanon et al. 1994; Hautier et al. 2010); and since (viii) the accumulation of 1 mM La yields an amount of energy equal to the consumption of about 3 ml O_2 kg^{-1} body mass (di Prampero 1981; di Prampero and Ferretti 1999), on the basis of an energetic equivalent of O_2 of 20.9 J ml^{-1}, this corresponds to a maximal energy release from La accumulation of 0.94–1.25 kJ kg^{-1}.

Hence the maximal capacity of the anaerobic stores, as given by the maximal amount of energy that can be obtained from PCr splitting and from La accumulation can be estimated in the range of 1.25 (0.31 + 0.94) to 1.64 (0.39 + 1.25) kJ kg^{-1} body mass, a value rather close to that calculated as described above (see Figs. 12.6 and 12.7).

12.8 Critique of Methods

The two procedures underlying the energetic analysis of the world records described above are based on several simplifying assumptions that have been discussed elsewhere (di Prampero et al. 1993, 2005, 2015 and by Rittweger et al., 2009) and that are briefly summarised below, the interested reader being referred to the original papers.

The assessment of the time course of the metabolic power in the three shorter distances, as based on the analogy between accelerated running on flat terrain and uphill running at constant speed (see Fig. 12.1), entails what follows.

(i) The overall mass of the runner is condensed in his/her centre of mass. This necessarily implies that the stride frequency, and hence the energy expenditure due to internal work performance (for moving the upper and lower limbs in respect to the centre of mass) is the same during accelerated running and during uphill running at constant speed up the same equivalent slope (ES).

(ii) For any given ES, the efficiency of metabolic to mechanical energy transformation during accelerated running is equal to that of constant speed running over the corresponding incline. This also implies that the biomechanics of running, in terms of joint angles and torques, is the same in the two conditions.

(iii) The highest ES values attained at the onset of the 100 and 200 m runs are substantially larger than the highest inclines actually studied by Minetti et al. (2002); hence the implicit assumption is also made that for ES > 0.45, the relationship between C_r and the incline is as described by Eq. (12.7). However, even for Bolt's 100 m world record, after about 6 m the actual ES values become <0.45, so that the assumption on which Eq. (12.7) was constructed is not likely to greatly affect the overall estimate of the energy expenditure, even if it may indeed affect the corresponding time course in the very initial phase of the run.

(iv) The calculated ES values are assumed to be in excess of those observed during constant speed running on flat terrain in which case the runner is lining slightly forward. This, however, cannot be expected to introduce large errors, since our reference value was the measured energy cost of constant speed running on flat terrain (C_0).

As concerns the simplified approach to estimate the average metabolic power, the energy spent in the acceleration phase over and above that for constant speed running was calculated as summarised below.

(v) The metabolic energy spent (per kg body mass) to accelerate the runner's body from zero to the mean (or peak speed) was estimated from the mechanical kinetic energy ($0.5 \cdot v^2$, where v is the mean, or peak, speed), assuming that the efficiency of converting metabolic into mechanical energy (η) amounts to 0.25 (Eqs. 12.10–12.12). This is consistent with the analogy between accelerated running on flat terrain and uphill running at constant speed, since in the latter case, for inclines between 20 and 40% η is on the order of 0.22–0.26 (see Fig. 12.2).

In addition, for all distances, the following two assumptions were also made, independently of the model utilised.

(vi) The energy expenditure against the air resistance, per unit body mass and distance, was assumed to be described by: $k' \cdot v^2$, where v (m s^{-1}) is the air velocity, the constant k' (J s^2 kg^{-1} m^{-3}) assumed throughout this study amounting to 0.01 (Pugh 1970; di Prampero 1986), a value lower than that which can be calculated from Arsac and Locatelli (2002) biomechanical data, and than that reported by Tam et al. (2012) which amount to 0.017 and 0.019, respectively.

(vii) The energy cost, per unit body mass and distance, for constant speed running on flat compact terrain neglecting the air resistance (C_0) was assumed to be =3.8 J kg^{-1} m^{-1}. The corresponding values reported in the literature range from 3.6 as determined on the treadmill by Minetti et al. (2002), to 4.32 ± 0.42 on the treadmill and to 4.18 ± 0.34 on the terrain, as determined more recently by Minetti et al. (2012) at 11 km h^{-1}, to 4.39 ± 0.43 (n = 65), as determined by Buglione and di Prampero (2013) during treadmill running at 10 km h^{-1}, the great majority of data clustering around a value of 4 J kg^{-1} m^{-1} (Lacour and Bourdin 2015). Thus, whereas on the one side it would be ideal to know the individual C_0 of each athlete, a rather unrealistic possibility when dealing with world records holders, on the other, the assumption of a common value of 3.8 J kg^{-1} m^{-1} does not seem to greatly bias the obtained results.

Finally the overall metabolic energy expenditure was partitioned into its aerobic and anaerobic fractions as follows.

(viii) The kinetics of O_2 consumption at work onset was calculated as described above, thus allowing us to estimate the mean $\dot{V}O_2$ throughout the run and hence the aerobic and anaerobic fractions of the total energy expenditure. This approach is supported by a recent series of experiments (di Prampero et al. 2015) in which the actual O_2 consumption was determined by means of a portable metabolic cart in 8 subjects during a series of shuttle runs over 25 m, each bout being performed in 5 s. The speed was continuously assessed by means of a radar system, thus allowing us to estimate the instantaneous energy cost and hence the metabolic power, as given by the product of this last and the speed. The actual O_2 consumption was then compared to that estimated as described above from the metabolic power curve and the subjects' $\dot{V}O_2$max. The two sets of data turned out to be rather close, thus supporting the approach used in this study to estimate the anaerobic and aerobic contribution to the overall energy expenditure.

12.9 Conclusions and Practical Remarks

Mathematical modelling of training and performance is becoming increasingly important in a number of professions ranging from health and physical training to rehabilitation from disease or injury, not to mention athletic performance (Clarke and Skiba 2013). Therefore, we deemed it important to condense the preceding analyses into a simple set of rules to estimate: (1) the overall energy expenditure, and (2) the aerobic and anaerobic contributions thereof, in running at speeds greater than the maximal aerobic one, provided that the subject's $\dot{V}O_2$max, the distance covered (d) and the running time (t_e) are known.

Equation (12.12) shows that the overall energy expenditure (E_{tot}) can be estimated with reasonable accuracy, if the energy cost of constant speed running (C_0), the constant relating the energy expenditure against the wind to the square of the speed (k') and the efficiency of metabolic to mechanical energy transformation in the acceleration phase (η) are known. Thus, replacing v_{mean} with the quantity d/t_e (where t_e is the performance time), Eq. (12.12) can be rewritten as:

$$E_{tot} = [C_0 + k' \cdot (d/t_e)^2] \cdot d + (d/t_e)^2 (2\eta)^{-1} \qquad (12.18)$$

In turn, the ratio of Eq. (12.18) to the performance time yields the average metabolic power ($\dot{E}$) throughout the distance in question:

$$\dot{E} = E_{tot}/t_e = [C_0 + k' \cdot (d/t_e)^2] \cdot d/t_e + [(d/t_e)^2 (2\eta)^{-1}]/t_e \qquad (12.19)$$

wherefrom the corresponding anaerobic energy yield (AnS) can be obtained thanks to Eq. (12.20), as derived from first principles in the Appendix:

$$AnS = (\dot{E} - \dot{V}O_2 max) \cdot t_e + \dot{V}O_2 max \cdot \tau - [-\ln(1 - \dot{V}O_2 max/\dot{E}) \cdot \tau \cdot (\dot{E} - \dot{V}O_2 max)] \qquad (12.20)$$

Thus, if $\dot{V}O_2 max$ and τ are known, it is an easy task to estimate both the aerobic and anaerobic components of E_{tot}.

As concerns the present study, the relevant quantities utilised in the calculations were as follows: $C_0 = 3.8$ J kg^{-1} m^{-1}, $k' = 0.01$ J s^2 kg^{-1} m^{-3}; $\eta = 0.25$; $\tau = 20$ s; d and t_e were the 100–5000 m running distances and the current world records times, and $\dot{V}O_2 max$ 22 or 27 W kg^{-1}, as indicated. It goes without saying that the choice of these quantities is somewhat arbitrary and can be replaced by more accurate estimates, if available.

Finally, whereas this simplified approach was constructed for running on the level on smooth terrain in the absence of wind, any other condition can be easily incorporated, assigning the appropriate values to C_0, k', η, etc., provided that, over the considered distance and time these quantities can be assumed to be constant.

In closing, and in spite of the inevitable limits of this brief review on sprint running and on world records, we do hope that the readers may consider it a stimulus and a challenge to further investigate this fascinating field.

Acknowledgements The financial support of the "Fondo Bianca e Chiara Badoglio" is gratefully acknowledged.

Appendix[2]

In a preceding section of this chapter (12.6) it was assumed that the overall energy (E_{tot}) spent during a supramaximal effort to exhaustion can be described by:

$$E_{tot} = AnS + \dot{V}O_2max \cdot t_e - \dot{V}O_2max \cdot (1 - e^{-t_e/\tau}) \cdot \tau \quad (12.21)$$

where AnS is the amount of energy derived from anaerobic sources, t_e is the time to exhaustion and $\tau (\approx 20 \text{ s})$ is the time constant of the $\dot{V}O_2$ kinetics at work onset. The third term of this equation takes into account the fact that $\dot{V}O_2max$ is not attained at the very onset of the exercise, but it is reached following an exponential function with the time constant τ (Wilkie 1980). Thus:

$$AnS = E_{tot} - \dot{V}O_2max \cdot t_e + \dot{V}O_2max \cdot (1 - e^{-t_e/\tau}) \cdot \tau \quad (12.22)$$

Furthermore, if t_e is sufficiently long (i.e. $\geq 4\tau$), the quantity $e^{-t_e/\tau}$ becomes vanishingly small and the third term of the equation reduces to $\dot{V}O_2max \cdot \tau$. In this range of exhaustion times therefore, AnS can be easily estimates as:

$$AnS = E_{tot} - \dot{V}O_2max \cdot t_e + \dot{V}O_2max \cdot \tau \quad (12.22')$$

Equations (12.21) and (12.22) are based on the implicit assumption that the $\dot{V}O_2$ kinetics is a continuous exponential function from the value prevailing at work onset to $\dot{V}O_2max$. However, as discussed in detail elsewhere (di Prampero et al. 2015), at the onset of supramaximal exercise [in which case the metabolic power requirement ($\dot{E}$) is greater than the subject's $\dot{V}O_2max$] $\dot{V}O_2$ increases exponentially towards $\dot{E}$ with the same time constant ($\tau \approx 20$ s), but stops abruptly at the very moment (t_1) when $\dot{V}O_2max$ is attained (Margaria et al. 1965). This is shown graphically in Fig. 12.8, where $\dot{E}$ (red horizontal line) and $\dot{V}O_2max$ (blue horizontal line) are indicated as a function of the exercise time, together with the time course of $\dot{V}O_2$ before the attainment of $\dot{V}O_2max$ (blue continuous curve) and with the hypothetical time course of $\dot{V}O_2$ above $\dot{V}O_2max$, were $\dot{E} = \dot{V}O_2max$ (green broken curve). Inspection of this figure makes it immediately apparent that, whereas Eq. (12.22) is correct for $\dot{E} = \dot{V}O_2max$, whenever $\dot{E} > \dot{V}O_2max$ it leads to an overestimate of AnS, the more so, the greater the difference between $\dot{E}$ and $\dot{V}O_2max$.

The aim of the paragraphs that follow is to describe an approach yielding a more accurate estimate of the anaerobic energy yield (AnS) in running at supramaximal constant metabolic power ($\dot{E}$), provided the subject's $\dot{V}O_2max$, the exercise duration (t_e) and τ are known. Indeed, on the one side, this set of data allows one to estimate $\dot{E}$, as described by Eq. (12.18):

---

[2]The equations appearing in this Appendix are numbered (12.21)–(12.33), even if some of them have previously been mentioned in the text.

$$\dot{E} = E_{tot}/t_e = \{[C_0 + k' \cdot (d/t_e)^2] \cdot d + (d/t_e)^2 (2\eta)^{-1}\}/t_e \qquad (12.23)$$

where all terms have been previously defined (see Sect. 12.6). On the other, if $\dot{V}O_2$max, τ and t_e are also known, Fig. 12.8 allows one to appreciate graphically that the anaerobic contribution to the overall energy expenditure is represented by the area delimited by $\dot{E}$ and the $\dot{V}O_2 - \dot{V}O_2$max curve, i.e. by the sum of the areas 1, 2, 3 and 3′, whereas the sum of the two areas 4 and 5, below the $\dot{V}O_2 - \dot{V}O_2$max curve, represents the aerobic energy yield.

What follows is devoted to assess quantitatively the anaerobic energy yield as given by the sum of the areas defined above, indicated numerically as in the Fig. 12.8.

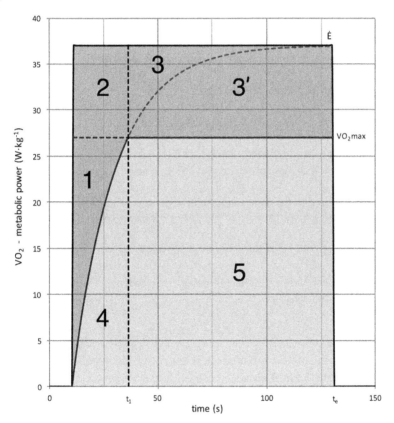

Fig. 12.8 Overall metabolic power requirement (red horizontal line, $\dot{E}$) as a function of time (t) during square wave exercise of constant intensity and duration t_e. Subject's maximal O_2 consumption ($\dot{V}O_2$max) is indicated by blue horizontal line. At work onset, $\dot{V}O_2$ increases exponentially towards $\dot{E}$, but stops abruptly at t_1, i.e. when $\dot{V}O_2$max is attained. Actual $\dot{V}O_2$ before t_1 is indicated by continuous blue curve, whereas after t_1 $\dot{V}O_2 = \dot{V}O_2$max. Green broken curve denotes hypothetical $\dot{V}O_2$ time course, were $\dot{E} \leq \dot{V}O_2$max. Anaerobic energy yield is given by the sum of areas $1 + 2 + 3 + 3'$ (red); aerobic yield by the sum of areas $4 + 5$ (blue). See text for details and calculations (Color figure online)

The $\dot{V}O_2$ kinetics at work onset is described by:

$$\dot{V}O_2(t) = \dot{E} \cdot (1 - e^{-t/\tau}) \tag{12.24}$$

However, as shown in Fig. 12.8, $\dot{V}O_2$ stops increasing abruptly at time t_1, i.e. when $\dot{V}O_2max$ is attained. Thus at t_1:

$$\dot{V}O_2(t) = \dot{E} \cdot (1 - e^{-t_1/\tau}) = \dot{V}O_2max \tag{12.25}$$

Rearranging Eq. (12.25), one obtains:

$$1 - \dot{V}O_2max/\dot{E} = e^{-t_1/\tau} \tag{12.26}$$

or:

$$\ln(1 - \dot{V}O_2max/\dot{E}) = -t_1/\tau \tag{12.27}$$

where from t_1 can be finally obtained as:

$$t_1 = -\ln(1 - \dot{V}O_2max/\dot{E}) \cdot \tau \tag{12.28}$$

It can therefore be concluded that Eq. (12.28) allows one to estimate the time necessary to attain $\dot{V}O_2max$, provided that $\dot{V}O_2max$ itself, together with $\dot{E}$ and τ are known.

It is now possible to estimate the anaerobic energy yield proceeding as follows. The area of the rectangle $(2+3+3')$ in Fig. 12.8 can be easily calculated as:

$$2 + 3 + 3' = (\dot{E} - \dot{V}O_2max) \cdot t_e \tag{12.29}$$

The area corresponding to the O_2 deficit incurred once $\dot{V}O_2max$ is attained $(1 + 2)$ can be estimated as:

$$1 + 2 = \dot{V}O_2max \cdot \tau \tag{12.30}$$

Finally the area of the rectangle 2 is given by the product of the time t_1 (Eq. 12.28) and the vertical distance between $\dot{E}$ and $\dot{V}O_2max$:

$$2 = t_1 \cdot (\dot{E} - \dot{V}O_2max) = -\ln(1 - \dot{V}O_2max/\dot{E}) \cdot \tau \cdot (\dot{E} - \dot{V}O_2max) \tag{12.31}$$

The overall amount of energy derived from anaerobic stores (AnS) is finally expressed by the algebraic sum of the Eqs. (12.29), (12.30) and (12.31):

$$AnS = 2 + 3 + 3' + 1 + 2 - 2$$
$$= (\dot{E} - \dot{V}O_2max) \cdot t_e + \dot{V}O_2max \cdot \tau - [-\ln(1 - \dot{V}O_2max/\dot{E}) \cdot \tau \cdot (\dot{E} - \dot{V}O_2max)] \quad (12.32)$$

It should finally be pointed out that Eq. (12.32) defines the anaerobic energy yield whenever the exercise duration is greater that that necessary for $\dot{V}O_2max$ to be attained ($t_e > t_1$). Whenever $t_e \leq t_1$, things become much simpler, since in this specific case the only energy derived from anaerobic stores (AnS') corresponds to the O_2 deficit incurred, as given by the product of the $\dot{V}O_2$ attained at the very end of the exercise, ($= \dot{E} \cdot (1 - e^{-t_e/\tau})$), Eq. (12.24) and the time constant of the $\dot{V}O_2$ kinetics (τ):

$$AnS' = \dot{E} \cdot (1 - e^{-t_e/\tau}) \cdot \tau \quad (12.33)$$

It should finally be pointed out that, whenever $\dot{E} = \dot{V}O_2max$, Eqs. (12.32) and (12.33) reduce to Eq. (12.22) or (12.22') depending whether, or not, t_e is sufficiently long for $\dot{V}O_2max$ to be attained.

It can be concluded that the anaerobic energy yield in running at supramaximal constant intensity can be easily estimated (Eqs. 12.32 and 12.33), provided that the metabolic power requirement (Eq. 12.23), together with the subject's VO_2max, the exercise duration (t_e) and the time constant of the $\dot{V}O_2$ kinetics at work onset (τ) are known.

References

Arcelli E, Cavaggioni L, Alberti G (2014) Il lattato ematico nelle corse dai 100 ai 1.500 metri. Confronto tra uomo e donna. Scienza & Sport 21:48–53

Arsac LM (2002) Effects of altitude on the energetics of human best performances in 100 m running: a theoretical analysis. Eur J Appl Physiol 87:78–84

Arsac LM, Locatelli E (2002) Modelling the energetics of 100-m running by using speed curves of world champions. J Appl Physiol 92:1781–1788

Beneke R, Taylor MJD (2010) What gives Bolt the edge—A.V. Hill knew it already. J Biomech 43:2241–2243

Buglione A, di Prampero PE (2013) The energy cost of shuttle running. Eur J Appl Physiol 113:1535–1543

Cavagna GA, Kaneko M (1977) Mechanical work and efficiency in level walking and running. J Physiol 268:467–481

Cavagna GA, Komarek L, Mazzoleni S (1971) The mechanics of sprint running. J Physiol 217:709–721

Charles JD, Bejan A (2009) The evolution of speed, size and shape in modern athletics. J Exp Biol 212:2419–2425

Clarke DC, Skiba PF (2013) Rationale and resources for teaching the mathematical modelling of athletic training and performance. Adv Physiol Educ 37:134–152

di Prampero PE (1981) Energetics of muscular exercise. Rev Physiol Biochem Pharmacol 89:143–222
di Prampero PE (1986) The energy cost of human locomotion on land and in water. Int J Sports Med 7:55–72
di Prampero PE (2003) Factors limiting maximal performance in humans. Eur J Appl Physiol 90:420–429
di Prampero PE, Ferretti G (1999) The energetics of anaerobic muscle metabolism: a reappraisal of older and recent concepts. Respir Physiol 118:103–115
di Prampero PE, Piiper J (2003) Effects of shortening velocity and of oxygen consumption on efficiency of contraction in dog gastrocnemius. Eur J Appl Physiol 90:270–274
di Prampero PE, Capelli C, Pagliaro P, Antonutto G, Girardis M, Zamparo P, Soule RG (1993) Energetics of best performances in middle-distance running. J Appl Physiol 74:2318–2324
di Prampero PE, Fusi S, Sepulcri L, Morin JB, Belli A, Antonutto G (2005) Sprint running: a new energetic approach. J Exp Biol 208:2809–2816
di Prampero PE, Botter A, Osgnach C (2015) The energy cost of sprint running and the role of metabolic power in setting top performances. Eur J Appl Physiol 115:451–469
Fenn WO (1930a) Frictional and kinetic factors in the work of sprint running. Am J Physiol 92:583–611
Fenn WO (1930b) Work against gravity and work due to velocity changes in running. Am J Physiol 93:433–462
Francescato MP, Cettolo V, di Prampero PE (2003) Relationship between mechanical power, O_2 consumption, O_2 deficit and high Energy phosphates during calf exercise in humans. Pflügers Arch 93:433–462
Francescato MP, Cettolo V, di Prampero PE (2008) Influence of phosphagen concentration on phosphocreatine breakdown kinetics. Data from human gastrocnemius muscle. J Appl Physiol 105:158–164
Graubner R, Nixdorf E (2011) Biomechanical analysis of the sprint and hurdles events at the 2009 IAAF World Championship in Athletics. NSA (New Stud Athletics) 26(1/2):19–53
Hanon C, Lepretre P-M, Bishop D, Thomas C (1994) Oxygen uptake and blood metabolic responses to a 400-m. Eur J Appl Physiol 109:233–240
Hautier CA, Wouassi D, Arsac LM, Bitanga E, Thiriet P, Lacour JR (2010) Relationship between postcompetition blood lactate concentration and average running velocity over 100-m and 200-m races. Eur J Appl Physiol 68:508–513
Hernandez Gomez JJ, Marquina V, Gomez RW (2013) On the performance of Usain Bolt in the 100 m sprint. Eur J Phys 34:1227–1233
Hill AV (1925) The physiological basis of athletic records. Nature 116:544–548
Kersting UG (1998) Biomechanical analysis of the sprinting events. In: Brüggemann G-P, Kszewski D, Müller H (eds) Biomechanical research project Athens 1997. Final report. Meyer & Meyer Sport, Oxford, pp 12–61
Lacour JR, Bourdin M (2015) Factors affecting the energy cost of level running at submaximal speed. Eur J Appl Physiol 115:651–673
Lloyd BB (1966) The energetics of running: an analysis of word records. Adv Sci 22:515–530
Maćkała K, Mero A (2013) A kinematics analysis of three best 100 m performances ever. J Hum Kinet 36(Section III—Sports Training):149–160
Margaria R (1938) Sulla fisiologia e specialmente sul consumo energetico della marcia e della corsa a varia velocità ed inclinazione del terreno. Atti Acc Naz Lincei 6:299–368
Margaria R, Mangili F, Cuttica F, Cerretelli P (1965) The kinetics of the oxygen consumption at the onset of muscular exercise in man. Ergonomics 8:49–54
Mero A, Komi PV, Gregor RJ (1992) Biomechanics of sprint running. A review. Sports Med 13:376–392
Minetti AE, Moia C, Roi GS, Susta D, Ferretti G (2002) Energy cost of walking and running at extreme uphill and downhill slopes. J Appl Physiol 93:1039–1046

Minetti AE, Gaudino P, Seminati E, Cazzola D (2012) The cost of transport of human running is not affected, as in walking, by wide acceleration/deceleration cycles. J Appl Physiol 114:498–503

Murase Y, Hoshikawa T, Yasuda N, Ikegami Y, Matsui H (1976) Analysis of the changes in progressive speed during 100-meter dash. In: Komi PV (ed) Biomechanics V-B. University Park Press, Baltimore, pp 200–207

Osgnach C, Poser S, Bernardini R, Rinaldo R, di Prampero PE (2010) Energy cost and metabolic power in elite soccer: a new match analysis approach. Med Sci Sports Exerc 42:170–178

Plamondon A, Roy B (1984) Cinématique et cinétique de la course accélérée. Can J Appl Sport Sci 9:42–52

Pugh LGCE (1970) Oxygen intake in track and treadmill running with observations on the effect of air resistance. J Physiol Lond 207:823–835

Rittweger J, di Prampero PE, Maffulli N, Narici MV (2009) Sprint and endurance power and ageing: an analysis of master athletic world records. Proc R Soc B 276:683–689

Scherrer J, Monod H (1960) Le travail musculaire local et la fatigue chez l'homme. J Physiol Paris 52:419–501

Summers RL (1997) Physiology and biophysics of 100-m sprint. News Physiol Sci 12:131–136

Tam E, Rossi H, Moia C, Berardelli C, Rosa G, Capelli C, Ferretti G (2012) Energetics of running in top-level marathon runners from Kenya. Eur J Appl Physiol. https://doi.org/10.1007/s00421-012-2357-1

Taylor MJD, Beneke R (2012) Spring mass characteristics of the fastest men on Earth. Int J Sports Med 33:667–670

van Ingen Schenau GJ, Jacobs R, de Koning JJ (1991) Can cycle power predict sprint running performance? Eur J Appl Physiol 445:622–628

van Ingen Schenau GJ, de Koning JJ, de Groot G (1994) Optimization of sprinting performance in running, cycling and speed skating. Sports Med 17:259–275

Ward-Smith AJ, Radford PF (2000) Investigation of the kinetics of anaerobic metabolism by analysis of the performance of elite sprinters. J Biomech 33:997–1004

Wilkie DR (1980). Equations describing power input by humans as a function of duration of exercise. In: Cerretelli P, Whipp BJ (eds) Exercise bioenergetics and gas exchange. Elsevier. Amsterdam, pp 75–80

Chapter 13
Metabolic Power and Oxygen Consumption in Soccer: Facts and Theories

Cristian Osgnach and Pietro E. di Prampero

Abstract After a brief overview of the principles underlying the assessment of the energy cost of accelerated/decelerated running on flat terrain, as obtained from the biomechanical equivalence between this last and uphill/downhill running at constant speed, metabolic power and actual O_2 consumption ($\dot{V}O_2$) will be estimated during a typical training drill and during a brief period of a soccer match. This exercise will show that: (i) $\dot{V}O_2$, as estimated from metabolic power assuming mono-exponential $\dot{V}O_2$ on/off responses with a time constant of 20 s at the muscle level, is essentially equal to that directly measured by means of a metabolic cart. Hence, (ii) the fractions of the overall energy expenditure derived from aerobic and anaerobic sources can also be estimated, provided that the individual $\dot{V}O_2$max is also known. Furthermore, the metabolic power approach will be updated to make it possible to estimate energy cost and metabolic power also during walking episodes. Indeed, whereas for any given incline the energy cost of running is independent of the speed, the energy cost of walking reaches a minimum at an optimal speed above and below which it increases. With increasing the incline of the terrain, the curve relating energy cost to walking speed retains this general U shape, but is shifted to a higher level, the optimum speed decreasing towards lower values. In addition, for any given incline there exists a "transition speed" at which the energy cost of walking becomes greater than that of running, and which corresponds to the speed at which the subject spontaneously adopts the running gait. Hence the updated algorithms detect this transition speed, as a function of the equivalent slope, and utilise the appropriate energy cost of walking or running in the subsequent calculations. The concluding paragraphs are devoted to a critical discussion of the principal assumptions underlying the metabolic power approach. Specifically, we will stress that the approach as such can be meaningfully applied only to forward walking or running. Thus, whereas other specific sport activities such as jumping,

C. Osgnach (✉) · P. E. di Prampero
Department of Sport Sciences, Exelio srl, 33100 Udine, Italy
e-mail: cristian.osgnach@gmail.com

P. E. di Prampero
Department of Medical Sciences, University of Udine, 33100 Udine, Italy

moving backwards, laterally, or with the ball, may be taken into account, at least in a broad sense, assuming appropriate correction factors (possibly on the bases of actual experimental data), other activities, such as those occurring in rugby, of American football, need entirely different approaches, a fact that, all too often, does not seem to be appropriately considered.

13.1 Introduction

Soccer, as well as many other team sports, is characterised by frequent episodes of accelerated and decelerated running. These bring about a substantial increase of the overall energy cost of any given fraction of a soccer drill or match, as compared to a similar period of running on flat terrain at an equal average speed. Nevertheless, the assessment of the work load in soccer has traditionally been based on speed and distance (Carling et al. 2008; Sarmento et al. 2014) regardless of the time course of the intervening speed changes, thus obviously neglecting the frequent occurrence of sudden transitions between low and high exercise intensities and vice versa (Osgnach et al. 2010). In recent years, however, thanks to the advances in GPS technology, on the one side, and in our understanding of the energetics of accelerated and decelerated running, on the other, it has become possible: (i) to estimate the time course of the instantaneous metabolic power requirement of any given player and (ii) to infer therefrom the time course of the actual O_2 consumption ($\dot{V}O_2$) (di Prampero et al. 2015).

The present chapter is devoted to review the basic assumptions underlying the estimate of metabolic power, as well as its physiological meaning and its relationships with the actual $\dot{V}O_2$, with the aim of arriving at a clear-cut picture of the overall energy expenditure, and of its partition into aerobic and anaerobic fractions, during any given time period of the match, or training session. In addition, we will also update our previous approach, with the aim of taking into account separately the running or walking episodes during any given time period of the match or training session.

However, before proceeding further into these matters it seems useful to summarise the main principles underlying the estimate of metabolic power, the reader being referred to Chap. 12 for a more detailed discussion.

13.2 Theory

Direct measurements of metabolic power during accelerated/decelerated running is rather problematic because of the short duration and high intensity of any such events that, on the one side, preclude the attainment of the steady state and, on the other, imply a metabolic power requirement that may greatly surpass the subjects' $\dot{V}O_2$max.

An alternative approach is to estimate metabolic power from the product of the energy cost of running (estimated) and the speed (measured) (see Sect. 13.3). In turn, the energy cost of running can be estimated assuming that accelerated/decelerated running on flat terrain is biomechanically equivalent to uphill/downhill running at constant speed.

The theory underlying the analogy mentioned above is summarised graphically in Fig. 13.1, as from the original papers by our group (di Prampero et al. 2005; Osgnach et al. 2010); it has been recently reviewed by di Prampero et al. (2015) and is briefly sketched below as well as in Chap. 12, the interested reader being referred to the original papers. Figure 13.1 shows that accelerated running on flat terrain and running uphill at constant speed can be considered biomechanically equivalent, provided that the angle (α) between the mean body axis and the terrain is equal in both conditions. It necessarily follows that the slope of the terrain, as given by the angle between the terrain and the horizontal, is proportional to the complement of α (90 − α) that, as shown in Fig. 13.1, increases with the forward acceleration (a_f). Furthermore, the incline of the terrain is generally expressed as the tangent of the angle 90 − α; as such it can easily be obtained from the ratio between the forward acceleration (a_f) and the acceleration of gravity (g) (Fig. 13.1):

$$a_f/g = tan(90-\alpha) = ES \qquad (13.1)$$

As calculated, ES is the tangent of the angle that makes accelerated running biomechanically equivalent to running at constant speed up a corresponding slope, hence the definition of "Equivalent Slope" (ES).

In addition to being equivalent to running uphill, accelerated running is characterised by yet another difference, as compared to constant speed running. Indeed, the average force that the runner must develop throughout a whole stride, in the former case is given by the product of the runner's body mass and the vectorial sum of the forward acceleration and the acceleration of gravity $\left(g' = \sqrt{(a_f^2 + g^2)}\right)$. As such it is greater than that exerted when running uphill at constant speed, in which case the force developed by the runner is given by the product of the body mass and the acceleration of gravity (see Fig. 13.1). Thus, to take into account this effect, the runner's body mass must be multiplied by a factor here defined "equivalent body mass" (EM) as described by:

$$EM = \sqrt{(a_f^2 + g^2)}/g = \sqrt{(a_f^2/g^2) + 1} \qquad (13.2)$$

Substituting Eq. (13.1) into Eq. (13.2):

$$EM = \sqrt{(a_f^2/g^2) + 1} = \sqrt{ES^2 + 1} \qquad (13.3)$$

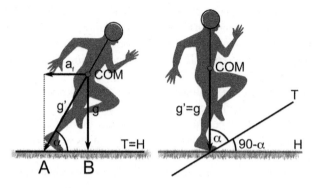

Fig. 13.1 The subject is accelerating forward while running on flat terrain (left panel) or running uphill at constant speed (right panel). COM, subject's centre of mass; a_f, forward acceleration; g, acceleration of gravity; $g' = \sqrt{(a_f^2 + g^2)}$, vectorial sum of a_f plus g; T, terrain; H, horizontal; α, angle between the runner's mean body axis throughout the stride and T; $90 - \alpha$, angle between T and H. Simple geometry shows that the angle A – COM – B is equal to $90 - \alpha$. Hence, since the length of the segment AB is equal to a_f, the tangent of the angle $90 - \alpha$ is given by the ratio AB/g = a_f/g. See text for details. (Modified after Osgnach et al. 2010)

It must also be pointed out that during decelerated running, which is equivalent to downhill running, and in which case the equivalent slope (ES) is negative, EM will nevertheless assume a positive value because ES in Eq. (13.3) is raised to the power of 2.

The Equivalent Slope (ES) and Mass (EM), as obtained from Eqs. (13.1) and (13.3), do not take into account the effects of the air resistance. Indeed, to overcome this last, during acceleration the runner must lean forward to a greater extent than required by the forward acceleration per se. Conversely, during deceleration the air resistance "helps" the runner's braking action, so that he/she leans backwards to a lesser extent that required by the (negative) forward acceleration. These effects, albeit small, can be taken into account as follows. The air resistance (R) increases with the square of the air velocity (v) as described by (di Prampero 2015):

$$R = k \cdot v^2 \quad (13.4)$$

where k is a constant, the value of which, per kg body mass and when expressing v in m s^{-1}, ranges from 0.0025 to 0.0048 (J s^2 kg^{-1} m^{-3}) (Pugh 1970; Tam et al. 2012; di Prampero et al. 2015). Dimensionally, R is the mechanical work per unit body mass and distance (J kg^{-1} m^{-1}) performed against the air drag. Since the numerator of this quantity is work (= mass · acceleration · distance), mass and distance cancel out; hence, numerically and dimensionally R = acceleration (m s^{-2}). Therefore, to take into account the (small) effects of the air resistance, both ES and EM can be calculated replacing a_f/g in Eqs. (13.1) and (13.3) with the quantity a_f + R. If this is so:

$$ES^* = (a_f + R)/g \qquad (13.5)$$

and

$$EM^* = \sqrt{(ES^{*2} + 1)} \qquad (13.6)$$

where the asterisk indicates that the appropriate quantities have been corrected for the air resistance. Incidentally, Eq. (13.5) allows one to estimate that running at 20 km h^{-1} (5.55 m s^{-1}) on flat terrain in the absence of wind is equivalent to running uphill on a treadmill at an incline of about 1% (ES* ≈ (0.0037 · 5.55^2) / 9.81 = 0.0115). This is consistent with the common practice of having the subjects run on the treadmill at a slope of 1% to simulate the air resistance encountered in the aerobic speed range.

It can be concluded that, if the time course of the velocity during accelerated/decelerated running is determined, and the corresponding instantaneous accelerations/decelerations calculated, Eqs. (13.5) and (13.6) allow one to obtain the appropriate ES* and EM* values, thus converting accelerated/decelerated running into the equivalent constant speed uphill/downhill running. Indeed, the energy cost of this last (C_r), as a function of the incline (i) in the range between i = −0.45 to i = + 0.45, is described by Minetti et al. (2002):

$$C_r = 155.4 \cdot i^5 - 30.4 \cdot i^4 - 43.3 \cdot i^3 + 46.3 \cdot i^2 + 19.5 \cdot i + 3.6 \qquad (13.7)$$

Substituting i with the equivalent slope (ES*), denoting C_0 the energy cost of constant speed level running, and multiplying by the equivalent body mass (EM*), the corresponding energy cost of accelerated/decelerated running can be easily obtained:

$$C_r = (155.4 \cdot ES^{*5} - 30.4 \cdot ES^{*4} - 43.3 \cdot ES^{*3} + 46.3 \cdot ES^{*2} + 19.5 \cdot ES^* + C_0) \cdot EM^*$$

$$(13.8)$$

For a more detailed discussion the reader is referred to Chap. 12.2, as well as to the original papers.

13.3 Metabolic Power and Oxygen Consumption

In running, as well as in any other form of locomotion, the product of the instantaneous velocity (v, m s^{-1}) and the corresponding energy cost per unit body mass and distance (C_r, J kg^{-1} m^{-1}) yields the instantaneous metabolic power (E, W kg^{-1}) necessary to proceed at the speed in question:

$$\dot{E} = v \cdot C_r \qquad (13.9)$$

As such, the metabolic power is a measure of the overall amount of energy required, per unit of time, to reconstitute the ATP utilised for work performance. We would like to point out also that C_r in Eq. (13.9) is the value applying at the very moment at which v is determined. Therefore, whereas at constant speed, be it on flat terrain or up (down) a constant slope, both C_r and v (and hence $\dot{E}$) are constant, this is not the case when either the slope of the terrain (and hence the corresponding C_r), and/or the speed, vary. It should be noted that, in this latter case, the speed changes have a twofold effect on metabolic power: on the one side because of the direct role of v in setting $\dot{E}$, on the other because in the acceleration (or deceleration) phase C_r is increased (or decreased) in direct proportion to the acceleration (deceleration) itself.

It should also be noted that the preceding considerations, as well as Eq. (13.9) as written, apply to running. In soccer, however, several walking periods are interspersed among running episodes. Since the energy cost of walking (C_w) is substantially less than that of running, in order to estimate correctly the metabolic power, these events must be clearly identified and the appropriate energy cost values applied, as discussed in detail below (Sect. 13.4).

At variance with the metabolic power, the actual oxygen consumption ($\dot{V}O_2$) is a measure of the amount of ATP actually resynthesized at the expense of the oxidative processes. As such, at any given time, $\dot{V}O_2$ may be equal, greater, or smaller than the metabolic power because:

i. The oxidative processes are rather sluggish as compared to the rate of change of the work intensity, in so far as they adapt to the required metabolic power following an exponential process, the time constant of which at the muscle level is on the order of 20 s;
ii. During strenuous exercise bouts of short duration, a common feature in soccer, the metabolic power requirement can attain values greatly surpassing the subject's maximal O_2 consumption ($\dot{V}O_2max$).

These considerations show that, whereas in a typical "square wave" aerobic exercise, after about 3 min, actual $\dot{V}O_2$ and metabolic power coincide, in soccer, as well as in many other team sports, because of the occurrence of several high intensity bouts of short duration, interspersed among low intensity periods, the time course of $\dot{V}O_2$ is markedly different than that of metabolic power requirement. Hence, at any given time during a match, or a training drill, in the course of which the intensity (metabolic power) changes rapidly (Fig. 13.2), the actual $\dot{V}O_2$ can be smaller, equal or greater that the instantaneous metabolic power, depending on: (i) the time profile of the metabolic power requirement and (ii) the subject's $\dot{V}O_2max$.

As discussed in detail elsewhere (di Prampero et al. 2015), knowledge of the time course of the metabolic power requirement allows one to estimate the

13 Metabolic Power and VO₂ in Soccer: Facts and Theories

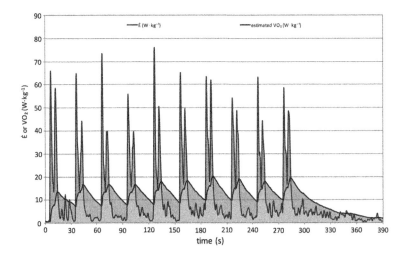

Fig. 13.2 Metabolic power (Ė, red line, sharp peaks) and O₂ consumption ($\dot{V}O_2$, blue line, smooth peaks) (W kg^{-1}) as a function of time (s) during a series of shuttle runs (25 m + 25 m in 5 s + 5 s) followed by a 20 s interval, in one subject. Each cycle was repeated 10 times for a total running distance of 500 m. Red areas (above $\dot{V}O_2$ and below Ė) and blue areas (below $\dot{V}O_2$) indicate the anaerobic and aerobic energy yield during the exercise bouts, respectively. Green areas (below $\dot{V}O_2$ and above Ė) indicate the amount of O₂ consumed in the recovery between bouts for phosphocreatine resynthesis (alactic O₂ debt payment). See text for details (Colour figure online)

corresponding time course of $\dot{V}O_2$, assuming a mean time constant of the $\dot{V}O_2$ kinetics during metabolic transients, as from literature data, on the basis of the individual $\dot{V}O_2$max. The so obtained estimated $\dot{V}O_2$ values turned out to be essentially equal to the actually measured ones in a group of 9 subjects who performed a series of shuttle runs over 25 m distance in 5 s. Each bout was immediately followed by an equal run in the opposite direction (again 25 m in 5 s). A 20 s interval was interposed between any two bouts, and the whole cycle was repeated 10 times (for a total running distance of 500 m). The running speed, the corresponding instantaneous acceleration, energy cost and metabolic power were then calculated by means of the same set of equations as implemented in a GPS system (GPEXE ®, Exelio, Udine, Italy[1]) and described in detail elsewhere (di Prampero et al. 2005, 2015; Osgnach et al. 2010). This allowed us to compare the overall amount of O₂ consumed, calculated from the integral of the $\dot{V}O_2$ time course estimated as described above, to the corresponding values obtained by means of a portable metabolic cart (K4, Cosmed, Rome, Italy) yielding the actual $\dot{V}O_2$ consumption on a single breath basis.

The so obtained data are represented for a typical subject in Fig. 13.3 that reports the time integral of:

---

[1]For information and further details on the system GPEXE ® (Udine, Italy) the reader is referred to www.gpexe.com.

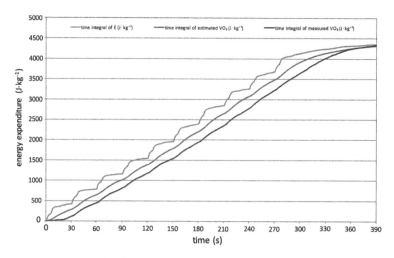

Fig. 13.3 Time integral (J kg^{-1}) as a function of time (s) of: (i) metabolic power requirement (red, upper), as determined by GPS (GPEXE ®, Exelio, Udine, Italy); (ii) $\dot{V}O_2$ estimated on the basis of a time constant of 20 s (blue, middle); and (iii) $\dot{V}O_2$ actually measured by means of a portable metabolic cart (K4, Cosmed, Rome, Italy) (green, lower), during the same training drill and in the same subject as in Fig. 13.2. See text for details. (Modified after di Prampero et al. 2015) (Colour figure online)

i. Metabolic power requirement (Ė, red line, upper);
ii. $\dot{V}O_2$ estimated on the basis of a time constant of 20 s at the muscle level (blue line, middle);
iii. Actually measured $\dot{V}O_2$ (green line, lower).

Inspection of this figure shows that:

i. Cumulated $\dot{V}O_2$ values (estimated or measured) are very close;
ii. They follow fairly well the time course of the total energy expenditure (i.e. of the time integral of the metabolic power requirement);
iii. The time constant of the $\dot{V}O_2$ kinetics, as given by the horizontal time difference between the two functions (measured or estimated $\dot{V}O_2$, and metabolic power) turns out longer ($\approx$35 s) at the measuring site (the upper airways) than that assumed to hold at the muscle level ($\approx$20 s).

These considerations lend experimental support to the approach briefly described above. Specifically, this approach allows one to assess: (i) the area below the $\dot{V}O_2$ curve (see Fig. 13.2), thus yielding the overall amount of energy actually derived from aerobic sources, as well as (ii) the area comprised between the curve reporting the time course of the metabolic power and that reporting the time course of $\dot{V}O_2$. It must be pointed out that the physiological significance of the area between these two curves depends crucially on their relative position. Indeed, when the metabolic power is greater than $\dot{V}O_2$ the corresponding area is a measure of the amount of energy derived from anaerobic sources. When, on the contrary the metabolic power is smaller than $\dot{V}O_2$, the corresponding area indicates the amount of energy derived from aerobic sources to "pay the alactic O_2 debt", i.e. to resynthesize a certain amount of phosphocreatine (PCr) split in the preceding phase of the exercise. This state of affairs allows one to draw a complete energy balance of any given period of a match or of a training drill. Indeed, if the overall time integral (recovery periods included) of the $\dot{V}O_2$ curve is equal to that of the metabolic power curve, it can be concluded that the energy sources utilised throughout the entire exercise cycle were completely aerobic, in so far that the amount of PCr split during the high intensity bouts was completely reconstituted in the recovery periods between bouts. On the contrary, if the time integral of the metabolic power curve is greater than that of the $\dot{V}O_2$ curve, it can be concluded that a certain amount of energy was derived from net lactate accumulation.

A practical example of the application of this approach to the first minutes of an actual soccer match is reported in Fig. 13.4. In this specific case, throughout the selected time window, the metabolic power was substantially greater than the player's $\dot{V}O_2$max and, in view of the relatively long duration of this high metabolic power phases, the actual $\dot{V}O_2$ did attain the player's $\dot{V}O_2$max. In addition in the subsequent period characterised by a milder metabolic power (i.e. below $\dot{V}O_2$max), the actual $\dot{V}O_2$ remained at the maximal level for several seconds. Because of this state of affairs, the actual overall anaerobic energy yield, as given by the sum of all red areas (above $\dot{V}O_2$ and below $\dot{E}$), was greater than the overall aerobic yield, as given by the areas below the actual $\dot{V}O_2$ during the high intensity exercise phases plus the areas corresponding to the alactic O_2 debt payment. The final outcome of this state of affairs is that a certain amount of Lactate must have been produced to cover the difference between the overall energy requirement (area below the metabolic power curve) and the overall aerobic energy yield (area underneath the $\dot{V}O_2$ curve).

It can be concluded that the approach briefly described above, and graphically illustrated in Fig. 13.4, allows one to draw a reasonably accurate overall energy balance of any defined time interval of a training period or of a match, provided that the subject's $\dot{V}O_2$max is also known.

13.4 The Role of the Energy Cost in Setting Metabolic Power Estimates

At variance with the data reported in Fig. 13.3, several recent studies reported substantial underestimates of the metabolic power obtained from GPS data, as compared to the actual O_2 consumption, as determined by portable metabolic carts (Buchheit et al. 2015; Brown et al. 2016; Highton et al. 2016; Stevens et al. 2015). However, apart from some inconsistencies in the data collection and analysis briefly discussed elsewhere (Osgnach et al. 2016), close inspection of these studies shows

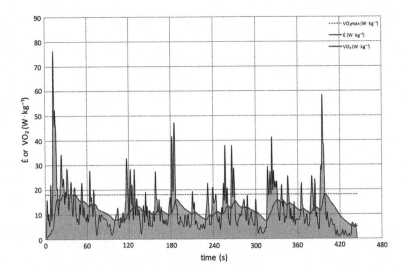

Fig. 13.4 Metabolic power (Ė, red line, sharp peaks) and estimated O_2 consumption ($\dot{V}O_2$, blue line, smooth peaks) (W kg^{-1}) as a function of time (s) during the first minutes of an actual soccer match in one player, whose $\dot{V}O_2$max above resting (18 W kg^{-1} ≈ 52 ml O_2 kg^{-1} min^{-1}) is indicated by the broken horizontal line. See text for detail and Fig. 13.3 for the meaning of the coloured areas (Colour figure online)

that all too often, the metabolic power approach was "stretched" to include activities such as training drills with the ball along winding paths, isometric efforts in rugby, or any other situations that cannot be easily reconciled with the theoretical pillars on which the metabolic power approach is based. In addition, in some of these studies the effects of the terrain and of the inter-individual variability in the energy cost of running are often neglected or only marginally discussed or are limited to rather low acceleration/deceleration values. Indeed, Stevens et al. (2015) recently reported that the average estimated energy cost of shuttle running turned out to be about 15% smaller than the directly measured values (i.e. 4.95–5.73, as compared to 5.73–6.71 J kg^{-1} m^{-1}). This difference, albeit not marginal, may well be due to individual difference of the shuttle running technique. In addition, this study investigated a relatively small range of accelerations/decelerations (which can be estimated to be <1 m s^{-2}) and hence of the corresponding ES* values lying in a range where the effects of the acceleration (upslope) are very nearly equal and opposite to those of the deceleration (downslope).

In view of these discrepancies, the aim of the paragraphs that follow is to review the basic assumptions underlying metabolic power estimates as obtained from GPS data, as well as to point out the limits of this approach.

As discussed at length in the original papers, to which the reader is referred for further details, the choice of the energy cost for constant speed running (C_0) on flat terrain is pivotal in setting the energy cost of accelerated/decelerated running, and, as such, the corresponding metabolic power. Its net value (above resting) (J kg^{-1} m^{-1}) ranges from 3.6 as determined on the treadmill by Minetti et al.

(2002), to 4.32 ± 0.42 on the treadmill and to 4.18 ± 0.34 on the terrain, as determined more recently by Minetti et al. (2012) at 11 km h^{-1}, to 4.39 ± 0.43 (n = 65), as determined by Buglione and di Prampero (2013) during treadmill running at 10 km h^{-1}, the great majority of data clustering around a value of 4 J kg^{-1} m^{-1} (Lacour and Bourdin 2015). Thus, whereas on the one side it would be advisable to determine C_0 on each subject, on the other it is often convenient to assume a unique value on the order of 4 J kg^{-1} m^{-1}.

It should also be pointed out that, apart from the inter-individual variability of C_0, its value depends also on the type of terrain (e.g.: artificial vs. natural grass, compact vs. soft surface, etc.), so that once again, its numerical value must be selected with a pinch of salt. Furthermore, it should also be stressed that the obtained estimates of energy cost and metabolic power apply to forwards running, so that the metabolic cost of other specific situations arising during a match or a training drill in soccer or in other sports, such as backwards running, running with or without the ball, sudden jumps, changes of direction, tackles, etc., inevitably introduces a certain degree of uncertainty and must be treated accordingly.

However, at the present stage, among the clear-cut scientific data available in the literature, none allow one to take into account with reasonable accuracy this state of affairs. As such, rather than relying on dubious corrections and assumptions, we deem it advisable to live with this uncertainty, inevitably built in the system's algorithms. This is particularly relevant in team sports such as rugby or American football in which the energetics of the numerous episodes in which the players collide violently or push forcefully against each other is not easily amenable to that of forwards running.

It should finally be pointed out that these general considerations apply also to the approaches based on speed and/or acceleration which cannot (and do not) yield any information on the sport specific activities mentioned above (Polgalze et al. 2015). Furthermore, even when applied to forward running these approaches overlook the inter-individual variability of the energy cost of running as well as the effects of the type of the terrain there-upon.

Another point to be considered when dealing with the energy cost and metabolic power of running is the fact that the equivalence between accelerated/decelerated and uphill/downhill running is based on the data obtained by Minetti et al. (2002) in a range of inclines between −0.45 and +0.45. For inclines outside this range, our approach relies on linear extrapolations of Minetti et al.'s data (see Giovanelli et al., 2016), rather than on the authors' polynomial equation which, obviously enough, cannot be extended beyond the experimental range of observations [see Eq. (13.7)]. However, the occurrence of such high values of acceleration/deceleration is rather infrequent so that this approximation cannot be expected to lead to any substantial errors.

These considerations apply to running; however, during a typical soccer match several walking episodes are interspersed among running bouts, a fact that may have indeed contributed to some of the above mentioned discrepancies between estimated metabolic power and measured O_2 consumption. Indeed, whereas for any given incline the energy cost of running at constant speed is independent of the speed itself, the energy cost of walking reaches a minimum at an optimal speed above and below which it increases. As reported in Fig. 13.5:

i. For level walking C_w attains a minimum of about 2 J kg^{-1} m^{-1} (i.e. about half the value of constant speed level running) at about 4.5 km h^{-1} (1.25 m s^{-1});
ii. At higher speeds C_w increases to reach a value about equal to that of level running (4 J kg^{-1} m^{-1}) at a speed on the order of 8 km h^{-1} (2.22 m s^{-1});
iii. With increasing the incline of the terrain, the curve C_w vs. speed retains a general U shape, but is shifted to a higher level, the optimum speed decreasing towards lower values;
iv. When walking downhill, until a slope of about −20%, the trend is reversed: at any given speed C_w is lower, whereas the optimum speed increases;
v. For slopes steeper than −20%, the C_w vs. speed curved is shifted again upwards and tends to become flatter so that the optimum speed is difficult to identify precisely.

As reported by Minetti et al. (2002), the relationship between the minimum energy cost of walking ($C_w$min, J kg^{-1} m^{-1}) and the incline of the terrain (i) can be appropriately described by the following 5th order polynomial equation:

$$C_w min = 280.5 \cdot i^5 - 58.7 \cdot i^4 - 76.8 \cdot i^3 + 51.9 \cdot i^2 + 19.6 \cdot i + 2.5 \quad (13.10)$$

where the last term (2.5 J kg^{-1} m^{-1}) is the energy cost of walking at the optimal speed on flat terrain. It should also be pointed out that for very steep slopes (> ± 0.50), this equation yields $C_w$min values rather close to those obtained from the 5th order polynomial equation describing the relationship between the energy cost of running and the incline [Eq. (13.7)]. According to the authors, this suggests that at these very high slopes walking and running "lose the pendulum like and bouncing ball mechanisms", so that at this steep slopes it may not be legitimate to speak of "walking" or "running" as different gaits. Be this as it may, it seems interesting to point out that, whereas the energy cost of running is independent of the speed, in walking (at least in the range of slopes between 0 and ± 0.40, where it is legitimate to speak of walking vs. running), as mentioned above, this is not the case. Indeed, within this range of slopes there exists a speed, here defined "transition speed", at which the subjects spontaneously change their gait from walking to running (dots in Fig. 13.5). On flat terrain, the transition speed is on the order of 8 km h^{-1} (2.22 m s^{-1}), essentially equal to that at which walking become more expensive than running, in terms of energy cost. Incidentally, this state of affairs is even more elaborate in dogs and horses, who, thanks to their quadrupedal locomotion, can adopt three types of gait: pace, trot and gallop. The spontaneously adopted gait corresponding always to the more economical one (in terms of energy cost). It can be reasonably assumed that also when walking or running uphill or downhill, the transition speed corresponds to that at which the energy cost of the two forms of locomotion become equal. The data reported in Fig. 13.5 allowed us to estimate the transition speed (v_{tr}, m s^{-1}) for any given incline (or ES*) between −0.30 and +0.40, as described by:

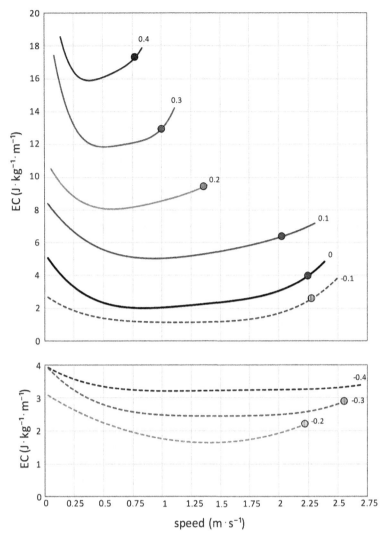

Fig. 13.5 Energy cost of walking (J kg^{-1} m^{-1}) as a function of the speed (m s^{-1}) at the indicated incline (i). The open dots indicate the "transition" speed, i.e. the speed at which the energy cost of walking and running are equal (see text). (Modified after Margaria 1938)

$$v_{tr} = -107.1 \cdot ES^{*5} + 113.1 \cdot ES^{*4} - 1.1 \cdot ES^{*3} - 15.8 \cdot ES^{*2} - 1.7 \cdot ES^{*} + 2.3 \tag{13.11}$$

It should be noted that this equation applies only within the range of inclines or ES* values indicated above. Indeed, for slopes steeper than +0.40, the minimum energy cost of walking is equal to that of running whereas, for slopes steeper that

−0.30, it becomes difficult to identify a clear-cut transition speed (see Fig. 13.5). Thus, for any given ES* within the range from −0.30 to +0.40 the energy cost utilised in the metabolic power calculations was that pertaining to walking or running depending on whether the speed was below or above the transition threshold described by the above equation. For steeper slopes the selected energy cost was, in any case, that pertaining to running. It should also be pointed out that this set of calculations is based on the implicit assumption that the analogy between accelerated/decelerated running on flat terrain, on the one side, and constant speed uphill/downhill running, on the other, can be applied to walking, and hence that the procedures to estimate the equivalent slope and mass are the same in the two forms of locomotion.

13.5 The Limits of Metabolic Power Assessment

The model summarized above is based on several additional assumptions briefly reported below, the interested reader being referred to the original papers for further details (di Prampero et al. 2005, 2015; Osgnach et al. 2010):

i. The overall mass of the runner is condensed in his/her centre of mass. This necessarily implies that the energy expenditure due to internal work performance (such as that required for moving the upper and lower limbs in respect to the centre of mass) is the same during accelerated running and during uphill running at constant speed up the same equivalent slope (ES*);
ii. Assumption (i) implies also that the stride frequency of accelerated running, for any given ES* value, is equal to that of constant speed (uphill/downhill running) over the corresponding incline;
iii. For any given ES*, the efficiency of metabolic to mechanical energy transformation during accelerated running is equal to that of constant speed running over the corresponding incline. This also implies that the biomechanics of running, in terms of joint angles and torques, etc. is the same in the two conditions;
iv. The calculate ES* values are assumed to be in excess of those observed during constant speed running on flat terrain in which case the runner is lining slightly forwards. This, however, cannot be expected to introduce large errors, since in all cases the reference value is the energy cost of constant speed running on flat terrain (C_0).

13.6 Conclusions

We would like to stress the fact that the technology for data collection must be appropriately selected. Indeed, since we deal with accelerations, any sampling frequency below 10 Hz is highly questionable. In addition, also the filtering of the signal to smooth the accelerations/decelerations of the centre of mass synchronous with the stride frequency must be appropriately considered in order to reduce the noise, without losing information. Finally, the often proposed utilisation of mechanical quantities obtained by inertial sensors is dubious, because of the absence, so far, of any convincing experimental data linking the observed mechanical variables with the corresponding energy expenditure.

In concluding, we strongly support the general validity of the proposed model, albeit within the limits discussed above, and provided that the actual investigated sport activities consist mainly of (or are reasonably amenable to) forward running or walking.

Acknowledgements Financial support of the "Lions Club Udine Duomo" is gratefully acknowledged.

References

Brown DM, Dwyer DB, Robertson SJ, Gastin PB (2016) Metabolic power method underestimates energy expenditure in field sport movements using a GPS tracking system. Int J Sports Physiol Perform 11:1067–1073

Buchheit M, Manouvrier C, Cassirame J, Morin JB (2015) Monitoring locomotor load in soccer: is metabolic power, powerful? Int J Sports Med 36:1149–1155

Buglione A, di Prampero PE (2013) The energy cost of shuttle running. Eur J Appl Physiol 113:1535–1543

Carling C, Bloomfield J, Nelsen L, Reilly T (2008) The role of motion analysis in elite soccer: contemporary performance measurements techniques and work rate data. Sports Med 38:839–862

di Prampero PE (2015) La Locomozione umana su Terra, in Acqua, in Aria. Fatti e Teorie. Edi-Ermes, Milano, Italy, 221 + IX. pp 108–112

di Prampero PE, Fusi S, Sepulcri L, Morin JB, Belli A, Antonutto G (2005) Sprint running: a new energetic approach. J Exp Biol 208:2809–2816

di Prampero PE, Botter A, Osgnach C (2015) The energy cost of sprint running and the role of metabolic power in setting top performances. Eur J Appl Physiol 115:451–469

Giovanelli N, Ryan Ortiz AL, Henninger K, Kram R (2016) Energetics of vertical kilometer foot races; is steeper cheaper? J Appl Physiol 120:370–375

Highton J, Mullen T, Norris J, Oxendale C, Twist C (2016) Energy expenditure derived from micro-technology is not suitable for assessing internal load in collision-based activities. Int J Sports Physiol Perform 20:957–961

Lacour JR, Bourdin M (2015) Factors affecting the energy cost of level running at submaximal speed. Eur J Appl Physiol 115:651–673

Margaria R (1938) Sulla fisiologia e specialmente sul consumo energetico della marcia e della corsa a varia velocità ed inclinazione del terreno. Atti Acc Naz Lincei. 6:299–368

Minetti AE, Moia C, Roi GS, Susta D, Ferretti G (2002) Energy cost of walking and running at extreme uphill and downhill slopes. J Appl Physiol 93:1039–1046

Minetti AE, Gaudino P, Seminati E, Cazzola D (2012) The cost of transport of human running is not affected, as in walking, by wide acceleration/deceleration cycles. J Appl Physiol 114:498–503

Osgnach C, Poser S, Bernardini R, Rinaldo R, di Prampero PE (2010) Energy cost and metabolic power in elite soccer: a new match analysis approach. Med Sci Sports Exerc 42:170–178

Osgnach C, Paolini E, Roberti V, Vettor M, di Prampero PE (2016) Metabolic power and oxygen consumption in team sports: a brief response to Buchheit et al.. Int J Sports Med. 37:77–81

Polglaze T, Dawson B, Peeling P (2015) Gold standard or fool's gold? The efficacy of displacement variables as indicators of energy expenditure in team sports. Sports Med 46:657–670

Pugh LG (1970) Oxygen intake in track and treadmill running with observations on the effect of air resistance. J Physiol 207:823–835

Sarmento H, Marcelino R, Anguera MT, Campaniço J, Matos N, Leitão JC (2014) Match analysis in football: a systematic review. J Sports Sci 32:1831–1843

Stevens TG, De Ruiter CJ, Van Maurik D, Van Lierop CJ, Savelsbergh GJ, Beek PJ (2015) Measured and estimated energy cost of constant and shuttle running in soccer players. Med Sci Sports Exerc 47:1219–1224

Tam E, Rossi H, Moia C, Berardelli C, Rosa G, Capelli C, Ferretti G (2012) Energetics of running in top-level marathon runners from Kenya. Eur J Appl Physiol 112:3797–3806

Printed by Printforce, the Netherlands